ISLAMIC CIVILIZATION IN SOUTH ASIA

Muslims have been present in South Asia for fourteen centuries. Nearly 40 per cent of the people of this vast land mass follow the religion of Islam, and Muslim contribution to the cultural heritage of the subcontinent has been extensive. This textbook provides both undergraduate and post-graduate students with a comprehensive account of the history of Islam in India, encompassing political, socio-economic, cultural and intellectual aspects.

Using a chronological framework, the book discusses the main events in each period between *c.* 600 CE and the present day, along with the key social and cultural themes. It discusses a range of topics, including:

- how power was secured, and how it was exercised;
- the crisis of confidence caused by the arrival of the West in the sub-continent; and
- how the Indo-Islamic synthesis in various facets of life and culture came about.

Excerpts at the end of each chapter allow for further discussion, and detailed maps alongside the text help visualize the changes through each time period. Introducing the reader to the issues concerning the Islamic past of South Asia, the book is a useful text for students and scholars of South Asian History and Religious Studies.

Burjor Avari is Honorary Research Fellow at Manchester Metropolitan University, UK. His research interests include Indian and South Asian History, and he is the author of *India: The Ancient Past – The History of the Indian Subcontinent from 7000 BC to AD 1200* (Routledge, 2007).

‘. . . the author’s story is extremely well told. This will be a most accessible book [. . .] most pleasingly his exposition, working over some heavily contested areas of history, is very well-balanced.’

Francis Robinson, *Professor of South Asian History, Royal Holloway, University of London, UK*

ISLAMIC CIVILIZATION IN SOUTH ASIA

A history of Muslim power and presence in the Indian subcontinent

Burjor Avari

LONDON AND NEW YORK

First published 2013
by Routledge
2 Park Square, Milton Park, Abingdon, Oxon OX14 4RN

Simultaneously published in the USA and Canada
by Routledge
711 Third Avenue, New York, NY 10017

Routledge is an imprint of the Taylor & Francis Group, an informa business

British Library Cataloguing in Publication Data
A catalogue record for this book is available from the British Library

Library of Congress Cataloging in Publication Data
Avari, Burjor.
Islamic civilization in South Asia: a history of Muslim power and presence in the Indian subcontinent / Burjor Avari.
p. cm.
Includes bibliographical references and index.
1. Islam–India–History. 2. India–Civilization–Islamic influences.
I. Title.
BP63.I4A946 2012
954.02–dc23
2012011479

ISBN: 978–0–415–58061–8 (hbk)
ISBN: 978–0–415–58062–5 (pbk)
ISBN: 978–0–203–09522–5 (ebk)

Typeset in Sabon
by Florence Production Ltd, Stoodleigh, Devon, UK

Printed and bound in Great Britain by the MPG Books Group

FOR ZARIN

CONTENTS

PLATES AND ACKNOWLEDGMENTS

Plate section can be found between pages 142–3

Front cover: A section of the Taj Mahal gardens
Source: Professor Ebba Koch, University of Vienna, author: *The Complete Taj Mahal*, Thames and Hudson, 2006

1 Islamic architectural styles

Source: Mr Ismail Lorgat, Blackburn, UK, private album

2 An artist's model of the Qutb complex in Delhi

Source: Sir Wolsey Haig, editor: *The Cambridge History of India,* Vol. 3, Cambridge University Press, 1928

3 Two styles of royal tombs: Sher Shah and Jahangir

3.1 Source: Sir Richard Burn, editor: *The Cambridge History of India,* Vol. 4, Cambridge University Press, 1937

3.2 Source: Mr Akhtar Hussain, Manchester, UK, private album

4 Two Mughal emperors: Babur and Akbar

4.1 Source: Dr Peggy Woodford Aylen, *Rise of the Raj*, Midas Press, 1978

4.2 Source: Dr Peggy Woodford Aylen, *Rise of the Raj*, Midas Press, 1978

5 Two Mughal forts: Lahore and Agra

5.1 Source: Mr Akhtar Hussain, Manchester, UK, private album

5.2 Source: Mrs Maharukh Desai, Goa, India, private album

6 Two gateways: Teen Darwaza (Ahmadabad) and Char Minar (Hyderabad)

6.1 Source: Miss Azzmin Mehta, Ahmadabad, India

6.2 Source: Dr George Michell, co-author: *Architecture and Art of the Deccan Sultanates*, Cambridge University Press, 1990

7 . . . 'space fit for royalty' – a Mughal palace and a garden

7.1 Source: Mr Imtiaz Patel, Blackburn, UK, private album

7.2 Source: Mrs Maharukh Desai, Goa, India, private album

8 An artist's view of Mughal Delhi
Source: Archaeological Service of India

9 Examples of Mughal art and aesthetics
9.1 Source: Victoria and Albert Museum, London, UK
9.2 Source: Mrs Maharukh Desai, Goa, India, private album
9.3 Source: Mrs Maharukh Desai, Goa, India, private album

10 Two Mughal mosques: the Badshahi (Lahore) and Moti Masjid (Agra)
10.1 Source: Mr Akhtar Hussain, Manchester, UK, private album
10.2 Source: Mrs Maharukh Desai, Goa, India, private album

11 Sultan Ibrahim Adil Shah II of Bijapur, and a Bijapur painting: *A Prince Reading a Book*
11.1 Source: The British Museum, London
11.2 Source: Chester Beatty Museum, Dublin, Ireland

12 Three nineteenth-century Muslim educators: Sir Sayyid Ahmad Khan and his colleagues, and a twentieth-century feminist: Jahanara Shah Nawaz Begum with her husband at the London Round Table Conference, 1932
12.1 Source: Professor Nizami, Aligarh Muslim University, India, *The History of the Aligarh Muslim University*, Vol. 1, Delhi, 1995
12.2 Source: Rozina Visram, *Women in India and Pakistan: The Struggle for Independence from British Rule*, Cambridge University Press, 1992

MAPS

EXCERPTS

PREFACE

The year was 1947; the month was August; and the date was either the fourteenth or the fifteenth. I was then a young boy growing up in the Stone Town of the island of Zanzibar, once a busy trading centre and a slave market off the east coast of Africa. Most people in the island were, and are, Muslims, divided into distinct ethnicities: the predominant Swahili, Arab, Persian and South Asians, or Indians as they were called. The Indian population grew substantially during British colonial rule from the late nineteenth century onwards. While a majority of them were Muslims, there were also numerous castes of Hindus and other non-Muslim Indians such as Sikhs, Goans and Parsees. Both 14 and 15 August 1947, were days of joy and celebration among South Asians worldwide, as they marked the independence of Pakistan and India. It was on one of those two days that I remember watching a procession of men marching along the main street that was indeed called Main Road; they were shouting slogans like 'Hindustan Zindabad' (Victory to India) and 'Pakistan Murdabad' (Death to Pakistan), or perhaps vice versa. Even today I can hear the echoes of those slogans in my ears. My historical and political education into my own South Asian/Indian origins began on that day, and has continued ever since.

My parents had a great regard for the British Raj and all that it had stood for: law and order, peace and security, education and progress for the many diverse peoples of India, but particularly for our own ethnic community of the Parsees. Yet, curiously, in our household we also held in great respect a set of heroes who had been waging a struggle for the independence of India from British rule; they were the Indian nationalist leaders like Dadabhai Naoroji, Sir Phirozshah Mehta, Mahatma Gandhi, Jawaharlal Nehru and Subhas Chandra Bose. In our sitting room, the picture that took the pride of place on the wall along with two others – the Persian prophet Zoroaster and King George VI and Queen Elizabeth – was the one that showed Gandhi and Nehru sitting on the ground, smiling at each other in a friendly embrace. It was taken for granted that Britain and India could never be enemies; Gandhi's humane and non-violent struggle had ensured that. What was, however, difficult for people like my parents to feel sympathy for was the demand for Pakistan by many of the subcontinent's Muslims. The founding of that nation on the eve of British departure was most unwelcome news to us and our other Indian friends. There was therefore no question of hanging the picture of Mohammed Ali Jinnah, the founder of Pakistan, or that of Sir Sayyid Ahmad Khan, the great Indian Muslim leader of the nineteenth century, in our house. During the years of my secondary schooling, before leaving home in the morning, I used to listen to the overseas news of the All India Radio in our Gujarati language; and I distinctly remember that the principal news item always seemed to begin in a strident tone with Pakistan and her misdeeds. I dare say that Pakistan Radio must have announced anti-Indian news in a similar tone, although we never listened to that. The drip-drip effect of biased news could not but have a deleterious effect on a young person's mind.

Old prejudices die hard. All my life since 1947 I have heard about the wars and animosities between India and Pakistan and, latterly, about the widespread incidents of Islamist terrorism against India and many Western countries. The result has been a great chasm of suspicion and hostility between Muslims and non-Muslims. My own childhood prejudices could have easily turned into hatred, were it not for two fortuitous circumstances of my life. The first was the decision of my parents to send me to Britain for higher education. Diverse peoples from the British Empire have made their homes in Britain over the last two centuries, and my encounters with them during my student days broadened my cultural and intellectual horizons. Since then, over nearly five decades of residence and teaching in Britain, I have made many Muslim friends; and I have a greater appreciation of their faith, cultural background and their hopes and aspirations. The fears and suspicions of childhood years were thus dissolved. The second big event for me was coming to read for a History degree at Manchester University. Since its foundation in the Victorian era, the Manchester History School has remained a robust centre of historical learning; its early pioneers were, in the memorable words of the historian Sir Maurice Powicke, 'big men in a big place'.[1] Eminent historians, like Professors Tout and Tait, Sir Lewis Namier and A.J.P. Taylor, had taught there and set the standards of scholarship that many generations of students have attempted to emulate. The lecturers I remember with great affection – the French Revolution specialist Albert Goodwin, the medievalists Michael Wallace-Hadrill and Gordon Leff, the Czech-born historian of sociology Werner Stark, economic historians W.H. Chaloner and T.S. Willan – helped to instil in me a lifelong love for my subject. Most of them were steeped in Anglocentric and Eurocentric knowledge, with somewhat limited appreciation of the history of India before British rule. That did not matter, however, because their greatest gift was the training they provided in historical skills and methodologies. That mattered more because, many years later, I self-studied Indian and South Asian History with the tools I had acquired at Manchester.

It was through my self-study and many years of teaching at both school and university levels that I came to explore the history of the land of my birth with fresh eyes. While writing my book *India: The Ancient Past – The History of the Indian Subcontinent, from 7000 BC to AD 1200* (Routledge, 2007), I had encountered varieties of bias and misconceptions that affect our understanding of events of the past. History has the potential either to inflame or modify existing prejudices in any people. With God's grace it turned out to be the latter for me. My initial knowledge of the history of the Muslims in South Asia was garbled in a haze of myths. One powerful, but dangerous myth, was that the glory of ancient India was destroyed by the coming of Islam and that medieval North India, particularly then, was a ruinous place. Many of the British imperial writings of the Victorian era had deliberately nurtured that myth, in the interest of maintaining British control over India. A more critical study of the subject, however, brought home to me that medieval India was in fact enriched, not destroyed, by the Muslim presence; and gradually, over a period of time, I also came to appreciate what may be called the Muslim perspective in South Asian history. For over 800 years, before the East India Company's domination from the mid- to late eighteenth century, Muslim sultans, emperors and princes ruled over extensive regions of the subcontinent. Their rule was often tyrannical and oppressive, yet many of them also governed their vast populations fairly and equitably, with panache and dignity. Under Muslim rule, India came to be introduced to the wider world of trade and knowledge; and society in India went through profound changes as a result of interaction between Hinduism and Islam. The study of Islamic or Indo-Islamic civilization is therefore as vital in our understanding of Indian and South Asian history as that of the ancient Hindu–Buddhist culture or the colonial period.

This book is my modest attempt to understand and articulate dispassionately the issues concerning the Islamic past of South Asia. It aims to present as comprehensively as possible, in one manageable and accessible volume, this history since the arrival of the first Muslim migrant traders and mariners from Arabia and Persia in the seventh century CE. I have attempted to provide the reader with information and a critical analysis of selected aspects of Indo-Muslim political, socio-economic and socio-cultural history in a chronological way. The narrative of the book and the interpretations arrived at are based upon numerous and valuable research works, written in English, by scholars over the last few decades. While many primary sources are in Persian or Urdu, languages which I do not read, much of their literature has fortunately been translated into English. A few selected excerpts from different sources are appended at the end of each chapter, as stimulus for further research in particular themes that the reader may be interested to pursue. For the reader's convenience, diacritics and accents have been omitted; and the vast majority of Muslim names in the text are spelt as widely pronounced today.

Among a number of people to whom I owe a great debt of gratitude I would first like to thank Professor Francis Robinson of Royal Holloway, University of London and Brasenose College, University of Oxford, for twice reading my manuscript and gently setting me on the right path with instructive comments and suggestions. His immense scholarship and wide empathies in Islamic Studies have been a source of inspiration for students and scholars in the field over many years, and I feel very privileged to have been guided by him. My good friend, Akhtar Hussain, a short-story writer, a community historian and an essayist on current issues in the English edition of the well known Pakistani newspaper, *Daily Jang*, has spent many hours reading the whole manuscript with great care, correcting errors and sensitizing me with nuances in matters Islamic; his encouraging advice has been very fruitful for the progress of my work. Other good friends and colleagues – the Islamic scholar Dr Mahmood Chandia of the University of Central Lancashire, the historian of mathematics Dr George Gheverghese Joseph, the philosopher Professor Lord Bhikhu Parekh, the sociologist Dr Mary Searle-Chatterjee and the Indo-Persian scholar at the Royal Asiatic Society Farrokh Vajifdar – have read sections of the book and provided most valuable comments. To them I am truly thankful; the shortcomings of the book, however, are solely my responsibility. Thanks are also due to my daughters Rushna and Anahita, and the IT specialist Gary Hullock at my university, who swiftly came to my assistance when I felt frustrated by my limited computer skills. The busy staff at the John Rylands Library of the University of Manchester and Sir Kenneth Green Library of the Manchester Metropolitan University always found time to speedily trace and order books for me. The friendly and congenial atmosphere of the Department of History at the Manchester Metropolitan University has nurtured my work for nearly two decades, for which I am ever so grateful. And, finally, a thank you to Dorothea Schaefter, South Asia Editor at Routledge, who gave me the opportunity to write the book.

Burjor Avari
Manchester
February 2012

1

INTRODUCTION

The first Muslims

The term 'Muslim' is of Arabic origin, and it means a person who submits to the will of God. More specifically, this person is a member of a community (*Ummah*) of people who profess to believe, however nominally, in the credo of the religion of Islam, again an Arabic term meaning 'submission' or 'peace'.[1] This religion was founded in Arabia in the early seventh century CE by a remarkable person, Muhammad (570–632 CE), who combined in him the qualities of a mystic, visionary, soldier, statesman and humanitarian. Muslims believe that he was the last prophet of God in the Abrahamic tradition that encompasses Judaism, Christianity and Islam, despite the fact that Christians and Jews do not accept this claim.

The Arabian society of Muhammad's time was both tribal and polytheistic in nature. The tribes were divided into warring clans, but a strict code of honour, common to many nomadic societies, governed their relationships during war and peace. Muhammad himself belonged to the powerful and influential tribe of the Quraysh who controlled the city of Mecca, his birthplace. This city, located on the trade route between Yemen in the south and Syria in the north, was a great trading metropolis, attracting to its bustling markets numerous merchants and buyers from the Arabian hinterland.[2] Congregating in the city annually were also many pilgrims who gathered to worship many pagan idols within the Kaaba, a monument reputedly built by Abraham (Ibrahim in Arabic) and his son Ishmael (Ismail in Arabic).[3] Despite the presence of Jews and Christians in Mecca, most Arabs then had only a dim understanding of the concept of monotheism, a belief in one supreme God (Allah).

In 610 CE, at the age of forty, Muhammad received his first prophetic revelation in the form of a command from God to recite, at a retreat on Mount Hira, just outside Mecca.[4] Many other revelations followed during the next twenty years, in response to the circumstances that he faced. The huge corpus of revelations that Muhammad recited verbatim verse by verse and preached to those who followed him, is what constitutes the Quran which, in Arabic, means 'recitation'.[5] Muslims consider the Quran as the word of God as revealed to Muhammad, and their sensibilities are gravely offended when the book is gratuitously mocked, insulted or abused. To his very small number of followers, initially consisting of his family members and friends, Muhammad preached the message of equity, justice and compassion; but it was not long before the most powerful people of the Quraysh in Mecca considered Muhammad's ideas to be highly dangerous both for the stability of the city and for their own power and position. Ultimately, in 622 CE, their hostility forced him and his small band of Muslim followers to migrate north to Yathrib, thenceforth known as Medina (Madinat al-Nabi, the City of the Prophet). This migration, called the *hijrah*, marks the beginning of the Muslim era, as it is from then on that an embryonic Muslim community was shaped by Muhammad.[6]

The people of Medina proved more receptive to Muhammad's ideas, but nonetheless he faced opposition from certain Jewish and Arab tribes. By tact and diplomacy, he kept the majority of people on his side. He and the Muslims, however, faced continuing hostility from the Quraysh in Mecca; and at least one of the Jewish tribes, although being warned twice, behaved treacherously by linking up with the Meccans. Muhammad therefore felt that he had no alternative but to prepare his followers to fight for survival. The Quran instructs the believers to work for peace, but Muhammad legitimized the idea of a just war and righteous revenge by his successful wars against Mecca and even approval of the mass slaughter of Jewish soldiers of Medina.[7] If he had not done so, his mission would have ended in failure. By the time he died in 632 CE Mecca was in Muslim hands, the Kaaba was cleared of its pagan idols and vast numbers of Arabs were converting to Islam.

It was the inspiration of the Quran that guided Muhammad to develop, over a number of years, a simple and effective programme of daily living for Muslims, in order to bind them together. This consisted of five practices or the Five Pillars of Islam.[8] First, a Muslim must for ever continue to declare that there is only one God and that Muhammad is His messenger (*Shahada*); second, a Muslim must prostrate before God and pray five times a day (*Salat*); third, a Muslim must donate alms for the benefit of the poor, traditionally set at 2.5 per cent annual levy on assets and capital (*Zakat*); fourth, with exceptions on the grounds of age or illness (to be re-compensated eventually), a Muslim must fast from sunrise to sunset in the ninth month, *Ramadan*, of the Muslim calendar (*Sawm*); fifth, a Muslim must endeavour to make a pilgrimage to Mecca at least once in a lifetime, without incurring debts (*Hajj*). These practices are a testament both to Muhammad's faith in God and his understanding of human nature. They not only foster a sense of inexplicable unity and bond among the believers but also appeal to the better nature and altruism of the followers. The rule on fasting, for example, is an effective way of promoting empathy for those suffering from hunger and famine. There are certain other Quranic injunctions on a number of subjects, such as dress, diet, marriage, inheritance, etc., but they were modified in later centuries. The bedrock of Muslim consensus lay in the five practices that constitute a comprehensive social and ideological programme of belief that is followed even today by over a billion Muslims across the world.

The Prophet's experiences and revelations in Mecca, his flight to Medina, his wars with the Meccans and the Jews, his final triumphal return to Mecca and the conversion of Meccans to Islam all mark the key stages of his life's work and achievements. With his death in 632 CE Muslim history took a new turn. Three major developments now shaped the early history of Islam and Muslims: territorial expansion, sectarian dissension, and legal sophistication. Most of the Arab territorial expansion took place during the period of the first three caliphs, Muhammad's successors, Abu Bakr (632–4), Umar (634–4) and Uthman (644–56), and later under the Umayyad dynasty (661–750).[9] The Persian/Iranian Sasanian Empire was destroyed and the Eastern Christian Byzantine Empire gravely threatened. Muslim Arab power spread throughout most of the Middle East and North Africa. From a small and struggling Muslim community with its two outposts at Mecca and Medina there arose a mighty Islamic Empire that extended as far as Sind, Spain, Central Asia and the Sahara desert. The message of Islam reached out to the conquered Zoroastrians, Jews, Christians and others; and, except within Persia/Iran, Arabic became the main language of communication and culture across the empire. How and why the vast masses of people were converted to Islam remains a controversial subject, but it would be utterly naïve to think that people were forced to become Muslim at the point of a sword.[10] The Arabs dominated the early Islamic Empire that reached its zenith under the Abbasid dynasty during the ninth and the tenth centuries; but with the decline of that dynasty and the fragmentation of the empire, other Muslim peoples and nations came to the forefront of Islamic imperialism. Today, nearly 1.5 billion people profess to be Muslims, and great centres of Islamic civilization are spread across the globe.

Muslim sectarian dissensions broke out over the question of succession shortly after the death of Muhammad. A faction that strongly believed in the idea of community leadership being vested in a member of his family claimed a place for Ali, Muhammad's cousin and son-in-law, as the caliph. This faction came to be known as Shia, or the partisans of Ali. Another larger faction, called the Sunni, disagreed and wanted a leader chosen by consensus and in accordance with the *Sunnah* or the established custom. The first three caliphs, Abu Bakr, Umar and Uthman were the Prophet's companions, not his relations. Ali had to wait his turn for twenty-four years before he became the caliph. Although his caliphate was not particularly noteworthy in its achievements and lasted only four years (656–60), the Shia continued to press the claim for his descendants too. In fact, the caliphate passed on to Muawiyah, a senior member of the Umayyad family in Mecca, who established his capital at Damascus. In 680 the Shia supported Ali's son Husain against Yazid who succeeded his father Muawiyah in that year. Husain's defeat and martyrdom at Karbala, Iraq, therefore entrenched among the Shia a feeling of being a beleaguered minority.[11] Later, in the eighth century, the Shia, too, split over the matter of rightful succession among their *imams*, with the Ismailis (the followers of seven *imams*) forming a separate branch from the majority Shia (the followers of twelve *imams*, the twelfth one being hidden).[12] In succeeding centuries, many other splits in the Muslim community arose over ideological or theological issues. The group called the Kharijites, for example, opposed absolute monarchy.[13] The Sufis were not a separate sect, but their eclectic interpretation of the Quran and other aspects of Islamic law marked them out as distinct from the traditional theologians, the *ulama*.[14]

The early Islamic Empire was a melting-pot of ideas exchanged between the dominant Arabs and the subject people. A high level of cultural and intellectual sophistication prevailed at the Abbasid court in Baghdad; disciplines as diverse as mathematics, chemistry, medicine, art, commerce, law and aesthetics were greatly enriched by ideas carried by scholars drawn from across cultures.[15] Europe's cultural transition from the Dark Ages to the Renaissance, for example, was due greatly to what she learnt from the Islamic civilization of that era.

An aspect of Muslim intellectual rigour may be seen in the way Islamic law developed during that period. At first, only the five basic practices and other injunctions from the Quran formed the basis of this law. In the seventh and eighth centuries, however, there began a systematic process of compiling, codifying and narrating all the events of Muhammad's life, his ideas, deeds and conduct, from a variety of historical sources and from the accounts provided by his companions. It was a highly complex process, involving checks, verifications, elucidations, interpretations and commentary; and ultimately all the reports were called the *ahadith* (or *hadith* in singular). The authenticity of each *hadith*'s narration had to be reliably supported and scrutinized by a chain of devout Muslims leading to the Prophet himself. The *ahadith* are the basis of the *Sunnah* of the Prophet, which has come to mean the behaviour of Muhammad.[16] For Muslims, therefore, Muhammad is the perfect and heroic ideal.[17] Islamic law encompasses both the Quran and the *Sunnah*, and the combination of the two constitutes what is called the *Shariah* or the straight path of divine law.[18] But the complexity goes further than that. Over many centuries, Muslim jurists have also laid stress on two other aspects of law: the idea of consensus (*ijma*) and reasoning by analogy (*qiyas*). These two are the foundation of what is known as *ijtihad* or independent reasoning. The science of jurisprudence (*fiqh*) thus came to be perfected through varied interpretations by jurists within different schools of law. Since a Muslim ruler could hardly be expected to be familiar with so many difficult issues of interpretation, an entire class of professional people, traditionally known as *ulama*, came to be involved with the interpretation, dissemination and enforcement of Islamic law, as expressed by the *Shariah* or the *Fiqh*. The *Shariah* and the *Fiqh* are the two great pillars of Islamic law, as it has developed over fourteen centuries; but the important point to remember is that the two are not necessarily opposed to each other.[19] Wise and learned human beings can interpret the law in line with customs and conventions of a prevailing age or a particular part of the world, but they must do so without infringing the limits set by the divine law. While the West has now abandoned any notion of divine law, Islamic law holds fast to it: and that is at the heart of the tension between the West and Islam. Within the Muslim world itself the absence of a central authority with powers to adjudicate has led to many disputes and conflicts.

The South Asian context

Clarifying the terms

In the context of this book, the term 'South Asia' refers to the three modern states of India, Pakistan and Bangladesh. There are other states in South Asia, but this book deals specifically with the history of Muslims in these three states. The term 'Indian subcontinent' has no political implication; it is simply the description of the more specific geographical zone in South Asia, covering the three states referred to. The term 'India' is widely used in this book in two particular senses. First, it refers to a pervasive culture and civilization within the region over many millennia, which the many diverse peoples of the subcontinent have been aware of through religion, art, literature or customs. Political India is a new nation, but cultural India is an ancient regional civilization. Second, with the rise of British power from the mid-eighteenth century onwards, 'India' also became a well recognized and acknowledged term in historical and political literature throughout the world. Additionally, for many centuries, the term 'Hindustan' was used by Muslims to denote the region we are talking about.

The term 'Islamic civilization' generally refers to the political, social and cultural advances achieved by the early Muslims in the heyday of the Islamic Empire under the Umayyad and Abbasid dynasties between the eighth and the tenth centuries CE. Yet, as we mentioned earlier, in each of the many regions of the world where Muslim influence spread during that period or subsequently, such advances only became possible through a fruitful interaction of Islamic ideas, values, aesthetics and imagination with those of the indigenous inhabitants. The resulting product was always hybrid. This phenomenon can also be observed in the case of the Islamic civilization that developed in South Asia over many centuries. There, too, a uniquely localized model developed, which may justifiably be called Indo-Islamic civilization.

The term 'power' in the subtitle refers to the rule exercised by Muslim monarchs and their governors over numerous regions of South Asia for many centuries. The high noon period of their sway was between the mid-sixteenth and late seventeenth centuries when the four great Mughal emperors ruled over virtually the whole of the subcontinent. The monarchs and their associates were essentially an elite group of people with a Central Asian warrior tradition behind them, but the vast majority of Muslims had no more power than any other South Asian groups. This lack of power did not mean a lack of presence, however. We use the term 'presence' in the subtitle to refer to the condition of Muslims in South Asia in general, the challenge of religious and social tensions from within and without that they faced, and how some of them reached the heights of cultural or intellectual achievements. The Muslim presence is pervasive throughout the subcontinent, as can be witnessed in different regions by varied sites of historic significance (Map 1.1). In each chapter of this book, therefore, an attempt

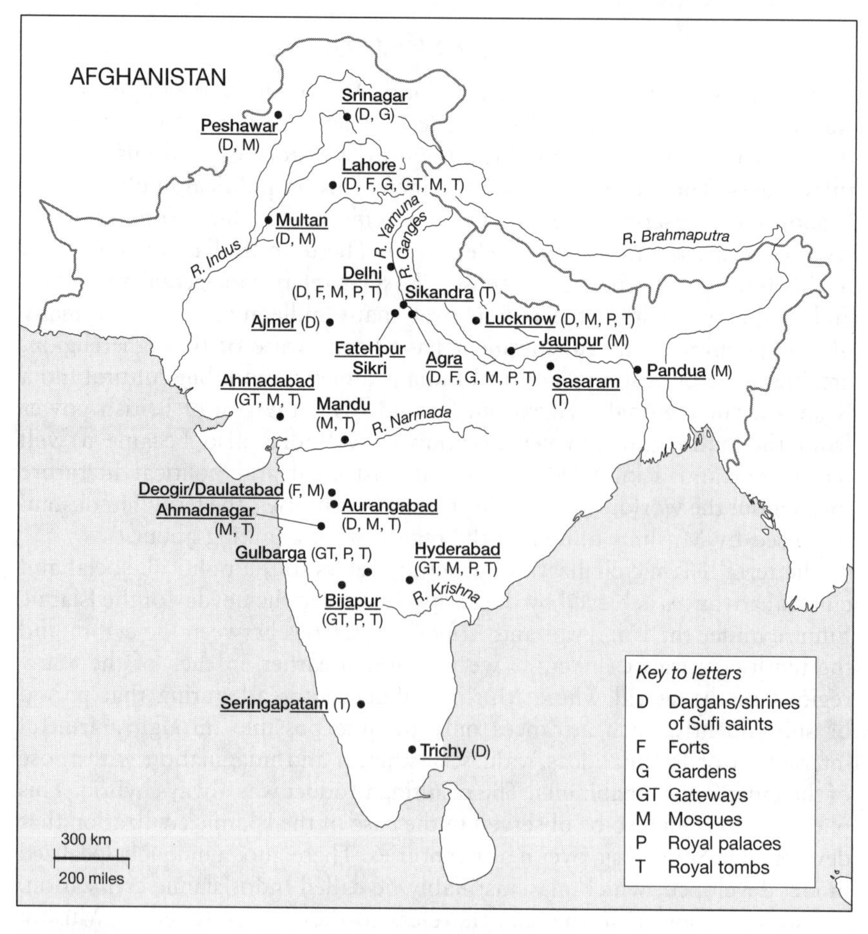

Map 1.1 Major Islamic sites of historic importance in South Asia

is made to balance the discussion between Muslim power (or the lack of it) and Muslim presence.

Correcting some popular misconceptions

During this present period of time, when the tension between Muslims and non-Muslims in many parts of the world has accelerated to a dangerous level, it may be instructive to examine some of the popular misconceptions and fallacies about South Asian Muslims, arising out of either general ignorance of historical and current facts or sheer prejudice. Here we present four of them. The first concerns their origins. They are often traced to Turkish and Mughal invaders, and vaguely dated to the Middle Ages. What is often forgotten is that the Muslims have been familiar with the subcontinent ever since the beginnings of Islam. Long before the Turks arrived, Islam had already been introduced, indirectly, into India by Arab and Persian sailors and traders. The first invasions by the Turco-Afghans did not begin until the early eleventh century; and their power did not become fully manifest until the early thirteenth century. The Mughal conquest followed three centuries later. What is also important to understand is that not all Muslims were invaders. The subcontinent attracted many Muslim traders, economic migrants, scholars, artists and artisans who made it their permanent home. On the other hand, it would be quite erroneous to claim that the present day Muslims in South Asia are mostly descendants of immigrants; the vast majority of them have their ancestry deeply rooted within the native gene-pool of the subcontinent.

Another misconception is about population statistics. Anti-Muslim voices, particularly in modern India, often exaggerate Muslim population figures and impute their increase to polygamy. The combined Muslim population of India, Pakistan and Bangladesh, in 2009, was estimated by the Pew Research Forum of the USA to be 480,339,000, or 30.5 per cent of the total world Muslim population of 1,571,198,000. After Indonesia, Pakistan (174 million Muslims), India (161 million Muslims) and Bangladesh (145.3 million) have the second, third and the fourth largest Muslim populations in the world today.[20] Post-independence India is not strictly a Muslim nation, but with 13.4 per cent of its vast population of 1.2 billion people being Muslim, the Muslim presence is too ubiquitous to be ignored. In the last few decades, in line with other peoples in the developing world, the Muslim population throughout South Asia has grown exponentially owing to high birth rate and low death rate through advances in modern medicine. This growth cannot be explained by polygamy, as the polygamy rate among Muslims of India is no higher than the rate for other communities.[21] Despite the huge increase, the fact of the matter is that most South Asians are, and have always been, non-Muslim; there has never been any period during the last 1,400 years when there was the slightest possibility of South Asia becoming a majority Muslim region, like somewhere in the Middle East.

A third misconception concerns the question of conversion of the non-Muslim population. Quite often, in popular discussions, there is an unspoken assumption that the Afghan or Turkish invaders cruelly put to death all those who refused to convert to Islam. This is not borne out by evidence. There were a number of Muslim rulers and their agents who were cruel, but they inflicted their cruelties on anyone they disliked, whether Muslim or non-Muslim (Chapters 3 and 4). Conversion by force is too simplistic an explanation for the growth of Islam in the subcontinent. There are other explanations. Muslim royal power, for example, could influence in two subtle ways. First, some of the most intelligent and ambitious non-Muslims could enhance their prospects and life chances by becoming Muslim (Chapters 2, 5 and 6). Their services were invaluable to Muslim rulers. Second, Muslim rulers often coerced non-Muslims into accepting Islam by exempting them from paying the poll tax or the land tax, although it is worth bearing in mind that in practice most rulers valued the income from these taxes more than the satisfaction of converting non-Muslims to Islam. The major explanation for large scale conversion lies in the efforts of three groups of people: mariners, merchants and missionaries. The first two, mostly Arab and Iranian, helped to spread and increase Muslim influence all along the coastline of India during the 400 years before the Turks (Chapter 2). The missionaries, mostly Sufis, played a major role, too, all over India; their compassion and common sense drew into the Islamic fold great numbers of non-Hindu forest and nomadic people and the lower-caste Hindus. The graves and mausoleums of Sufi *pir*s are revered places of pilgrimage for many non-Muslims even today (Chapters 3, 4, 6 and 8).

Finally, there is the vexed question of unity and diversity within the Muslim community in South Asia. Many non-Muslims often assume that Muslims are a tightly knit homogeneous community. This is correct only to the extent that they feel that they are guided by the singular example of Prophet Muhammad's life and the practices that he preached. In all other ways they are a very diverse people. They are divided by ethnicity, language, sectarian differences, class, education, occupations and wealth (Chapters 5 and 8). A version of the untruth of homogeneity was also propagated by the followers of the Pakistan Movement who insisted upon the idea of Muslims being a separate nation and the need for them to have their own homeland (Chapter 10). It was essentially a misconceived idea, as it became clear in 1971–2 when a violent division among the Muslims in Pakistan split that state into two (Chapter 11).

Phases of Muslim authority

In the early eighth century, India was a patchwork of big and small regional kingdoms. The northern kingdoms had emerged out of the ashes of a former great Hindu state, known as the Empire of the Imperial Guptas (320 to 550 CE). The Hindu–Buddhist kingdom of Sind was the first and the only kingdom to be conquered by Arab invaders (711). From then on, for nearly 300 years, the rulers of Sind were the Arab governors appointed first by the rulers of the Umayyad dynasty (661–750) based in Damascus and later by the Abbasid caliphs (from 750 onwards) based in Baghdad. The political instability in the Abbasid Caliphate during the ninth and the tenth centuries resulted in the passing of power in Sind to the Shia Ismaili sect and other local Sindhi Arabs who ignored the caliph in Baghdad (Chapter 2).

The next stage of immigrant invasion was initiated in 1000 CE by Turco-Afghan warlords, the descendants of Central Asian Turkish slaves who were in great demand at both the Abbasid and other courts of the various Islamic kingdoms in Iran and Afghanistan. Many of these slaves went on to become commanders of armies and sometimes to capture power themselves. A particular warlord who began attacking India was Mahmud, king of Ghazna in Afghanistan. His Ghaznavid successors carved out a territorial base that covered the frontier areas straddling today's Afghanistan and Pakistan, Baluchistan, Punjab and Sind. Eventually, however, it was the forces of a rival dynasty, the Ghurids from Ghur, also in Afghanistan, who conquered extensive tracts of North India by the end of the twelfth century (Chapter 3).

The key decisions affecting the Ghaznavids and Ghurids were essentially taken in Afghanistan; but a critical change came in 1206 when, on the death of the Ghurid leader, his principal lieutenants decided to continue to operate in India without further consultation with the authorities at Ghazna or Ghur. The so-called Slave dynasty that was established in 1206 was followed by four other dynasties; and together they constitute what is generally known as the Delhi Sultanate (1206–1526). The Delhi monarchs were of Turco-Afghan origin, but nearly all had been born and brought up in India. They were not a foreign or a colonial ruling class. The descendants of immigrants cease to be immigrants after one or two generations. The Turco-Afghans remained a principal elite force within the Sultanate, but increasingly, with the passage of time, Afghanistan's interests began to be separated from India's. The Sultanate went on to become a great imperial power in India and, by the early fourteenth century, its influence stretched beyond the Indo-Gangetic heartland into South India. Its demise came through a series of regional revolts in the late fourteenth and fifteenth centuries (Chapters 4 and 5).

The Mughal imperium, established by Babur, the Uzbek warlord from Afghanistan, enjoyed its golden period of authority during the sixteenth and seventeenth centuries (Chapter 6). Its long decline, however, began from the 1680s onwards, reaching its tragic end with the last emperor being forced into exile by the British. The regional post-Mughal Muslim and non-Muslim kingdoms were able to survive in the first half of the eighteenth century by playing off the two great European powers, France and Britain, which sought colonial dominance in the subcontinent. After the defeat of France in the 1750s, the British were the sole European power that mattered; and from then onwards, gradually over a number of decades, the East India Company emerged as the hegemonic power (Chapters 7 and 8).

Through all the phases of the feudal and pre-modern era, the effectiveness of Muslim monarchy in the subcontinent depended upon five vital factors. The first was the personality and charisma of the individual ruler. Weak, indecisive or wayward rulers brought destruction unto themselves as well as the kingdom that they ruled (Chapter 4). Constitutional monarchy of the modern variety was not an option for them. The second was the army and its commanders. Turkish horsemanship and hardy soldiers from Afghanistan and Central Asia had won India for the Turco-Afghan rulers; and the maintenance of correct balance of forces between the infantry and cavalry depended upon the quality and skills of the monarch's closest commanders. A priority for the monarch was to ensure that the commanders' interests were served and their concerns addressed. Although artillery was used by the army from at least the sixteenth century onwards, the lack of investment in research into military technology put all Indian armies at a disadvantage in comparison with European forces by the early nineteenth century. The third factor was the efficient bureaucracy. Muslim monarchs used the Persian model that had been perfected in Sasanian Iran and first introduced into India by the Ghaznavid rulers (Chapter 3). The Mughal Emperor, Akbar, refined it through a graded system of civil and military service, called the *mansabdari* system (Chapter 6), whose effectiveness lasted for over a century. The fourth factor, upon which the first three depended, was sound finance. The efficient collection of land revenue was the critical element in this, as this was essentially a pre-industrial society (Chapters 4 and 6). The final, fifth factor that was critical to the success of Muslim monarchy over a long period of nearly eight centuries was its ability to draw a positive response from the subcontinent's majority non-Muslim population by its adherence to certain established norms of justice that both the rulers and the ruled understood.

Inter-cultural understanding

During the first millennium CE India stood at the forefront of world civilizations. By the time that the Prophet Muhammad received his first revelation from God in the form of a command to recite, India had numerous linguists, grammarians, mathematicians, astronomers, doctors and architects. The first Muslim Arabs, full of curiosity, were keen to learn from everywhere; and at the court of the Abbasid caliphs in Baghdad, during the Golden Age of Islam between the eighth and tenth centuries, Indians were welcome visitors. Sind was the conduit through which the intellectual traffic between

Baghdad and India flowed; and the first Arab governor of Sind had no hesitation in declaring Hindus and Buddhists as *dhimmi* or 'People of the Book' (Chapter 2).

The Turkish officers and soldiers who began to arrive in North India from 1000 CE onwards were essentially a nomadic people whose appreciation of India and her culture was at first limited. This changed once they began to respect the immutable ways of the sedentary Indian peasantry whose toil on the land produced such abundant wealth. They also came to appreciate the rich variety of Indian trades and occupations which were flourishing concerns made possible by complex networks of kinship and caste arrangements. While militarily triumphant, the Turks nevertheless had to make compromises with Indian realities, and over time they became integrated in Indian society. The process was made easier by the fact that the Ghaznavids and Ghurids were much influenced by Persian culture, and Persia's connections with India went a long way back into history.

Royal patronage was one of the critical factors in the promotion of inter-cultural understanding, particularly in urban centres where Muslim power and presence were highly visible. The wisest of the Turco-Afghan sultans, the regional sultans and the Mughal emperors invited skilled people from Iran, Central Asia and other parts of the Middle East to settle in India and contribute to the refinement of Indian culture in general. This was how an Indo-Islamic style developed in art and architecture, aesthetics, literature and music. New products and agricultural technologies were imported from other Islamic lands and they enriched India. Styles of dress and cuisine underwent great sophistication too. It should be noted that the best of the royal patronage was not religion-bound. Hindus and other non-Muslims with skills and abilities were employed in royal studios and workshops without any discrimination (Chapter 6). The intimate secrets of Mughal finances were often in the hands of Hindu accountants.

Another form of inter-cultural understanding was at work through the efforts of some of the Sufi missionaries in the countryside. We have mentioned their role in converting to Islam a great number of marginalized and low caste people in different parts of India. What is not always appreciated was their ability to empathize with the followers of the predominant Hindu faith, to engage in dialogue with them and to seek reconciliation between religious ideas. This empathy evoked a positive response from Hindu sages, and many of the inter-faith movements in medieval India arose out of the encounters of Sufis and Hindus (Chapter 5). In contrast to the more legalistic theologians of Islam, the Sufis showed tolerance towards those who had become Muslim but who could not shake off their original caste rituals and conventions (Chapter 7). Thus, we see even today in the subcontinent, Muslims who unconsciously follow many a practices of their Hindu ancestors. This diversity is an integral part of the subcontinental Muslim society and culture and cannot be wished away.

The challenge from the West

During the sixteenth and seventeenth centuries Mughal power was able to limit the ambitions of Western maritime nations like Portugal, the Netherlands, France and Britain to just trade. With the rapid decline of the Mughals in the eighteenth century, the territorial ambitions of France and Britain came to the fore; and by the third quarter of that century Britain was the sole European power that mattered. The dual role of the East India Company as both a trading concern and a major territorial power was something that the Muslim (and Hindu) ruling classes in India had to learn to get used to. The defeat of anti-British forces in the great rebellion of 1857 marked the final end of independent Muslim authority in the subcontinent (Chapter 8). For Muslims, the complexity of their situation went beyond the fortunes of their rulers, however. Besides the power game, there was also taking place a culture war in the subcontinent. As harbingers of Western knowledge, scientific ideas and new social and cultural attitudes, the British had begun both consciously and unconsciously a cultural revolution that was bound to be both traumatic and highly consequential. The Muslim response to this challenge assumed two forms. One consisted of a movement to reform Islam and to revive it (Chapters 7 and 8), while the other was about accepting Western modernity and advocating Western education (Chapters 8, 9 and 10). Both responses had their strengths and weaknesses, but there was also some overlapping of interests, in the sense that those who desired religious reforms were by no means uninterested in gaining political power. By the end of the nineteenth century, both Hindus and Muslims had recovered their self-confidence and were ready to challenge colonial rule. Unfortunately, the anti-colonial struggle became distracted by the Muslims' need to resolve the agonizing dilemma of unity or separatism during the final decades of colonial rule (Chapter 10).

After colonialism

The splintering of South Asian Muslims occurred in two stages: through the creation of Pakistan in 1947 and the break-up of Pakistan in 1971–2. Their fortunes have varied in the three countries of India, Pakistan and Bangladesh (Chapter 11). In India, while they face covert discrimination in employment and occasional hostility from the Hindu Right, they enjoy the full benefits of a well established parliamentary and democratic system. Active and fruitful Muslim participation in all walks of Indian life has given them a unique position of strength and a valued stake in the future development of the country. In contrast, the Muslim population of Pakistan has been worn down by all the disadvantages resulting from a lack of genuine democracy and unrealistic political and military ambitions of the country's rulers. The hopes of 1947 have been cruelly betrayed. Although Bangladesh is at present a formal democracy, its future remains uncertain owing to its multiple and daunting economic problems. Millions of peasants in the Ganges delta face a bleak future owing to the rising seas.

Historiography and sources of study

Our primary source of knowledge for learning about the pre-Mughal Muslim history in South Asia lies in a variety of historical surveys and accounts left by medieval historians, such as al-Baladhuri (died *c.* 892 CE), Abu-Fazl Bayhaqi (995–1077 CE), Ali al-Kufi, Minhaj al-Din bin Siraj al-Din Juzjani and Zia ud-Din Barani (1285–1357).[22] A large number of Persian manuscripts left behind by these and other writers were collated and translated into English in eight comprehensive volumes by two English scholars of the nineteenth century, Sir Henry Miers Elliot (1808–53) and Professor John Dowson (1820–81). These two men lived in the heyday of supreme British power in India, and a careful reading of Elliot's original preface in the first volume betrays an imperial arrogance in his withering analysis of the weaknesses of medieval historians.[23] These weaknesses, however, cannot be ignored, and must be taken into account in our attempt to make sense of the medieval Muslim history of South Asia. Long and tedious narration of deeds and wars of individual monarchs, absence of analysis of social or cultural aspects of life, didacticism and extreme bias towards Islam, eulogistic obsequiousness, the generally hostile and triumphalist tone towards non-Muslims – all these weaknesses do diminish the value of the medieval works. Having said that, we nevertheless have to acknowledge the fact that without these histories we would have no source to rely on, except some histories by outsiders or fragments of state papers and documents. Elliot and Dowson's laborious efforts are a testimony to the usefulness, however limited, of medieval histories.

In contrast to this earlier period, diverse sources of information are available from the Mughal period.[24] The personal and intimate autobiographies of Emperors Babur and Jahangir, and the *Akbarnama* chronicle and its companion volume, the *Ain-i Akbari*, written by Abul Fazl, Akbar's chief adviser and court historian, give us exhaustive information about the Mughal court, administration and economy. When we add to these the Mughal state records, the biographies and memoirs of nobles and *mansabdar*s, and the works of two well known scholars, Ferishta and Badauni, we are in possession of a comprehensive range of primary historical material for the Mughal age as a whole (Chapter 6). Although this material was all written in Persian or occasionally in Turkish, the core works have been translated into English during the last two centuries. Our range of sources extends further with the varieties of fascinating accounts and travelogues written by European visitors to India in the sixteenth and seventeenth centuries. Diversity of evidence thus makes historical corroboration relatively easier.

Two distinct schools of thought influenced British works on India from the late eighteenth century onwards. The so-called Orientalist school, whose pioneer was Sir William Jones (1746–94), was Indophile in its approach to Indian history and civilization. Much of the ancient history of India was carefully unearthed and systematized by academics of this school. They did a great service to Indian historiography in general; and their work stands as a corrective to the more negative view of Orientalism popularized in the second half of the twentieth century. The more assertive Anglicist school, whose ideologues included such imperial historians and administrators as James Mill (1773–1836), John Malcolm and Lord Macaulay (1800–59), lacked empathy with Indian cultural and intellectual traditions. A common factor that bound the representatives of this school together was their immutable belief that India's salvation lay in British rule. They scorned in equal measure both the Hindu and Muslim cultural legacies, but they expressed a particular disdain of Muslim governance of India and Muslim historiography. In their eyes the nearly 800 years of Muslim power and presence before the British were nothing more than an exercise in what came to be termed Oriental Despotism (Excerpt 1.1). Muslims, in the words of Sir John Malcolm, the East India Company's Governor of Bombay, were 'the scourge of the human race'.[25] Notwithstanding the differences of opinion between the Orientalists and Anglicists, we need to acknowledge the fact that the systematic study of history in schools and colleges of India only began under the British. A number of fine Indian scholars were able to proceed to Britain for further research, and Muslim history of India came to be enriched by their works.[26]

British imperial and sometimes racist attitudes were bitterly resented by the resurgent Hindu elite in the nineteenth and early twentieth centuries. At the same time, the specifically anti-Muslim remarks by the imperialists gave some members of that elite, particularly the nationalist historians, the opportunity to condemn the Muslim record in Indian history. Hindu-oriented organizations such as the Bharatiya Vidya Bhavan (Institute of Indian Culture) and historical projects like the multi-volume *History and Culture of the Indian People*, edited by the eminent Hindu historian, R.C. Majumdar, have rightly promoted Hindu pride and identity, but in the process they have also exaggeratedly drawn the image of Hindus as being helpless victims of Muslim oppression over many centuries.[27] It is a short step from that to the fanatic anti-Muslim tone of those who today champion an unthinking form of Hindu assertion and offer a version of Indian history completely different from the well recognized historical data (Excerpt 1.2). Hindu revisionists have their counterparts among Muslims too; attempts to re-shape history textbooks in Pakistani schools are a good example of historical subversion that has been carefully investigated[28] (Excerpt 1.3).

Fortunately, owing to the very rigorous standards of historical scholarship originally developed and perfected in the West, we can now rely on an increasing number of good and serious scholars for excellent interpretations of the Muslim past in South Asia. They include both South Asian and foreign historians. This book would not have been written without the inspiration that the author received from reading the works of such writers as Muzaffar Alam, Justice Ameer Ali, Catherine Asher, K.K. Aziz, Christopher Bayly, C.E. Bosworth, R.M. Eaton, Irfan Habib, Professor Muhammd Habib, Peter Hardy, Mushirul Hasan, S.M. Ikram, Peter Jackson, Ayesha Jalal, Humayun Kabir, Sayyid Ahmad Khan, Ebba Koch, Sunil Kumar, Donald Maclean, Hafeez Malik, Barbara Metcalf, Muhammd Mujeeb, K.A. Nizami, Francis Robinson, Annemarie Schimmel, Sanjay Subramanyam, Andre Wink and many others listed in the bibliography.

Select excerpts

1.1 An early British view of a Muslim ruler

The writers of the Anglicist school attacked both Hindu and Muslim systems of government, but their primary target remained the Muslim ruler, since the form of government most familiar to them was Mughal. In his well known work, *The History of Hindostan*, published in London, 1772, the historian Alexander Dow, for example, castigates all Muslim government as wicked and primitive.

> The (Islamic) legislator . . . derived his success from the sword, more than from his eloquence and address. The tyranny which he established was of the most extensive kind. He enslaved the mind as well as the body. The abrupt argument of the sword brought conviction, when persuasion and delusion failed.
>
> (Chatterjee 1998: 162)

1.2 Irrational hatred of Muslims

A Hindu chauvinist, P.N. Oak, gained much notoriety with the publication of his book on the Taj Mahal, in which he asserts that the great monument was a Hindu temple and that the story of Emperor Shah Jahan's undying love for his wife Mumtaz Mahal is pure fiction. The following piece is an example of his style. Unfortunately, his ideas have in recent years gained traction among a large number of otherwise sensible Indians.

> Islam transformed all Arabs and other conquered peoples into illiterate brutes who perpetrated rape, ravage, massacre and plunder wherever they went . . . On studying Islamic history of India, I have

> concluded that 1) every Muslim is a descendant of a captured Hindu; 2) every mosque or mausoleum is a captured Hindu building; 3) at historic sites the construction is all Hindu and destruction all Muslim . . .
>
> (Oak 1969: xx)

1.3 Fairy tales about Urdu

Nearly twenty years ago the distinguished Pakistani historian, K.K. Aziz, wrote about the crisis of history teaching in Pakistani schools and colleges after surveying sixty-six textbooks on Social Studies, Pakistani Studies and History. He identified numerous errors of fact, emphasis and interpretation, often deliberately committed by respectable academics and professors, many of whom wished to be on the right side of their superiors in the government and the military. The effect of their texts on young minds can only be retrogressive. The extract below describes the glory of Urdu, as quoted from a widely used textbook.

> The special feature of Urdu is that it is spoken not only in every nook and corner of South Asia but people who know and understand it are found in the whole World . . . Gradually Urdu has developed to a stage where now it is considered one of the more developed languages of the world. Not only that, but next to Arabic, Urdu is the only language which has no equal in the world. The fact is that even English and French languages are losing their popularity and importance before the Urdu language.

Aziz describes the above as a fairy tale, and responds thus:

> Two claims deserve notice and then ridicule. First, Arabic as a language has no equal in the world (we are not told in what sense). Secondly, Urdu comes next in the order of this distinction. But the author . . . proceeds recklessly to enter another title on behalf of Urdu, and on his way in this hazardous journey gives us the great and heart-warming news that in the world of today Urdu is leaving English and French behind in popularity and importance. He should have gone the whole hog and told us that the British and the French are giving up their languages and adopting Urdu.
>
> (Aziz 1998: 39–40)

2

THE ERA OF ARAB PREDOMINANCE (*c.* 600 TO 1000 CE)

Forged early in pre-Islamic times, the Arab connection with India encompassed an ocean-going trade of global importance and a brief period of domination in Sind, a vital frontier region of India[1] (Map 2.1). Trade and power initially propelled the Arab advance, but their momentum slackened through a lack of clear strategy at the Abbasid court. After the tenth century, Arab political control of Sind was at an end, although Arab ships continued to ply the Indian Ocean lanes. It was the Arabs, however, who laid the foundations of a Muslim civilization in India and whose wisest minds engaged in a dialogue with the intellectuals of India.[2]

Early Arab commerce and settlement in India

The Arabs were no strangers to the Indian coast. Long before the rise of Islam, their traders and seafarers, along with the Persians, had been engaged in the Indian Ocean commerce. Numerous small ports, dotted along the coastlines of the Persian-Arabian Gulf and the Red Sea, Yemen and Hadhramaut, East Africa and Western India, were interconnected by *dhows* of varied sizes.[3] Along with mariners from different nations, the Arabs had a major share in sea transport. Indian commodities were in great demand throughout the Middle East, the Levant and Mediterranean lands; they included pepper, spices, ointments, perfumes, leathers, precious woods and stones, iron, dyes and medicines. India imported such goods as silks, copper, lead, paper, glass, horses and slaves. The Indians received gold and silver bullion that made up for the difference between the values of their exports and imports.[4]

The first Muslim traders and mariners in India were those same Arabs and Persians who had become Islamized. They were particularly attracted to the western Konkan coast on account of the valuable teakwood from the forests of the Hindu Rashtrakuta kingdom; this wood was in great demand in the Iraqi shipbuilding industry. Appreciating the profits to be made from this trade, the Rashtrakuta kings treated the Muslims honourably

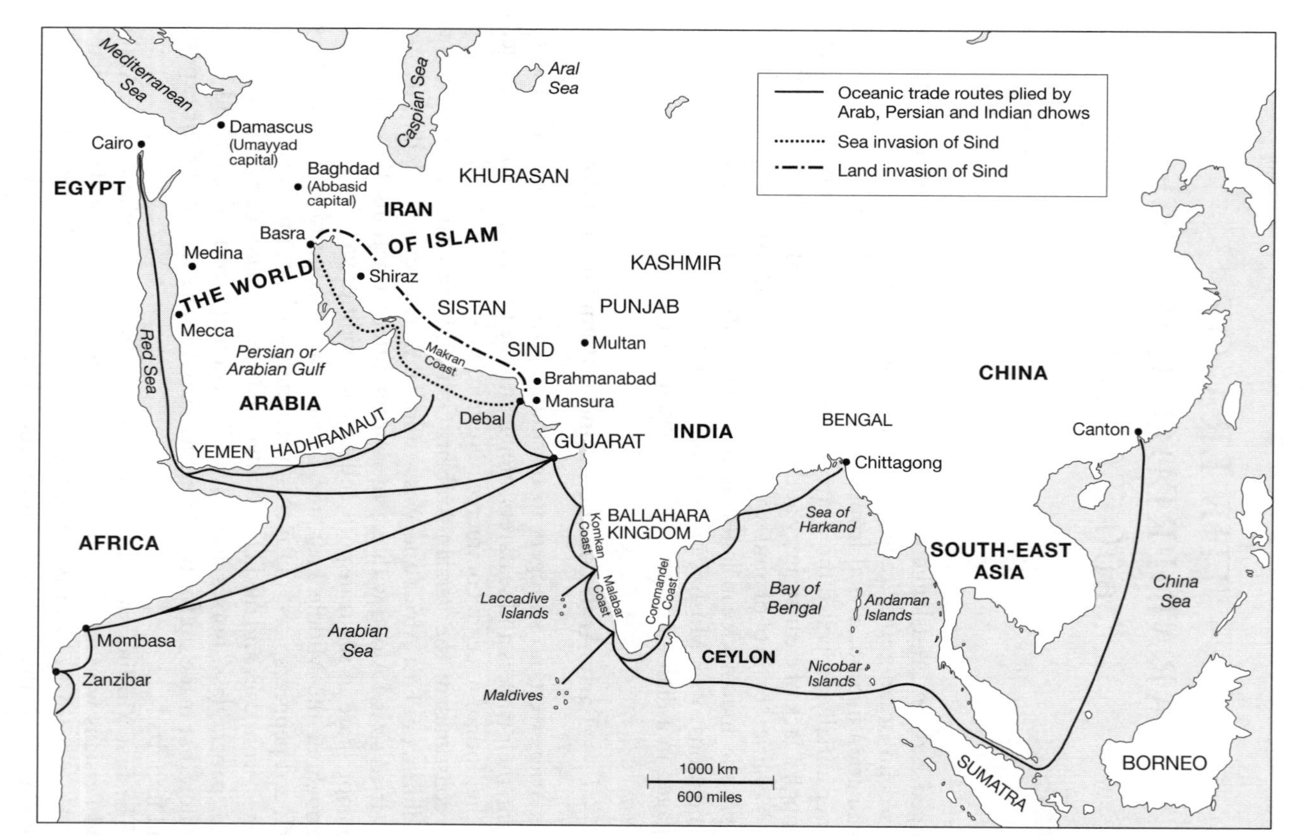

Map 2.1 South Asia, the Islamic Empire and the Indian Ocean: eighth to ninth centuries CE

and permitted them to build mosques. The Arabs called them the Ballahara kings; and, according to the great Muslim traveller and observer, al-Masudi, 'amongst the kings of Sind and Hind none treats the Muslims who are established in their domain with more distinction than the Ballahara'.[5] According to Masudi, the 10,000 or so Muslims who lived in the district of Saymur, were the *bayasira*, or 'Muslims born in Al-Hind of Muslim parents'.[6]

In the heyday of Arab/Muslim control of the Indian Ocean, a flourishing and sophisticated trade network developed in Malabar, the Keralan coast and Sri Lanka. Most interestingly, the Jewish merchants and brokers of the Middle East and the Levant were at the forefront of this Arab maritime enterprise. Although the history of Indian Jews goes back to pre-Christian times, it was in the era of the early Islamic Empire that they became prominent in Kerala.[7] The Muslims, both Arab and Persian, were also able to establish their respective coastal settlements. A large number of South Indian Muslims in particular trace their descent from the Hadramauti Arabs of south Arabia. The custom, popular among the Arabian tribes there, of arranging 'temporary marriages', or the *muta*, facilitated the increase in the Muslim population in Malabar, because many of the Arab sailors married the women from the marginalized caste of Hindu fisher-folk.[8] The offspring, although brought up as Muslims, stayed with their mothers in conformity with the matriarchy of Keralan society. The Mappillas ('the big children', in Malayali language), as they came to be known, remained a fairly invisible minority for a number of centuries, partly because they confined themselves to the coast and partly for caste reasons. Although Muslim, and theoretically out of caste structures, they could not escape the societal conventions of their Hindu neighbours. When, however, caste rules of purity and pollution became excessively rigid in Hindu social life in Kerala, to the extent that the Hindus shunned all maritime association, the Mappillas came to be valued for their specialization in maritime matters.[9] Through their intimate ties with Hindu kings, the Mappillas secured privileges and carved out their space in the wider society. They built mosques and tombs and followed customary rules of the Arab Shafiite school of law, but they also leant towards integration with the Hindus by speaking Malayali, dressing like the Hindu Nayars and following the matrilineal system of social organization.[10] The last Chera emperor of Kerala, Cheraman Perumal, who went on a pilgrimage to Mecca in 822 CE after converting to Islam and who died there, is revered among the Mappillas as the Zamorin, meaning the mariner.

On the Tamil/Coromandel coast, too, from the earliest periods of the eighth and the ninth centuries, an Arab-Islamic element had generally permeated into Tamil society. The common name for Tamil Muslims is *Ilappai* or Labbai, a corrupted form of the word *arabi*. The coastal Arabs were mostly engaged in maritime occupations; most of them were fishermen, but the more enterprising among them traded with Sri Lanka and South-East Asia, reaping great profits from dealings in gems and pearls. In due course the term 'Labbai' meant a jeweller or a merchant. Their wealth enabled them to build elegant mosques, strengthening their own perception of being the custodians of a pure Arabic Islam. Nevertheless, they were always prepared to accept the patronage of Tamil Hindu kings who protected them.[11]

The islands of the Laccadives and the Maldives in the central Indian Ocean lie across the routes from south Arabia to South India, Sri Lanka and the Far East. The Arab sailors recognized their importance, and a ninth-century Arab account refers to these islands as the *Dibayat*. While the Laccadivians are ethnically closer to the Malabar Mappillas, the Maldivians are related to the Sinhala people of Sri Lanka. The Laccadivians were originally Hindus with their own sense of caste hierarchy, following the traditional Keralan matrilineal kinship system. The Maldivians were Buddhists who had eschewed caste and followed a patrilineal system. Conversion to Islam among both groups started in the Arab period, although full scale Islamization was not achieved for many more centuries after that. Quite remarkable is the transcending of Arabic Islam and its Shafiite school of law across two completely different cultural patterns.[12]

The earliest Arab sailors and traders had also reached the river delta of Bengal, Chittagong and the Andaman and Nicobar Islands in the eastern Indian Ocean. Al-Masudi describes these places as part of *Al-Hind* which 'extends over land and sea'; the Arabs called the Bay of Bengal the 'sea of Harkand'.[13] Islam reached both mainland Bengal and the islands very early in the Arab period, and it continues to maintain its cultural and religious dominance among the peoples of these lands.

The Arab conquest of Sind

One of the objectives of Arab expansion was to secure effective naval control of Indian Ocean trade routes. A fleet sent by Caliph Umar in 637 to capture Thana, near Mumbai, failed,[14] but a bigger prize was secured a little later when, as a result of expeditions into Persia, the Makran coast of Baluchistan fell to his army. The province of Baluchistan straddled (as it still does) both the Persian and the Indian domains; and the control of its coast was invaluable for the Arab navy to patrol the sea lane that ran from the Persian Gulf to the Arabian Sea. Following this, during Caliph Uthman's reign, the Arab armies arrived on the western fringes of the River Indus delta, where they defeated the army of a local Hindu potentate, Raja Rasil, who then retreated to the river's east bank. On being informed about the forbidding and uncongenial nature of Makran geography and the ferocity of its people,[15] Uthman ultimately decreed that the west bank of the River Indus was to be the easternmost limit of his new Islamic empire.

The next phase of expansion in India took place in the reign of the Umayyad Caliph Walid I (705–15).[16] Among the many governors who served the Umayyads, none was more able or focused in his aims than al-Hajjaj, the governor of Iraq (694–714).[17] His authority also extended over the conquered Persian territories, including far flung places like the Makran coast. Walid I and al-Hajjaj took the decision in 711 to revoke Uthman's earlier decision and invade Sind, the part-fertile, semi-arid territory of the lower Indus basin and gateway to inner India or, as the Arabs called it, *al-Hind*. It was not an unprovoked attack, but it was still part of the Arab strategy of expansion. A conveniently plausible excuse was the protection of Arab trade in the Arabian Sea.

A particular danger that lurked along the Persian-Arabian Gulf route was that of piracy by a semi-nomadic and pastoralist ethnic group, the *Mids*,[18] whose boats lay concealed in the confusing criss-cross of waters in the Indus delta surrounding the main port of Debal (near modern Karachi). They presented a constant threat to a flourishing inter-regional trading network extending from Persia to Sri Lanka built up by Buddhist merchants of Sind,[19] and which was profitable for Arab and Persian traders. Before Islam, the Persians had taken steps to protect this valuable India trade; and the Arabs now did likewise.[20] As Governor al-Hajjaj saw it, piracy could only be ended by Arab invasion of Sind; and he found a perfect pretext in 711. When a group of Arabs settled in Sri Lanka died, the local king thought it prudent to send their widows and children back to Arabia, along with gifts and letters of good will for al-Hajjaj. Unfavourable winds, however, drove the ships carrying the survivors and the gifts close to Debal, where the pirates struck, taking the women and children captive and confiscating the gifts. Hajjaj asked the king of Sind, Dahar, to take action against the pirates and free the captives, but received only an evasive reply.[21] This, therefore, became the *casus belli* for him, and he successfully persuaded his master, Caliph Walid, to authorize a punitive expedition against Sind and to rescind Caliph Uthman's injunction to the Arabs not to proceed further than the west bank of the Indus Delta.[22]

With a force of 6,000 cavalry, 6,000 camel riders and a naval force with five catapults, the Arab commander, Muhammad bin Qasim, a seventeen-year-old nephew and son-in-law of al-Hajjaj, successfully took the port of Debal. In accordance with Hajjaj's instructions, Muhammad gave no quarter to the people of Debal, many of whom were massacred. The Buddhist stupa, or the *budd*, as the Arabs called it, was destroyed; and an area of the town was marked out for the settlement of all who already were Muslims or converting to Islam. A mosque was built, which must be considered as the first officially sanctioned mosque in the subcontinent. The women prisoners were freed; and, in accordance with Umayyad military custom, one-fifth of the booty, cash and slaves were sent to al-Hajjaj. Having fulfilled his brief, Muhammad could have concentrated his efforts on destroying the pirates; he went on instead to undertake further conquests inland. King Dahar had escaped to the town of Alor but, having been deserted by his mercenaries, lost the final battle. Gruesomely, his head and those of his other commanders, along with one-fifth of the booty and the Hindu royal princesses, were sent to al-Hajjaj as victory trophies.[23] With Dahar's death, it was relatively easy for Muhammad bin Qasim to capture the main cities of Alor, Brahmanabad and Multan in the far north.

Two factors may account for the Arab success. From a military point of view, the use of siege artillery, including a massive ballista, was highly effective in the initial attack on the port of Debal. A fleet of keel-less, nailed and bitumen-smeared boats also proved revolutionary.[24] And swift Arab cavalrymen, with baggage provision on camels, were more than a match against bulky Indian infantries supported by elephants. The other factor was the prevailing tension between Hindus and Buddhists in Sind. The Buddhists felt aggrieved that Dahar's father, Chach, had earlier usurped the throne from the Buddhist Rai dynasty; also, they and the Hindus had divergent socio-economic interests in Sind.[25] The Hindus were mostly villagers and farmers, while the Buddhists wielded financial control through their domination of the inter-regional and maritime trade. The Buddhists found little support from the Hindu political establishment. A group of Buddhist envoys had also been informed by no less a person than Governor al-Hajjaj himself that the Arabs intended to invade Sind 'up to the border of China'.[26] Since Buddhist traders had long had flourishing contacts with Central Asia and China, they could clearly anticipate the commercial advantages they might secure under the Arabs. It is quite likely therefore that some form of Buddhist collaboration with the Arabs may have begun even before the Arab invasion.

Arab rule in Sind

The Arab governors and soldiers who ruled Sind (now part of Pakistan) for nearly 300 years, from 711 to 1025, were the first Muslims to wield effective power and authority over a part of India. With very rare exceptions, none of the future Muslim rulers were Arabs; but they retained a certain consciousness of the importance of those early centuries, whose events came to be etched in the collective memories of the Muslim rulers and the ruled.

Shifts in Arab authority

Muhammad bin Qasim's military appetite grew with each victory, and he harboured ambitions to lead his forces into China. His plans, however, were quashed by the successive deaths, first of his immediate overlord, al-Hajjaj, and shortly thereafter of Caliph Walid in Damascus. The new Umayyad caliph, Suleiman (715–17), nursed grievances against Hajjaj and his extended family, including Muhammad bin Qasim. Muhammad was recalled by Suleiman and was then executed. There thus began a period of turmoil in Sind in the first half of the eighth century when various Hindu potentates alternately cooperated with and fought against the Arabs. The Umayyad caliphs attempted to expand the Arab dominion into Gujarat but, apart from destroying the fine Gujarati city of Vallabhi, decisively failed. They were reduced to merely consolidating their remaining hold on Sind.

A significant milestone in Islamic history was reached in 750 when, except for their continuing authority in Spain, the Umayyads ceded the caliphate to a rising power, the Abbasids. This was not just a reflection of inter-Arab ethnic tensions; there were far-reaching geo-political implications. The Abbasids were of course Arabs, but their leaders drew their strength from Yemen and Muslim Persia; their power base was at Baghdad. The Persian connection became crucially important with the transfer of the caliphate to that city. The Persian model of bureaucratic government, imitated from the earlier non-Islamic Sasanian rulers' administrations, replaced the more open and aristocratic leadership of the Umayyads.[27] The Abbasid Caliphate of Baghdad lasted for over 500 years until 1258, and then continued as a minor sideshow in Cairo for another three centuries. Its control over Sind, however, became increasingly tenuous, as local Arab chiefs, called *mutaghalliba* (those having seized power) increasingly exercised their autonomy by deliberately withholding annual tributes to Baghdad. By the end of the ninth century, the Abbasid caliph had lost all power in Sind, although his name was formally and ritually mentioned in the Friday *khutba* sermons at mosques in Sind.[28]

It was not long before another power shift occurred. Challenging the might of the Abbasids from the end of the ninth century onwards were a group of Yemeni Shia, called Ismailis, 'the infallible ones' (with reference to their imams), who helped to set up what came to be known as the Fatimid dynasty of Egypt. This dynasty, named after the Prophet's daughter Fatima, lasted until 1171. Egypt came under its control in 910 and eventually its power became extended across North Africa, Sicily, Palestine, Lebanon, Syria, the Red Sea coast of Africa, Yemen and Hejaz. The Fatimids too had their imam-caliphs, and their expansion was rapid under Imam-Caliph al-Muizz (953–75).[29] As the Fatimid Empire spread, its trading abilities increased greatly. The control of Egypt and the Red Sea coast meant that the Indian Ocean trade arena lay more in their hands rather than those of the Baghdad Abbasids. While Sind remained outside the immediate range of Fatimid maritime interest in the Indian Ocean, the commercial influence of Sindhi merchants and mariners in the Arab–Indian trade could not be ignored by the Fatimids, if they were to challenge the Abbasid maritime monopoly.[30] More important than trade, however, the Fatimid interest in Sind came to be represented by the Ismailis who, through secret societies and missions (*dawa*s), became influential in the cities of Multan and

Mansura. In the late tenth century, Multan particularly became the epicentre of their authority and the seat of Fatimid power in Sind.

A decisive shift of authority in Sind occurred when both Multan and Mansura, the two great Fatimid strongholds, fell to the armies of the Sunni Turco-Afghan warlord, Mahmud of Ghazna, in the years 1010 and 1025 CE respectively. Despite these setbacks the Ismailis, with their missionary zeal, continued to retain a cultural presence in Sind over many centuries. The end of Arab dominance in Sind is generally dated from 1025 when the region was incorporated into the Ghaznavid Empire of Mahmud and his successors. Over time, of course, the Arabs intermarried with indigenous Muslims and non-Muslims and immigrants from Afghanistan, Iran and Central Asia, becoming an indistinguishable part of the general population of Sind.

Administration and economy

The Arabs established a highly pragmatic administrative system both in their garrison towns and in civilian areas. The military forces, generally under Muslim command, were let loose on the populace only in the most extreme circumstances of rebellions or mass protests.[31] For the most part, civilian authorities, whether Hindu, Buddhist or local Muslim, carried on with traditional administrative and judicial processes and collecting taxes from both rural and urban people. At least a fifth of the proceeds had to be sent to the caliphal treasuries in Damascus or Baghdad,[32] while the rest was used for army upkeep, government machinery and the maintenance of law and order. Islamic interests were given a priority over those of other faiths, but there is no evidence of mass coercion. For the welfare of poor Muslims, the Islamic state relied on the mosque and the *zakat* donations, but the building of new mosques was a state project. The mosque at Debal, for example, 'contained the largest number of inscriptions so far known of all Arab mosques of the early caliphate'.[33]

The Arab rulers subdued the nomadic tribes and encouraged more cultivation of land.[34] The geography of Sind, however, hindered agriculture becoming the prime economic motivator for the rulers, because vast areas beyond the fertile Indus banks and delta were unsuitable for cultivation. It was trade that was the real source of large revenues for the state. Through Sind crossed three major land trade routes: from Khurasan in Iran to Multan, from Sistan, also in Iran, to Debal, and from Debal to Kashmir and beyond into Central Asia and China. Sugar cane, bamboo and other woods, camels and salt were some of the chief traded commodities, while slaves were the human exports. Slave raiding and slave trading occurred on a substantial scale; and the crushing of those nomadic tribes that offered resistance also made available a large supply of slaves for sale in the markets of the Umayyad and Abbasid empires, yielding much wealth for the Sind rulers.[35] Another source of wealth lay in maritime trade. The Buddhist merchants of Sind had, before the Arab intrusion, acquired a vast share of this trade, with their influence extending to China and the Malay Peninsula. As the overland route from the Levant to China was not always safe, the Chinese preferred to send their export goods first to Debal, from where they could be re-exported by sea.[36] This was a great money-earner. With the arrival of the Arabs the maritime trade further expanded and customs revenues were generated on a large scale. Urbanism received a boost from increasing trade, as attested by the building of the new city of Mansura.

Religious policy

The Islamic Empire aspired to spread Islam among the peoples of the world. In practice, however, this depended upon the circumstances within each invaded country and the attitude of the victorious Muslim commander or ruler. In the case of Sind, the religious policy was settled in the correspondence between Muhammad bin Qasim and his patron al-Hajjaj. Muhammad was guided by the latter's advice that was based on the Quranic interpretation of military policies towards conquered peoples. In his letter al-Hajjaj instructed thus: 'My ruling is given: Kill anyone belonging to the combatants; arrest their sons and daughters for hostages and imprison them. Whoever submits . . . grant them *aman* and settle their *amwal* as *dhimmah*'.[37] The term *aman* signifies protection, while *amwal* was the tribute that non-Muslims had to pay, of which the most significant were the *mal* (tax in general) and *jizya* (poll-tax). The *dhimmah/dhimmi*, were the People of the Book (i.e. Christians and Jews in the first place, then Zoroastrians and, for Muhammad bin Qasim, Hindus too). As far as Sind was concerned, the enemy was to be killed, not because he was a non-Muslim but because he was offering military resistance. Once his resistance was broken, he could become a Muslim or remain a non-Muslim. The choice was not between Islam and the sword, but Islam and *jizya*, as an eminent authority on Sind's history has reminded us.[38] The price of remaining a non-Muslim was to pay the *amwal* or *jizya*, which was like an extra *mal* that was supposed to be a fixed contract of honour, binding, and not to be lightly changed.[39]

Muhammad bin Qasim interpreted this ruling in a consistently liberal manner. Nearly 60 per cent of Arab success in Sind was secured through *aman* or *sulh* (treaty), much higher than in Egypt (8 per cent) or Syria (36 per cent).[40] Muhammad granted the *amans* most generously, but particularly to the defeated towns where the merchants, artisans and agriculturists were important elements in the population. He remained truly magnanimous in the application of Islamic injunctions. In the so-called Brahmanabad Settlement, agreed with the Hindus in that town, Muhammad followed al-Hajjaj's good advice on religious toleration (Excerpt 2.1); he permitted the Brahmans to continue as revenue officers in rural areas, as had been the custom in pre-Islamic Sind; and he withheld 3 per cent of the principal of the revenue assessment, for charity to Brahmans and Buddhist monks.[41] He also offered high positions to talented non-Muslims in his administration.[42] The Arab administration in Sind remained non-sectarian for a considerable time – a legacy of Muhammad bin Qasim's benevolence – and most of the later Arab rulers of Sind respected that legacy. Small wonder that a distinguished Indian historian, Mohammed Habib, has said: 'Alone among the many Muslim invaders of India Muhammad Qasim is a character of whom a conscientious Mussalman need not be ashamed'.[43] It has also been claimed that a great historical narrative, the *Fathnama/Chachnama*, which is of primary importance to the historians of Sind, might have been translated from Arabic into Persian in the early thirteenth century to remind the Islamic rulers of the Delhi Sultanate (established in 1206) of the standards set by Muhammad bin Qasim.[44] We shall examine this work at the end of the chapter.

Final decline of Buddhism in Sind

Before the Arab conquest, the Buddhists were in a majority in south Sind and in many urban centres. They were also a wealthy community, owing to their control of both the inland and maritime trade of Sind. Yet, by the end of the tenth century, there was hardly any trace of Buddhism left. It was as if the community had never existed. Three sets of explanations are usually given for this.[45] One theory is that the Buddhists might have migrated to other parts of India; but there is little evidence of population movement. Another explanation could be that the Buddhists were assimilated into the Hindu fold. Since Hinduism was strong in the villages, it is possible that rural Buddhists might have moved over to Hinduism in the face of Arab and Muslim advance. The weakness of this theory is that the majority of the Sind Buddhists belonged to the Hinayana tradition. This is the quintessential anti-Brahmanic tradition; and it is doubtful whether such Buddhists would have so easily given up their cherished beliefs. Finally, there is the theory that the Buddhists converted to Islam in massive numbers. At first sight this seems a curious theory, because Islam was more alien to the Buddhists than Hinduism. Strangely, however, it is a plausible theory.

The explanation may lie in the subtle way that economics can influence one's sense of identity.[46] As merchants and traders who had offered least resistance to Arab forces, the Buddhists had cherished high expectations of prosperity under Arab rule. And yet it was the Arabs and not them who were gaining commercially. The Buddhist monasteries, which were also like industrial units, were being superseded by Arab production centres and

Arab caravanserai. The Buddhists were indeed facing a covert form of discrimination by Arab merchants against non-Muslims. Feeling a sense of relative deprivation in these circumstances, they more than likely concluded that conversion to Islam was the most convenient and practical route to maintain their families' fortunes and well-being. It did not have to be an arduous process. Conversion to Islam was not exactly the same thing as Islamization.[47] While conversion is about following the not too onerous five key practices of Islam, along with a few customs including male circumcision and some external modes of behaviour like modesty in dress for women, Islamization is about a much more demanding mental and spiritual adherence to the Islamic belief systems and philosophy. By converting to Islam the urban Buddhists escaped the *jizya* and the petty restrictions and discriminations to which they could easily fall victim: that was what immediately mattered to them. The conversion was a steady protracted process rather than a single event. But eventually the urban Buddhists became urban Muslims. Then, as contributions to their monasteries shrank and these institutions fell into disuse and disrepair, there remained no further social structure for the survival of the Buddhists and Buddhism in Sind.

Hindu resilience

In contrast to the decay and decline of Buddhism, Hindu society in Sind survived mainly intact under Arab rule. In the towns the same pressures that applied to the Buddhists had affected the Hindus; and indeed a large number of urban Hindus found conversion a convenient route to follow. But the vast majority of Hindus lived not in towns but in rural heartlands. The main psychological strength of Hindu village society lay in closely knit networks of relationships marked out by caste gradations. The Arabs were disinclined to interfere in village life; and Muhammad bin Qasim had wisely left revenue collection in the hands of the Brahmans who controlled the strings of rural power. This tradition continued under his successors. Rural Hindu life in the villages was free from the vicissitudes felt by the urban Buddhists within a highly competitive arena. Hindus felt relatively less stressed than Buddhists. It is also worth bearing in mind that Arab interest in Sind was for long focused on the southern region. Apart from Multan, Arab interest in the north remained marginal and peripheral. In Multan there was a great Hindu sun-temple that withstood all local Islamic hostility and continued to attract Hindu pilgrims from far and wide, even after the iconoclastic Ghaznavids had captured the city – a certain indication of continuing Hindu influence.

Another interesting theory about the persistence of Hinduism has to do with caste and caste rules. Purity and pollution are the criteria by which caste lines and boundaries have been drawn in the Hindu world over a period of some five millennia. The Hindu religious literature, consisting of *Dharmashastras* and *Dharmasmritis*, clearly distinguishes which groups are pure and which ones polluted. The *Brahmans*, the *Kshatriyas* and the *Vaishyas* have to take special care not to be polluted by the impure *Shudras*. Foreigners who came to India were generally called the *mlecchas* or non-Aryan barbarians, and mixed-caste persons were referred to as *chandalas*. Both *mlecchas* and the *chandala*s were impure according to caste rules; and caste contamination had to be avoided.[48] However, if an upper-caste Hindu felt contaminated or polluted, he or she could be redeemed through performance of elaborate purification rites legitimized in the scriptures. This meant that there was already an established procedure within Hindu society for addressing issues such as the pollution caused by intermixture with a *mleccha* or a *chandala*. This procedure was re-visited in a Hindu legal text called the *Devala Smriti*, written significantly between *c.* 800 and 1000 CE in Arab-controlled Sind. Devala, the sage, laid down certain injunctions about how a re-purification ceremony can be performed for both men and women in circumstances of caste pollution caused by conversion or marriage outside the caste, even after a delay of twenty years. Re-purification became a flexible tool for Hindus to re-convert those who had been drawn to Islam. Whatever the extent of its influence might have been, the *Devala Smriti* helped to staunch the haemorrhage of the Hindu population in Sind.[49]

Religious functionaries among Muslims

The very first Muslims in Sind were Arab soldiers, merchants, governors and travellers. Their ranks were, in due course, swelled through the conversions mostly of Buddhists but also of Hindus. A commonly used method for acquiring information about Muslims is by examination of their *nisbahs* or names of attributions.[50] A *nisbah* is generally a geographical attribute, which not only tells us the place where the person comes from but can also reveal his occupation if that place is well known for a particular craft or skill. An Iraqi Arab whose name ended with al-Kufah, for example, was quite likely to be connected with the silk industry, because Kufah was a centre of silk manufacture. The Muslims of Sind, too, follow the practice of attaching the *nisbah* to their names. Historians of Sind differ as to the value of information gleaned from Sindhi *nisbahs*, but it is recognized that such *nisbahs* do help to identify the religious occupations of Sindhi Muslims. It is through the *nisbahs* that we learn about those persons and families who engaged in the study and transmission of the great Islamic traditions based on the Quran and the *hadith*. The so-called 'traditionists' of Sind in the late eighth and ninth centuries greatly increased their influence when their ranks were augmented by converted Buddhists. The discipline of studying the Quran and the *hadith* was quite similar to the study of the

Vinaya-Pitaka, the great Pali canon of Buddhism covering several traditions concerning the Buddha.[51] Both studies require of their followers a very high degree of intellectual acumen. Converted Buddhists possessing that level of cerebral rigour brought their own individual genius, rooted in their pre-Islamic past, to bear on the interpretation of Islamic traditions. From the ranks of the traditionists were drawn the great and the good of conventional Muslim society, such as the *qadis* (judges). Their judgments in disputes of varied nature were based on the work of scholars, theologians and philosophers of Islam who likewise were held in high esteem.

From the *nisbahs* we can also learn about the first Sufis who came to Sind. These mystics and wandering saints attempted to understand Hindu and Buddhist ideas and reconcile them to Islam. A legend says that the great Persian Sufi, Abu Yazid al-Bistami (d. 874), had been introduced to the Sufi science by a mystic of Sindhi origin, Abu Ali Sindi.[52] The eclectic philosophical ideas of a Hindu sect known as *Pashupata* bore great similarities to those espoused by the Sufis. The often unconventional behaviour of the Sufis had its parallel among the *Pashupata* followers; and sometimes both groups held eccentric or even illicit ideas.[53] The famous Arab mystic, al-Hallaj, who came to Sind in 905 and might have been influenced by eclectic Indian ideas drawn from Hinduism, was executed in Baghdad in 922, after he fell victim to the wrath of the more dogmatic traditionists.[54] Today, he is venerated at numerous Sufi shrines in Sind and Punjab (Excerpt 2.2).

Although the vast majority of Sindhi Muslims in the eighth and ninth centuries were Sunni, some *nisbah* names indicate Shia origins. As a minority, the Shia were present in Sind from the earliest period of Arab intrusion. As we have noted, from the mid-tenth century, the charismatic Shia Ismailis enjoyed a brief period of power and influence through their connections with the Fatimids. In Multan, which was under the Fatimid-Ismaili condominium, the Ismaili *dai* missionaries were highly proactive in their drive to bring Islam to as many village Hindus as possible. Their zeal for proselytization was markedly different to the Sunni reluctance to meddle in the religious life of the village Hindus. The Ismailis employed three particular methods to win over the Hindus.[55] First, they acted amidst great secrecy, which must have been appreciated by those who wished to convert. Second, their missionaries did not engage in the mêlée of mass conversions; rather, they approached individuals and set up a variety of programmes to suit the special needs of those wishing conversion. Third, they did not hurt the identity or self-esteem of the proselytes, as the latter were permitted to retain much of their original culture. The Ismaili missionary raised no objection to the converted Hindus following their age-old customs in such matters as food and dress. Consequently, the converted Hindus of Sind felt relaxed in their new faith. Of all the Muslims of South Asia today, the Ismaili Muslims show the greatest empathy with their Hindu past. They are very multicultural.[56]

Early Islamic thirst for knowledge

Abbasid Baghdad

Islam became the dominant ideology between Spain and Sind under the first four caliphs and the Umayyads, but Islamic civilization as a force for creativity and a centre of universal knowledge began with the Abbasid Caliphate of Baghdad. In the early centuries of this caliphate, the Arabs remained the major political and economic power, but in matters concerning culture and knowledge, they perforce had to acknowledge the abilities and contributions of non-Arabs and non-Muslims. Just as the modern USA attracts skilled people from across the world to its laboratories and institutes of science and engineering, the libraries and schools of learning in Abbasid Baghdad invited scholars from all over the then known world, providing them with facilities and hospitality to conduct research into a multitude of scientific disciplines. Caliph Harun al-Rashid (r. 786–809) established a great library for scholars from far and wide, the *Khizanat al-Hikmah*, the 'Storehouse of Wisdom'. His successor, Caliph Al-Mamun (r. 813–33) instituted in 830 an official translating institution, the *Bayt al-Hikmah*, on the pattern of the Persian intellectual centre of Jundishapur.[57] While Arabic remained the *lingua franca*, many of the works were also written in Persian, Greek or Syriac. Original works in these languages and others such as Sanskrit or Chinese were also translated and distributed for further study. Although the spirit of Islam remained an inspiring guide for all Muslim scientists, academics and knowledge-seekers, Islamic civilization encompassed a world broader than Islam. Every branch of knowledge – philosophy, spiritualism, cosmology, geography, geology, history, botany, mineralogy, physics, mathematics, astronomy, astrology, medicine, pharmacology, anatomy, optics, agriculture, irrigation, zoology – came under the scrutiny of scientists and scholars, Muslim and non-Muslim, Arab and non-Arab, of the Islamic civilization.[58]

Early travellers

Knowledge can be mutually exchanged. The first Muslims to seek knowledge of and about India were the geographers and travellers of this period. With Arab and Persian sailors having opened up the sea routes as far as China, and with most scientific knowledge translated from the Greek, Persian, Syriac and Indian treatises, the age of the curious and observant traveller had arrived. European historians have accredited the fifteenth and sixteenth centuries as the age of the great voyages and geographical explorations. This Eurocentric understanding of the maritime history of the world underestimates the ambitious scale and vision behind the journeys that the Muslim travellers undertook five or six centuries before Columbus and

Vasco da Gama.[59] As early as the middle of the ninth century a traveller, simply named Suleiman the Merchant, sailed from the Persian Gulf to India and further east. He has left us both popular and scientific accounts in his *Akhbar al-Sin* (travelogue on China) and *Akhbar al-Hind* (travelogue on India)[60] (Excerpt 2.3). A ninth-century traveller, Ibn Khordadbih, who wrote *Kitab ul Masalik wa-l Mamalik* or the 'Book of Roads and Kingdoms', provides us with much information about the classes of people in India.[61] The best known of the tenth-century travellers was al-Masudi, known as the 'Muslim Pliny',[62] whose travelogue *Muruj al-Zahab* or 'The Meadows of Gold' is still a highly readable piece of geographical description.[63] Speaking of himself, al-Masudi said that 'he traveled so far to the west (Morocco and Spain) that he forgot the east, and so far to the east (China) that he forgot the west'.[64]

On reading the works of these voyagers and wanderers, we are struck by their vivid observations on India. All were impressed by the immensity of the land which they record (Excerpt 2.4), providing useful information about the physical features, urban settlements and great rivers, along with their own itineraries. As Sind was Arab and Muslim-controlled, it was natural that they were much preoccupied with this region. The charting by them of the River Indus was relatively accurate, and they had a clear understanding of the locations of such centres as Multan, Mansura and Debal and their geographical relationship with places like Makran, Punjab and Rajasthan. Although naturally biased towards Islam, these early Muslim travellers admired the varied aspects of the Hindu religion, its emphasis on metempsychosis or transmigration of souls, and the generally mild nature and abstemious habits of the Hindus. They also correctly understood and noted the historical fact that the Chinese received Buddhism from India. They observed with great sensitivity the behaviour and conduct of Indian kings, both in peace and in war, and were greatly impressed by them. Their unanimous admiration was, however, reserved for the Rashtrakuta monarchs, whom they called the *Ballahara*.

Indian intellectual input

Whereas the travelogues of Muslim visitors to India are extant, no direct written evidence exists from Indian travellers to the Arabian lands. Scholarly evidence, however, does point to Indian intellectual influence in the institutes of Baghdad. From the beginnings of Islam, Muslims were interested in determining accurately the times of day and the direction of prayers; and Arab and Persian travellers needed to calculate longitudes for their extensive voyages. Astronomy and navigational geography were therefore fields of significant investigation by the scientists. While Persian astronomical knowledge from the Sasanian period, partly based on the Indian tradition, was available, a major advance was made during the reign of the Abbasid

Caliph al-Mansur (r. 754–75) with the arrival of a set of mathematical and astronomical texts from India, known as *siddhanta*s, which provided useful planetary tables and texts for the calculation of eclipses, zodiacal ascensions and other astronomical data of great use at prayer times or on the high seas.[65] An astronomer at al-Mansur's court, Muhammad bin Ibrahim al-Fazari (died 806) translated the Indian works and created a book called the *Sindhind*, the original of which is lost.[66] Through the *Sindhind* the scientists and mathematicians in Baghdad also came to learn about another Indian treasure, the so-called Hindu system of notation and numeration. Although outside India, as early as 662, Severus Sebokht, a Syrian bishop, had made a reference to Indian numerals,[67] it was in the eighth and the ninth centuries that they truly came to the attention of the intellectuals at Baghdad. Abu Jafar Muhammad bin Musa al-Khwarazmi (*c.* 800–*c.* 850), a mathematician at the court of Caliph al-Mamun, produced an arithmetic textbook which incorporated both the Indian decimal place-value idea and numerals.[68] His work, called *al-Jam wal-tafriq bi hisab-al-Hind* or 'Addition and Subtraction in Indian Arithmetic', survives in a thirteenth-century Latin manuscript called *Algoritmi de numero indorum* or 'The notation of Indian numbers'. Nearly 100 years later, in 952, another scholar, Abul Hasan Ahmad bin Ibrahim al-Uqlidisi, wrote a more comprehensive work on the subject, called *Kitab al-fusul fil-hisab al-hindi* or 'The Book of Chapters concerning Indian Arithmetic'.[69]

The spread of the Indian system of medicine, *Ayurveda*, within the early Islamic Empire, is well documented. Some of the Ayurvedic knowledge filtered through the Greek and Persian sources. Both Greece and Persia had, over the centuries, received Indian physicians with their ideas.[70] An important channel of communication was established through Muslim trading centres on the coastline of India. A large amount of Indian exports consisted of spices, dyes, drugs and fragrances; and both Arab and Persian traders would have noticed the skill with which the Indians used these commodities in Ayurvedic preparations. The control of Sind gave the Abbasid Caliphate a vital route to tap into Indian ideas and expertise.[71] Caliph Harun al-Rashid strongly encouraged the learning of the Ayurvedic corpus and arranged its translation into Arabic. His own physician was an Indian doctor who helped to translate the Sanskrit work *Susruta Samhita* into Arabic under the title of *Kilal-Samural-hind-i* of 'Susrud'. The translation activity was much expanded in the reign of the next caliph al-Mamun who set up at Baghdad the official translating institution, the *Bayt al-Hikmah*. The important Indian treatises on poisons were translated under the title *Shanaq*, while the work known as *Saharik-al-hindi* was a translation of the great Indian corpus of *Charaka Samhita*.[72] It is beyond speculation that all these works would have adorned the shelves of the great Baghdad library, the *Khizanat al-Hikmah*.

Arabic historical writing and the first important works of Indo-Muslim history

Among the varied forms of Islamic historiography that developed after the rise of Islam, historical narratives and biographical dictionaries constituted what is generally known as *tarikh*. Some of the greatest of Islamic historical works are *tarikhs*.[73] One of the early achievements of Muslim historiography was the idea of historical chronology, set out year by year, by month and even day to day. This was the first step towards a scientific approach to historical events. Part of the rationale for the narratives was to memorialize the great Arab conquests after the Prophet's death. The purpose of writing the accounts of conquests was not merely to rejoice in the Arab victories; its objective was also to understand the practical outcomes of their conquests. That meant making a detailed study of the affected areas of the world and understanding their histories and political economies for the ultimate benefit of the newly emerging Islamic empire.[74] The invasion of Sind could be considered a subject for both remembrance and study, thus meriting its own history. In Abbasid Baghdad of the ninth century, there lived at the court of Caliph al-Mutawakkil (r. 847–61) the Persian historian al-Baladhuri (died *c.* 892).[75] His encyclopaedic history, known as *Futuh al-Buldan* (The Book of Conquests), covers comprehensively the story of all the great early Islamic Arab victories, with evidence drawn from oral history, campaign reports, genealogies of rulers and information drawn from other writers. As al-Baladhuri had not gone to Sind to collect his evidence at first hand, an author on whom he relied for information for his chapter on the Sind campaign was an early Arab historian, al-Madaini (*c.* 753–840).[76]

A chapter on Sind in al-Baladhuri's *Futuh al-Buldan* is only a shadow of a more exhaustive work in Arabic on pre-Islamic Sind, the Arab conquest and also its aftermath: the *Chachnama*, originally known as the *Fathnama*. The title *Chachnama*, popular since the sixteenth century, refers to the story of Chach, the Brahman minister who usurped the throne of the Rai dynasty in 632, while the earlier title of *Fathnama*, meaning 'victory account', alludes to the Arab victory. The work's author remains uncertain, but there is strong conjecture that the author was in fact al-Madaini. Whoever wrote the *Fathnama*, the fact is that the work lay ignored for many centuries. In the year 1216, however, during the reign of Sultan Nasir ud-Din Qabacha, the ruler of Sind and South-West Punjab (r. 1205–28), a diligent scholar named Ali al-Kufi came across a copy of the Arabic manuscript which had been preserved with a well known Sindhi family which traced its descent from the earliest soldier-clans of the Arabs. Recognizing the significance of the manuscript, he translated it into Persian, the language of culture among the thirteenth-century Muslims of India. The Arabic account, according to al-Kufi, consisted of many acts of heroism and much advice to those in power. There was also an element of historical romance in the narrative; and al-Kufi used the power and beauty of language in Persian expressions to embellish the contents and give linguistic lustre to the *Fathnama*. Perhaps he meant his translation to be a memorial to what had been achieved in Sind by the Arabs and the early Muslims.[77]

Apart from a very detailed narrative commencing from the pre-Islamic Rai dynasty down to Muhammad bin Qasim's conquests and his generally benevolent rule, the *Fathnama* contains oblique nuggets of advice to a Muslim ruler, very much in the tradition of a great Indian manual of political theory: the *Arthashastra*.[78] The ruler should maintain a harmonious balance among all the various people who surround him: women, chiefs, ascetics, merchants, artisans, cultivators, soldiers, moneyed men, etc. None should be especially favoured or ignored. The ruler must show religious tolerance towards people of faith and must heed advice from those learned in faith. Battles must be fought hard, but the enemy must not be forced into a humiliating surrender; and the productive classes of the defeated side must be respected and nurtured. The chiefs of the defeated parties must accept the paramountcy of the Muslim ruler, but they should be allowed to retain their faith as long as they accepted the terms of payment of tributes, etc. The idea of having a wise counsellor giving advice to the ruler is also strongly inculcated.

Select excerpts

2.1 The Brahmanabad Settlement: an example of early Islamic liberalism

This is brought out in the exchange of correspondence between al-Hajjaj, the reputedly cruel Umayyad governor of Iraq, and Muhammad bin Qasim, the first Muslim invader of India.

> The Hindus said: 'In this country, the existence and prosperity of towns depends on the Brahmins, who are our learned men and philosophers. All our affairs, on occasions of mirth or mourning, are conducted and completed through their medium, and we agreed to pay the tax, and to subject ourselves to scorn, in the hope that everyone will be permitted to continue in his own religion. Now, our temples are lying desolate and in ruins, and we have no opportunity to worship our idols. We pray that our just commander will permit us to repair and construct our . . . temples . . .'
>
> Muhammd Kasim wrote about this to Hajjaj, and in a few days he received a reply, which ran as follows: '. . . With regard to the

> request of the chiefs of Brahmanabad about the building of . . . temples, and toleration to religious matters, I do not see (when they have done homage to us by placing their heads in the yoke of submission, and have undertaken to pay the fixed tribute for the Khalifah and guaranteed its payment) what further rights we have over them beyond the usual tax. Because after they have become dhimmis we have no right whatever to interfere with their lives or their property. Do therefore permit them to build the temples of those they worship. No one is prohibited from or punished for following his own religion, and let no one prevent them from doing so, so that they may live happy in their own homes'.
>
> (Fredunbeg 1900/1985 reprint: 168–9)

2.2 *The independent outlook of Sufis*

Conservative theologians, disturbed by eclectic ideas, have historically shown much harshness towards the Sufis. What happened to the mystic al-Hallaj is a case in point. Although a teacher in the royal household of the Abbasid caliph, he was executed for suggesting an alternative to pilgrimage to Mecca.

> If a man would go on pilgrimage and cannot, let him set apart in his house some square construction, to be touched by no unclean thing and let no one have access to it. When the day for the rites of pilgrimage comes, let him make his circuit round it, and perform all the same sacraments as he would perform at Mecca. Then let him gather thirty orphans for whom he has prepared the most exquisite feast he can get; let him bring them to his house, and serve them that feast; and after waiting on them himself, and washing their hands as a servant himself, let him present each of them with a new frock, and give them each seven dirhems. This will be a substitute for Pilgrimage.
>
> (Falconer 1991: 108–11)

Caliph Muqtadir signed the execution warrant after the judges and the *wazir* found that what he suggested was heretical.

2.3 *Travellers' fascination with unusual behaviour*

Early Arab travellers recorded many factual details concerning the places they visited, but they were also keen observers of lifestyles they were unfamiliar with. The following account from the travelogue of Suleiman the Merchant is an amusing example.

> In India there are persons who, in accordance with their profession, wander in the woods and mountains, and rarely communicate with

the rest of mankind . . . Some of them go about naked. Others stand naked with the face turned to the sun, having nothing on but a panther's skin. In my travels I saw a man in the position I have described; sixteen years afterwards I returned to that country and found him in the same posture. What astonished me was that he was not melted by the heat of the sun.

(Elliot & Dowson 1867/2001 reprint, Vol. 1: 6)

2.4 Al-Masudi's understanding of the extent of India

India is a vast country, having many seas and mountains, and borders on the empire of ez-Zanij (modern Sumatra-Java), which is the kingdom of the Maharaj, the king of the islands, whose dominions form the frontier between India and China, and are considered as part of India. The Hindu nation extends from the mountains of Khorasan, and of ez-Sind as far as et-Tubbet (Tibet). But there prevails a great difference of feelings, language and religion, in these empires; and they are frequently at war with each other.

(Sprenger 1851, Vol. 1: 176–8)

3

TURKISH POWER AND PERSIAN CULTURE IN THE AGE OF TRANSITION (1000–1206)

For nearly thirty years after 1000 CE, Mahmud, the Turco-Afghan sultan of Ghazna (now Ghazni) in eastern Afghanistan, mounted lightning raids on urban settlements of northern and north-western India on seventeen different occasions. The whirlwind rapidity with which his swift horsemen arrived at the gates of palaces, forts and temples and executed their plan of loot and destruction of sacred Hindu, Buddhist or Jain sites and the massacres of those to whom they took a dislike, was a terrifying experience for the sedentary Indian populace of traders, artisans, bureaucrats and priests. In the absence of any form of international law there could be no redress from a higher authority that would conceivably restrain this marauding ruler. The Indians accepted their fate with stoicism but their hearts were filled with fear of the Turk.

Mahmud's death in 1030 brought only a temporary respite for the Indians. On a smaller scale, and less frequently, the raids continued under his Ghaznavid successors who firmly planted themselves in Punjab and tenuously in Sind. In 1186 the Ghaznavids lost their Indian lands to another Turco-Afghan dynasty, the Ghurids; but, by then, many Hindu rulers of North India had suffered a permanent loss of their lands and influence. From Punjab to Bengal, the vast swathe of Indian territory was run from Ghazna, not from an Indian metropolis. Only when, in 1211, Delhi was proclaimed by the 'slave-king', Shams ud-Din Iltutmish, as the capital of his new Indian sultanate that the affairs of North India again came to be conducted from within India. The sultanate, however, remained in Turkish hands; and it would take some centuries before it was to become Indianized. Moreover, the official faith protected by the sultanate was to be Islam, not Hinduism. Under the Ghaznavids and Ghurids the Muslim population, hitherto confined to Sind, Gujarat and the south, expanded in the north too. By the time Iltutmish died in 1236, a new Muslim empire in North India was secured.

We might, with some justification, perceive the Ghaznavid and Ghurid monarchs as aggressive bandits and barbarians, but such an overly Indo-centric view needs to be tempered by some empathy towards them. Originally descended from the Central Asian Turkish slaves, theirs had been a culture of nomadism; raiding their neighbours' lands, far and near, was for them almost a natural order of things.[1] The rapacious instinct for conquest and control of land remained with them well after their arrival into the new habitats of Persia and Afghanistan. They carried it into India too. But conquests and aggression are only part of the story. Through their migration into Persia and Afghanistan, the Central Asian Turks had come to imbibe both Persian culture and the Islamic religion. The power of the Ghaznavids and Ghurids was exercised within the cultural milieu of Islamic Iran. Turkish power in India spread the influence of Persian rather than Turkish culture; and it was to be some centuries before Turkish culture could rival the Persian. At the same time, with the decline of the Abbasid Empire, there began the diminution in the influence of Arabic culture; and, while Arabic remained for all Muslims the primary language of their faith, it was the Persian style that triumphed in India in the areas of government, literature and art.[2] By being the harbingers of Persian culture into India, the Ghaznavids and Ghurids provided an exciting stimulus to the multi-cultural society of India.

Ghaznavids and Ghurids: their wars and conquests

After their initial conquest of Sind in the early eighth century, the Arabs lacked men and resources to extend their dominion further into India. Although they continued to wield power in Sind, that region remained the limit of Muslim authority in India until the early eleventh century. From then on, however, Muslim power and presence in India continued to accelerate for many centuries under the tutelage and might of the Central Asian Turco-Afghan dynasties. The first of these dynasties was that of the Ghaznavids, whose style of rule can be appreciated first by exploring the background of the Turks within the eastern Iranian province of Khurasan[3] and the nearby region of Transoxania traversed by the River Oxus (now known as Amu Darya).[4]

Military slavery and the Turks

The institution of slavery was a well entrenched feature of Arab society. Islam regularized it but did not ban it. Most domestic slaves were brought from Africa, but the Turkish slaves from Central Asia proved to be an ideal military resource for the Islamic rulers. The caliphs required them as bodyguards or mercenary soldiers. After invading pre-Islamic Iran in the mid-seventh century, the Arab armies had penetrated deep into the adjoining lands to the north-east in Central Asia, in the ancestral homeland of the various ethnic groups of Turks. The Turks of Central Asia were a people familiar to the Persians of Khurasan and the Transoxanians.[5] Although basically a nomadic people they were divided into many ethnic groups. These groups entered the Iranian zone quite often as nomadic migrants, through the pressure from other Central Asian people like the Mongols. Both Mongols and Turks vied for grazing land on the vast plains of Inner Asia: and whenever the surges from the Mongols to their east accelerated, the Turks pushed towards the Iranian borders and northern Afghanistan, most often with aggressive intent. Nomadism even led some of them into being itinerant traders and skilled workers; and, for centuries, there continued a sustained trade in various commodities across Iranian and Transoxanian frontiers.[6] Some Turkish traders sold their own lowly and dispossessed fellow Turks as slaves to the Persians and the Arabs. Turkish migration and slave trade both went hand in hand. While Turkish slaves never made good domestic servants, their soldiering qualities soon came to be appreciated by powerful families in Khurasan and Transoxania. The slaves with true military prowess, called *ghulams* or *mamluks*, were sought after by all who needed to protect their persons, lands or titles. The higher the status of the families, the higher the corresponding status of the military slave. A great honour for a Turkish slave was to serve as bodyguard or retainer in a royal family or as a soldier in that family's army;[7] and an even greater honour was to become a slave-commander in the royal Islamic armies. Brought up in the challenging geographical environment of Central Asia, the Turks had earned a reputation for three qualities: their prodigious strength, their skilled horsemanship and their intense personal and tribal loyalty, all of which could be useful to any monarch. Additionally, the Turkish slaves soon embraced Islam and, in time, they became defenders of Islam as ardent as the Arabs had been earlier. Throughout the Islamic world, the ruling dynasties demanded their share of the Turkish military slaves. The unquestioned loyalty of these soldiers was generally both expected and secured. Inevitably there were occasions when either the adverse circumstances of their masters or their exceptional abilities tempted the highest ranking slaves to turn upon their masters and seize power. Such indeed were the beginnings and rise of the Ghaznavids.

The Ghaznavids: 977 to 1186

By the tenth century, the weaknesses of the Abbasid Empire had led to the proliferation of smaller kingdoms in the Iranian world which, while showing nominal respect to the caliph in Baghdad, became politically autonomous. The Samanid kingdom, with its capital at Bukhara, was one such autonomous unit that extended across Khurasan, Transoxania and Afghanistan.[8] The Samanids maintained a benign level of law and order, without recourse to an accustomed oppression of subject peoples, but they too recruited Turkish slave-soldiers into their army, and on whom they relied for protection. One of these Turks, Alptigin, while paying homage to the Samanid rulers, had seized control of the eastern Afghan city of Ghazna. A few years after his death, in 977, yet another Turkish slave-commander, Sabuktigin, seized control of Ghazna and ruled over it until 997. Most of Afghanistan by now was in the hands of the Muslims, but a large slab of territory between Kabul and the land of five rivers – the Punjab – was controlled by the Indian Hindu Shahi dynasty.[9] By 991, Sabuktigin had destroyed their authority west of the River Indus. On his death in 997, his son Mahmud took control of not just the Ghazna throne but of the rich province of Khurasan and all Afghanistan not controlled by the Hindu Shahi dynasty.

An experienced warrior who had gained his spurs in his father's military campaigns, Mahmud now began his infamous raids into India. While keeping a watchful eye on his Khurasanian possessions, his attention was focused on India for two basic reasons: first, the opportunity to loot its wealth; second, the popularity to be gained along with honours and accolades in the Islamic world by desecrating Hindu temples and deities. Mahmud did not hate the Hindus, but once he became aware of their military weakness, he could not resist the temptation to invade repeatedly. He campaigned in different regions for different purposes. The first principal target, Punjab, the last bastion of the Hindu Shahi kingdom, was taken in 1026; and it eventually became the headquarters of the Ghaznavid Empire in India.[10] The second zone that Mahmud attacked was Sind itself, which was anyway in Muslim hands. As a Sunni, however, Mahmud resented the dominant Shia Ismaili influence there (Chapter 2); and he viewed an attack on the Ismailis as a way of gaining favours from the Abbasid caliph.[11] Then he went further east and attacked wealthy cities like Mathura and Kanauj in the Gangetic Doab (the fertile land between the rivers Ganges and Yamuna), the heartland of Hinduism. Finally, there was Gujarat, where the great Hindu temple of Somanatha, with its hoard of treasures, aroused Mahmud's avarice. The Gujarat campaign of 1025–6 was perhaps the riskiest that Mahmud undertook; and although, in the long run, it made him the most reviled Muslim in Indian history, in the short term it gained untold riches for Ghazna.[12]

Within ten years of Mahmud's death, his son and successor, Masud, had lost the prime Iranian estate of Khurasan to a new and powerful Turkish group, the Seljuqs, who defeated him at the famous Battle of Dandanquan in 1040. The Ghaznavids retained Ghazna in Afghanistan and Lahore in Punjab; even without Khurasan, they remained a formidable power for 100 years and more. In the mid-twelfth century, however, they faced an internal challenge from the Shansabani princes of Ghur who, although under their sway, were gaining strength while they and the Seljuqs were at strife. The Ghaznavid king, Bahram Shah, could not prevent the wanton destruction of Ghazna and its monumental buildings constructed from the ill-gotten wealth from India. For seven days in 1151, the Ghurid leader, Ala ud-Din Husain, torched the city; thousands were put to death and women and children carried off into slavery. Precious literary treasures stored in the libraries that Mahmud had been so proud of were destroyed. The loss of all the works of the Iranian philosopher and physician, Ibn Sina, was certainly a major blow to the world of culture. The only notable structure that escaped the wrath of Ala ud-Din was the tomb of Mahmud himself.[13] By 1160 the Ghaznavids had lost their capital of Ghazna and moved on to Lahore. But Lahore, too, was lost in 1186 and, with it, whatever Indian territories they had held.

The Ghurids: 1173 to 1206

The Ghurids were a power to be reckoned with for barely half a century but, in relation to India, their achievement was substantial and permanent. Possession of Punjab and Sind was all that Mahmud of Ghazna and his successors had secured territorially, while the Ghurid influence extended right up to Bengal (Map 3.1). The Ghurid conquests of India were to form the core lands of the future Sultanate of Delhi.

To understand the Ghurid background, we have to begin once again with Afghanistan and the eastern Iranian world. The local people of the Ghur mountains which are located between Ghazna and Herat were ruled by the Shansabani family whose members were Tajiks, not Turks.[14] The Turkish presence, however, could not be avoided because, as with all rulers in the Islamic Middle East, the Shansabanis also recruited Turkish slave-soldiers. These slave-soldiers, at the vanguard of Ghurid power, were to become the shapers of India's future destiny. Many of the Shansabani princes married Turkish slave-girls or possessed them as concubines. A notable admixture of Tajik, Persian, Turkish and indigenous Afghan ethnicities therefore characterized the Shansabanis. The unity of this family was remarkable; with deep family loyalty and mutual respect, the Ghurids avoided debilitating fratricidal feuds that broke out in other dynasties. They operated a system of *appanage*, which meant that every prince was assigned jurisdiction over a territory where he could exercise personal power and which he could bequeath to his successor, without any contestation from other family members.[15] At the same time, the head of the family, whose seat was at Firuzkoh (most probably modern Jam, 200 km east of Herat), was unquestionably obeyed by all the minors. It was this strong family allegiance that had led Ala ud-Din Husain to raze Ghazna in 1151 in revenge for his brother's death, ordered by Bahram Shah, the Ghaznavid king. The same family ties brought Ala ud-Din's nephew, Ghiyas ud-Din, to the Ghurid throne at Firuzkoh in 1163. The family unity was further strengthened when, in 1173,

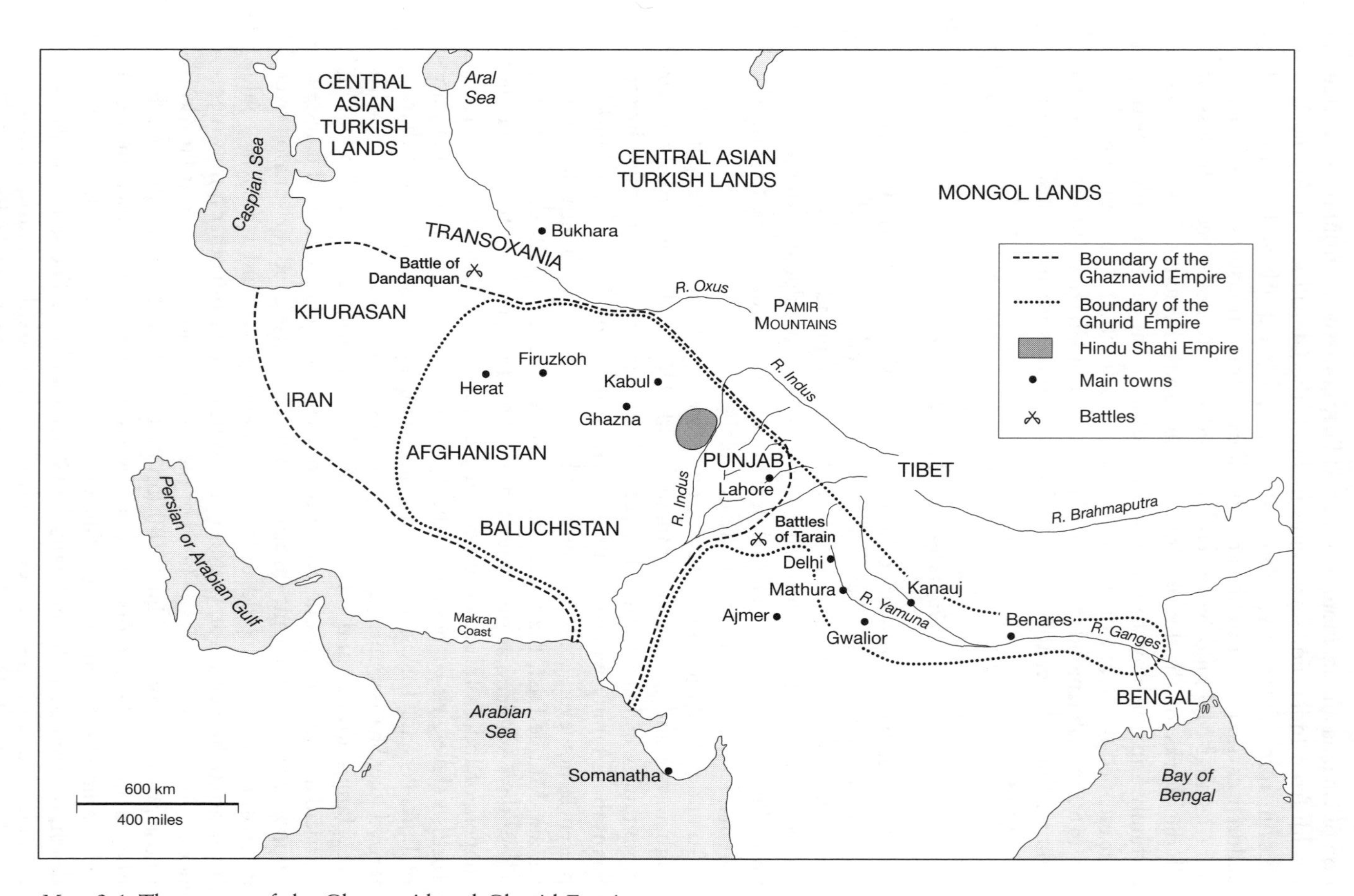

Map 3.1 The extent of the Ghaznavid and Ghurid Empires

Ghiyas ud-Din's younger brother, Shihab ud-Din (also known in history as Muizz ud-Din or Muhammad Ghuri), ejected the Seljuqs who had occupied Ghazna since 1160. Following customary practices, Ghiyas ud-Din gave away Ghazna along with Indian territories controlled from that city to this brother. Thenceforth, neither Ghiyas ud-Din nor any other Shansabani prince was to interfere in Shihab ud-Din's Ghazna; by the same token Shihab ud-Din unconditionally accepted the seniority and the continuing advice of his elder brother who remained head of the family.[16] Since Ghazna was the centre from where all raids into India had been initiated, and since the Ghaznavids still controlled Punjab from Lahore, it became Shihab ud-Din's responsibility to extend the Ghurid sway into India. In the context of all his Indian conquests, Shihab ud-Din/Muizz ud-Din is best known in Indian history as Muhammad Ghuri: and that is the name by which we shall call him henceforth.

For over three decades, Muhammad Ghuri fought hard to make his fiefdom a power to reckon with. Until the death in 1203 of his overlord and elder brother, Ghiyas ud-Din, Muhammad's principal area of operations was India. There he was served well by his loyal Turkish slave-commanders, the most formidable of whom was Qutb ud-Din Aybeg. His successes in India were also Aybeg's successes. By 1186, they had succeeded in ejecting the Ghaznavids from Punjab. Lahore, the prized capital, fell to the Ghurids. Next, it was the turn of the Indian kingdom of the Chauhans/Chahamanas whose ruler, Prithviraj Chauhan (sometimes called Rai Pithora), was defeated at the two Battles of Tarain 1191 and 1192. The Turks stormed the great cities like Kanauj, Benares and Gwalior. Each success led to further invasions. Delhi fell in 1193, the Upper Ganges district and Rajasthan in 1198 and the Chandella kingdom in 1202–3.[17] Aybeg's deputy, Muhammad bin Bakhtiyar Khalji, carried the Ghurid sword into Awadh, Bihar and Bengal. However, his senseless destruction of Nalanda Buddhist University in Bihar, one of the leading intellectual centres of Asia at that time, left a shameful blot on the record of the Ghurids and the early Turco-Afghan invaders. In 1204, the Sena kingdom of Bengal came to an end.[18]

Having established a definitive supremacy over North India, the Ghurids were now to face two major reverses which were to determine the shape of future events. A new Turkish power, the Khwarazm Shahs, in alliance with the Qara Khitai Turks, had dealt a blow to the Seljuq Empire in Iran; and after the death of his elder brother Ghiyas ud-Din in 1203, Muhammad Ghuri had to distract himself from Indian affairs. In 1205, he suffered a major defeat at the hands of the Qara Khitai.[19] Returning to Ghazna on 15 March 1206, he was assassinated, probably by a religious dissident.[20] Aybeg and the other slave-commanders were now the supreme power in India. The same Muhammad bin Bakhtiyar Khalji, the conqueror of Bengal, overreached himself by launching a major invasion of Tibet in 1206, but he and his army were decisively defeated.[21] This second Ghurid defeat was to mark out the northernmost limits of the new Muslim dominion in India. In the west, there was little prospect of tangible links with Afghanistan or eastern Iran because the Khwarazm Shahs, followed by the Mongols, were shortly to overwhelm those regions. Muhammad Ghuri's slave-commanders had now no choice but to make the best of their extensive Indian possessions.

Turkish military success in India

In less than thirty years the Ghurids had overwhelmed the various Indian kingdoms and extended their sway into all of North India. Yet, at first, there was nothing inevitable about their victories. The Indian armies were formidable and, on several occasions, proved their mettle against the Turks. Much earlier, Mahmud of Ghazna had himself twice failed to storm the great fort of Lohkot in Kashmir, and Muhammad Ghuri's expedition into Gujarat ended disastrously. Hindu and Rajput soldiers solemnly swore loyalty to their kings out of sincere religious conviction, and they generally fought to the death. The Muslim instinct to fight for the *jihad* was equally matched in battle by the Rajput code of honour. And on many occasions, when certain defeat loomed, the Rajputs preferred individual or mass suicide rather than surrender. Indian kings were adequately provided with various accoutrements and military materials. They deployed thousands of elephants protected by chain mail; and, from the capacious *howdah*s constructed on these elephants, the archers wrought general havoc on the enemy. Again, the supply lines of Indian armies were never really stretched because they were closely defending their forts and were near to sources of foodstuffs and other goods and procurements.

It was essentially the horse that brought victory to the Turks in India.[22] Although the chariot is frequently mentioned in the earlier pages of Hindu history, Indian horsemanship was, at best, inferior; and the Indian military machine, while thoroughly accustomed to the effective use of elephants, had a poor grasp of the deployment of the horse on the battlefield. The Turks not only possessed the best Ferghana horses, but knew exactly how to manage them. Their slave-soldiers were trained from childhood in mounted archery. Not an easy skill to master, it was to prove invaluable, for the mounted archers were more mobile than any numbers of infantry and therefore more effective militarily. The Turkish slave-commanders employed both the swift light cavalry, favoured by the early Arab invaders in the armies of Islam, and the heavy cavalry where mounted archers were effectively protected by coats of mail and other heavy armour. The careful juxtaposition of both types of cavalry on the battlefield and the precise ways in which the intensity of charges were adjudged, were all part of the Turkish military stratagem. On an Indian battlefield, the Turkish horsemen's first instinct was to disorient the enemy's phalanx of elephants; their infantry could then complete the manoeuvre. The Turkish infantry consisted of soldiers from various nationalities in the eastern Iranian world. Besides the Central Asian Turks themselves there were other militarized people from the Khurasani, Tajik, Arab and Afghan populations, all accustomed to horsemanship and archery.[23] A large number of Islamic volunteers, or *ghazis*, also formed part of the infantry: they enlisted for a variety of motives, particularly plunder, but were never permitted to be in the infantry vanguard.[24] That honour was reserved for the highly professional Turkish slave-soldiers. The Turkish commanders also made elaborate preparations for setting up supply depots at regular distances along the route of their expeditions.

With victory secured in India, Muhammad Ghuri's slave-commanders began combining the best of the military practices of both the Ghaznavids and the Indians. In all spheres of military life and activity – the administrative rules, the procurement of armaments, the training of cavalry and infantry, the use of horses and elephants, the recruitment of soldiers, the knowledge and use of new weaponry from foreign nations – the Ghaznavid and Indian traditions ultimately converged under the great sultans of Delhi.

The Ghaznavid model of authority

Authoritarianism

The administrative prototype under the Delhi sultans after 1211 was first tested by the Ghaznavids in Punjab – the only part of India that they had managed to fully incorporate within their empire. The governing principles of this administration were based on the Persian traditions of government and power that began with the pre-Islamic Sasanians and was continued by the Samanids.[25] A genre of Persian literature, known as *adab* (loosely translated as 'manners'), contained manuals of behaviour and protocols for all those involved in affairs of state, whether monarchs, *wazirs*, bureaucrats, army commanders or humble clerks.[26] The *adab* texts are collectively known as the 'Mirrors for Princes'; and, among the most famous of these texts, was the *Siyasatnama*, compiled at the end of the eleventh century by Nizam ul-Mulk, the great *wazir* of the Seljuq Sultan Malikshah.[27] In it, he portrays an idealistic vision of a model king based on the heroes of ancient Persia, and offers hard-headed advice for running an effective public administration. Good governance, in his view, was only possible when there was complete order in society. The king had to stamp his authority, transcend all social differences, and conduct himself righteously with decorum, honour and probity. For Nizam ul-Mulk, even the tyranny of a monarch was more acceptable than anarchy or disorder.[28]

The Ghaznavid king interacted with numerous subordinates. Among his five great ministers of state (for finance, war, correspondence, secret service and household departments) the two most important were the chief financial adviser to the king (*wazir*), and the head of the secretariat for correspondence (*Diwan-i Risalat*).[29] In the uneasy relationship within the system of monarchical autocracy, the *wazir's* position was frequently unsafe; he could fall foul of a capricious or unpredictable monarch, leading to his utter ruination or even death. Nizam ul-Mulk therefore advised all his peers to ensure they had specific contracts (*muwadaa*) of duty drawn up upon their appointment by the king.[30] The *muwadaa* provided greater legal protection to both parties than an undefined master–servant paternalist relationship. The *Diwan-i Risalat* oversaw all documents containing information ranging from the highest affairs of state to items of political, economic and financial information of varying value. An official working for the *Diwan-i Risalat* had to be thoroughly competent in Arabic and Persian and be able to prepare simultaneous translations of documents.[31] A hierarchical chain of command ran through each department of state, both at Ghazna and in the provinces. In Lahore, the capital of Punjab, the administrative system was more or less replicated by the Ghaznavids and maintained by the Ghurids.

The effective links between the metropolis and provinces were maintained by an efficient postal system (*barid*).[32] The Achaemenid rulers of Iran (550 to 330 BCE) had first pioneered the postal relay system for the purpose of effective communications to and from the governors (*satraps*) in the distant parts of the Persian Empire. The Samanids and the Ghaznavids continued this royal Persian tradition, more for the benefit of the state rather than the public. Along the postal routes were stationed postmasters (*sahib barids*), one of whose tasks was to report on disaffection or rebellion on the part of a provincial governor. A group of inspectors (*mushrifs*) ensured that no wrongdoing took place against the king. The *sahib barid* and the *mushrif* were the twin pillars of the highly sophisticated Ghaznavid espionage system which Nizam ul-Mulk thoroughly endorsed (Excerpt 3.1).

Financial administration

The heart of the Ghaznavid Empire had been in Khurasan and most of Afghanistan, and its financial machinery was originally perfected in the context of those lands. Indian lands were at first only a marginal possession. However, with the loss of Khurasan to the Seljuqs at the battle of Dandanquan in 1040, the Ghaznavids fell back on their Afghan lands and Punjab in India. Punjab, being almost as rich as Khurasan, was carefully nurtured, and its wealth helped maintain the Ghaznavids economically for the next 146 years, until they finally lost out to the Ghurids in 1186. Punjab was the sole patrimony that the Ghaznavids enjoyed during the last 26 years of the dynasty (from 1160 to 1186). The Ghurids, who occupied Punjab in 1186, made no particular changes to the Ghaznavid financial administration. In the thirteenth and fourteenth centuries, the Delhi sultans developed their own ideas on finance; but some of the key Ghaznavid principles of the eleventh and twelfth centuries remained effective. A study of the Ghaznavid financial administration is therefore essential to understanding the system that operated in later centuries.

There were three major heads of expenditure. First, there were the incessant wars requiring continual purchase, training and upkeep of slave-soldiers, along with all other necessary materials. Despite heavy cost, the Ghaznavids retained a strictly professional army that was exclusively loyal to them. Second, there was the luxurious lifestyle of the royal court and the nobility. Hundreds of cultured, and not so cultured, retainers and hangers-on, along with a retinue of nobles and commanders, were kept content with lavish gifts of lands or bullion. Third, there was the expense of a grand building programme in both Ghazna and Lahore. The revenue for such lavish expense was derived from many sources, through skilful application of strict procedures. The accountability was secured by means of punishments that we would now consider ferocious and excessive, but they were the norms then.

The first source of revenue lay in the produce of the monarch's own extensive estates that included the best of the agricultural and pastoral lands.[33] Since the Ghaznavids began their careers as slave-commanders of the Samanids, they would have received the privileges of revenue assignment (*iqta*). The *iqta* privileges were, in theory, strictly limited; but since the Ghaznavids eventually displaced the Samanids from power they became the sole owners of their lands, with the right of granting the *iqtas* to their subordinates.[34] A commander or a favoured noble receiving an *iqta* grant from the king had to temper his pleasure with the knowledge that it could be revoked at the king's whim. Generally, the escheat or the return of all property, including land, to the king occurred when a lord died without issue. Through a variety of escheats, the Ghaznavid kings continually replenished their treasuries. Further revenues arrived in the form of tributes and presents from lesser kings. From within India the most valued presents were gold and silver bullion, elephants and, curiously enough, dyestuff indigo, a product in great demand in the Islamic world. Hundreds and thousands of Indian slaves were a huge pool of cheap labour, another major economic resource for the Ghaznavids. Although the Turco-Persian historians of those centuries gave exaggerated figures of the bullion looted from India, it was nevertheless substantial in quantity and value. Another source of revenue was taxation, which included the *kharaj*, a tax on land and agricultural produce, the *jizya*, and assorted levies and tolls.[35] It was a harsh tax regime that the Ghaznavids operated. Each tax collector was charged with collecting current tax as well as the arrears; and each was made fully aware that no mercy would be shown him if he failed to collect his quota. The peasants, therefore, suffered great hardships, and very often they simply fled their land. After the Battle of Dandanquan in 1040, the people of Khurasan gladly swapped Ghaznavid rule for the Seljuq, partly because of the punitive taxation imposed by the former.[36] Owing to the military administration being in charge in Punjab, the burden of taxation would have been higher for the Indians.

Zealotry and realism in religious policy

Nearly all the Ghaznavids, Ghurids and Muslim Turks were Sunnis. The caliphate, first that of the Umayyads and then of the Abbasids, was the legitimization of Sunni doctrinal authority. The Shia, who had supported the claims of Ali, the son-in-law of Prophet Muhammad, in the seventh century CE, were considered dissidents. Among them were those like the Ismailis who actively sought the destruction of the caliphate. They achieved a measure of success in the tenth century with the establishment of the Fatimid dynasty of Egypt and North Africa, heavily influenced by Ismaili ideology.[37] Some of the Persian dynasties like the Buyids and the Dailamis supported Shiism out of a feeling of resurgent Persian nationalism.[38] With their political authority on the decline, it was therefore important for the Abbasid Caliphate to position itself as the premier authority in the Sunni world. The important Persian power inside the caliphate was that of the Sunni Samanids; and, since the Ghaznavids had been subordinate to them, they too followed the Sunni path. There was no innate Turkish preference for Sunnism; but for someone like Mahmud of Ghazna, carrying an inferiority complex born of his family's lowly origins, a show of humble submission to the caliph in Baghdad was an expedient necessity in the harsh world of eastern Iranian dynastic politics.[39] Mahmud therefore took upon himself the role of being the hammer of those whom he thought of as heretics. In 999 CE, in return for restoring the *khutba* in Khurasan, he received from the caliph two prestigious Islamic titles of honour.[40] The legal and moral confirmation of his authority by the caliph was of utmost importance to him. The Ghaznavid coins, for example, normally bore the imprint of the caliph's name or title.[41] Mahmud received further honours as he undertook campaigns against the Ismailis. According to somewhat exaggerated estimates put out by the Turco-Persian historians of India, he killed nearly 50,000 so-called heretics.[42]

The fanatical zeal with which Mahmud and his successors attacked their Muslim enemies was replicated in the destruction of Hindu sacred symbols and images. The Hindu devotee conceives of the images of gods and goddesses of the Hindu pantheon as icons possessing life-breathing and life-giving properties that facilitate the union of the human with the divine. The creation of images of deities, by drawing or sculpturing, is for such a Hindu an act of pious love and devotion to the god.[43] The Muslim attitude is a result of Islam's uncompromising championship of monotheism. Islam took its early guidance in this matter from Judaism. The Quran refers to Abraham as the smasher of idols and images,[44] and Moses issued the Second Commandment of Judaism and Christianity that expressly forbids the making of graven images. From the very beginnings of Islam, Muslim theologians had castigated India as the land of idolatry. Mahmud of Ghazna grew up in that tradition. During his Indian raids he took every opportunity to either smash the varieties of sculptures that adorned the sacred spaces of Hindu temples externally and within, or break what he considered as idols.

In many temples of North India, dedicated to the worship of Lord Shiva, the chief idol is the *Shiva-lingam*, the phallic symbol of nature's generative power. It was the *lingam* that irked Mahmud most and aroused his wrath. His iconoclasm reached new heights at the Gujarati temple of Somanatha that he attacked in 1025. He understood the history of this temple in the context of an old Muslim legend of *su-manat*, concerning a pre-Islamic goddess of Arabia, known as Manat. This goddess, along with her two sister goddesses, Lat and Uzza, was worshipped at her shrine by all members of the Prophet Muhammad's own Quraysh tribe. In his attempt to impose Islam among his people, Muhammad had ordered the destruction of the shrines of all three goddesses, but a general belief spread that the shrine of Manat somehow came to be saved and transported to the Gujarat coast in India. The identification of the pre-Islamic Manat with the Hindu temple of Somanatha became rooted in early Muslim imagination; and Mahmud of Ghazna was a naïve believer of the legend. For him, the black cylindrical-shaped *lingam* of the Somanatha Shiva was Manat herself.[45] He and his soldiers carried out much desecration inside the temple.

Later, Turco-Persian historians, giving vent to their own prejudices, wrote in triumphalist terms about Mahmud's actions. These historians of medieval India used hyperbole in their accounts of Mahmud's deeds and exploits, partly to please their royal masters and partly to inflame Islamic passions among the believers. They thus did a great disservice to future generations of Indians, because some of the venom in modern day Hindu–Muslim relationships stems from the tone and style of their writings. The reality was rather different. Mahmud never succeeded in wholly destroying the Somanatha temple because, according to both Hindu and Jaina accounts, the temple continued to operate for centuries afterwards.[46] Mahmud's actions did not result in any permanent estrangement of Hindus from Muslims in that part of Gujarat, because mutual commercial needs and interests sensibly took precedence over religious prejudices.[47] Mahmud certainly pillaged much treasure from the temple, but for contemporary Indians this would not have been a matter of great surprise; even until recent times the victors looted from the defeated.

The iconoclasm carried out by Mahmud and his Ghaznavid and Ghurid successors was a result of imbibing certain theological presumptions in Islam as indicated above; however, the Turkish monarchs were not inherently anti-Hindu or anti-Indian. Their attacks were ferocious in times of conflict, but they did not indulge in senseless genocidal massacres. They employed Indians (or *hunud*, *hindu*, *hindwan* in Turkic parlance) in various capacities: as high ranking officers and *diwan*s in state positions, as commanders in battles, as guards of the royal palace in Ghazna or as humble slave-soldiers. Separate living quarters were available in response to the religious needs and susceptibilities of non-Muslim soldiers. There was no attempt at forced or mass conversions of Indians into Islam.[48] There were many Sufi-led efforts at conversion in the Hindu Shahi kingdom, Kashmir and in rural Punjab from the early eleventh century onwards, but the bulk of conversions in North India occurred after the mid-thirteenth century. The vast majority of the Muslims of India in the eleventh and twelfth centuries were descended from those who had been converted earlier by the Arabs in Sind, Gujarat, the Malabar coast or Bengal. Equally, it is certain that the Turco-Persian historians would not have missed any opportunity to trumpet the number of conversions under the Turks if such indeed had taken place in huge numbers.

The dominance of Persian cultural tradition

With the conquest of Iran by the Arabs from the mid-seventh century onwards two sets of developments, apparently contradictory and yet inevitable in their outcome, can be noticed. On the one hand, there took place the gradual displacement of the Iranian state religion of Zoroastrianism by Islam over the following four or five centuries and the steady infusion of Arabic into the Persian linguistic and literary corpus. The *Pahlavi* language, known also as Middle Persian, absorbed much of the Arabic influence, culminating in New Persian.[49] On the other hand, owing to the respect and admiration evinced by Arab philosophers and intellectuals towards the pre-Islamic Persian civilization that had reached heights of refinement at the Sasanian courts, the Persian language became a powerful tool for dissemination of Islamic ideas relating to both government and literature. While Arabic, as the principal language of Islamic faith, retained the patronage of the Abbasid caliphs in Baghdad and other lesser Muslim rulers, it was Persian that dominated at the courts of the eastern vassals of the caliphate, particularly at the Samanid court.[50]

With the eclipse of the Samanids by the Ghaznavids, it was the latter who became the patrons of Persian. The linguistic map of the Ghaznavid territories was complex, with many languages and tongues vying for supremacy. First, the Ghaznavids themselves were Turks, and their Central Asian language was naturally popular among their people. Second, they had to cultivate Arabic as part of their Islamic obligation and in view of their Sunni connections and close links with the Caliphate. Then there were several other local dialects spoken in the Ghazna region of Afghanistan. Finally, there was New Persian that was developing as the language of high culture in the eastern caliphate. The general preference of the Ghaznavid court was for Persian; and the spread of Persian into India, too, owed much to its royal patronage.

Literature at the royal courts of Ghazna and Lahore

The royal court of Ghazna shone with great lustre in the reigns of Mahmud (998–1030) and his son Masud I (1030–41). Despite all his aggressiveness and fanaticism, Mahmud retained a love and passion for things of beauty and admiration for people of culture.[51] His raids on India had yielded him unimagined quantities of treasure which, to his credit, he put to good use by establishing a library and a museum at Ghazna. At his sumptuous royal palace, he attracted nearly 400 writers, poets and intellectuals who were all maintained by him. They included such luminaries as the great scientist, Ibn Sina, the scholar al-Biruni and Firdausi, the poet. Without Mahmud's patronage and support these men's learning might have gone unrecognized, and history would have been poorer. At the same time, without the eulogies and panegyrics by some of them, Mahmud's fame – and shame – would not have spread as widely and afar. Coming from 'the cities of the Oxus (River Amu Darya) and the shores of the Caspian Sea, from Persia and Khurasan',[52] as a writer has described their movement, these learned courtiers appreciated the mutual interdependence they found at Mahmud's court. The Victorian historian, Stanley Lane-Poole, compared Mahmud with Napoleon and wrote that while 'Napoleon imported the choicest works of art from the countries he subdued to adorn his Paris, Mahmud did better, he brought the artists and the poets themselves to illuminate his court'.[53] For the sake of balance, one should also note the remark of the scholar and traveller, Edward Browne, who described Mahmud as 'a great kidnapper of literary men whom . . . he often treated in the end scurvily enough'.[54] In the light of Mahmud's testy relationship with both al-Biruni and Firdausi, Browne's judgement has much probity. In contrast to Mahmud, his son, Masud, was himself a highly educated monarch and showed due reverence for the learned at his court, in appreciation of which al-Biruni dedicated to him his most substantial work, the *Qanun al-Masudi*.

While a majority of poets and writers at the Ghazna court were concerned with themes from the eastern Iranian world, there were also those who brought their Indian experience into their works. Thus the great poets, Minucheri and Farrukhi, for example, eulogized Mahmud's Indian victories in their odes.[55] The most outstanding name, however, in connection with India, is that of Abu Raihan Muhammad bin Ahmed al-Biruni (*c.* 973–1048). A native of Khwarazm, he had attained high office in the service of his king, before being captured by Mahmud during the latter's Khwarazmian campaign. Mahmud soon came to appreciate his genius, a man endowed with copious skills in a variety of arts and sciences. Forced to accompany

Mahmud on a number of Indian expeditions from 1016 onwards, al-Biruni made the best of his situation by engaging with the learned scholars in India and acquiring a most intimate knowledge of the country, her people and their customs and beliefs. Two of his mathematical works, *Risalah* (Book of Numbers) and *Rasum al-Hind* (Indian Arithmetic) contain an assessment of the Indian system of numeration.[56] His best known treatise is called *Kitab fi Tahqiq ma lil-Hind*, a truly innovative study of India, incorporating the disciplines of history, sociology, anthropology and geography, and arriving at conclusions of a highly compelling nature.[57] Although adopting an overall Muslim perspective in his elucidations, al-Biruni was candid yet sincere in describing the Hindu character. While he approached the study of their culture in a sympathetic manner, he found something lacking in their scholarship (Excerpt 3.2). Regarding his own patron, Mahmud, he was likewise objective and open. While complimentary towards the king, he roundly condemned him for his cruel and barbaric attacks on the sacred sites of Hindu culture, deeply regretting the fact that Mahmud's actions had scattered the Hindus like 'atoms in the dust'.[58] In this he was very different from the other luminaries who, mostly uncritically, sang Mahmud's praises.

Although al-Biruni wrote in both Arabic and Persian, it was the latter language that began to spread rapidly inside India. A major reason was that the Ghaznavid soldiers were mostly Persian-speaking; and their successive raids and marches into India made it possible for them to increasingly commingle with ordinary Indians.[59] In the wake of the soldiers came Islamic missionaries, teachers and Sufis, all helping to expand the Persian language deeper into India. After 1040, with the loss of Khurasan to the Seljuqs at the Battle of Dandanqan, the Ghaznavid monarchs paid greater attention to their one Indian territory: Punjab. Rich in resources and occupying a strategic frontier position, Punjab became the mainstay of Ghaznavid power; and its governor enjoyed a quasi-regal status. The court of Lahore came to be as splendid as the one at Ghazna, and it became a great magnet for Persian-speaking poets, writers and learned Sufis. In Lahore were established innumerable Persian *madrasa*s, and through them 'the study of the Persian language became wider and more popular day by day among the people of upper Hindustan'.[60] Lahore was the sole capital of the Ghaznavid Empire from 1160 to 1186; it retained its status as the second capital under the Ghurids; and Qutb ud-Din Aybeg, the founder of the Delhi Sultanate, was crowned not in Delhi but in Lahore. Such was the importance of the city at that time.

The Lahore court became famed throughout the Iranian cultural world for its patronage of Persian poetry. Persian historians have been somewhat reticent in their acknowledgement of the genuine literary advances on Persian poetic styles in India.[61] The main conventions of versification in Persian poetry, such as the *qasida* (long monorhymed panegyric poem), the *ghazal* (short lyrical poem), the *rubai* (a quatrain of four-line verses) and the *masnavi* (long narrative poem), were all well known to many poets, good and mediocre, who thronged the Lahore court.[62] Two Indian-born poets merit a special mention. Abul Faraj Runi (died 1091) was the master of *ghazals* that became models for compositions both in Iran and India.[63] Masud Saad Salman (1046–1131), a more famous poet, composed nearly 5,000 verses in Persian. Experiencing prison life twice, he composed prison poems, known as his *habsiyat* collection (Excerpt 3.3). Following a Sanskrit poetic tradition he also wrote month-poems (*baramasa* poems) which celebrated festivities associated with each month.[64]

The most well known Persian historian of this period was Abul-Fazl Bayhaqi (995–1077) who, after having worked as a secretary in the chancellery at Ghazna, wrote nearly thirty volumes of memoirs describing all the activities at the court and the workings of the Ghaznavid government. Only one volume, the *Tarikh-e Masudi*, covering the reign of Sultan Masud (1030–40), survives, but it is enough for us to appreciate Bayhaqi's intellectual curiosity and capacity for historical research.[65] Having had access to vast numbers of official documents, treaties and other state papers, some of which he retained or copied from the original versions during his period of office as secretary, he was able to construct a credible account, richly packed with solid evidence. He was also very critical in his judgements on people and issues: and this was perhaps because 'he began writing his history after many years had passed and he had achieved an emotional distance from earlier affections and animosities'.[66]

Two Sufi poet-saints

The great mystical tradition of Islam, Sufism, blossomed in India between the thirteenth and the sixteenth centuries, but its early harbingers had arrived in the country before the thirteenth century. The mystic, al-Hallaj, for example, was active in Arab Sind (Chapter 2).[67] The two prominent names of the eleventh and twelfth centuries are those of Ali Hujwiri and Muin ud-Din Chishti. Born and brought up in the eastern Iranian world, both eventually settled in India. They represented the Persian strand of Islamic mystical tradition. Ali Hujwiri (*c.* 990 to 1071) is today considered one of the great Sufi saints of Islam.[68] He imbibed his knowledge and understanding of his faith from the works of Abul Qasim al-Junayd, another Iranian mystic, who founded his own Junaida school of mysticism in Baghdad. By attempting to reconcile the legalism of the *Shariah* with mystical inspiration Hujwiri attracted ordinary Muslims towards Sufism. At Lahore he wrote a fine treatise in Persian called *Kashf al-mahjub* or the 'Unveiling of the Hidden', which is a review of all the Sufi works extant up to his time.[69] His shrine there, called that of Data Ganj Bakhsh, has been a revered place of pilgrimage ever since.[70]

One man who came to pay his respects at the shrine was Khwaja Muin ud-Din Chishti (*c.* 1139–1236), a Persian, born in Sistan.[71] After having visited famous Islamic cities like Mecca, Medina, Samarqand and Bukhara, he finally settled in Ajmer *c.* 1190, and began his life's work of drawing people towards Sufi ideas developed by the Chishti order to which he belonged. The followers of this order, founded by Abu Ishaq Shami of Syria, in the town of Chisht, near Herat in Afghanistan, in 930, believed in simple and unsophisticated preaching and practice of the love of God, without discrimination against any groups of people.[72] Khwaja Muin ud-Din was loved and respected by both Hindus and Muslims; and there can be little doubt that both his work and personality led many Hindus from unprivileged backgrounds to convert to Islam. Muin ud-Din wrote very fine Persian poetry but, unlike other Sufi-inspired Persian poets, like Hafiz or Rumi, did not indulge in poetic pleasures evoked by worldly pleasures of wine, tavern, roses, nightingale and the tresses of damsels; his poetry is about spiritual truth and compassion. This had a deep appeal to the masses in a country like India that is suffused with spiritual lore and literature from ages past.[73] Millions of Muslims and countless non-Muslims, particularly those without children, money or robust health, have continued to visit the *dargah* of Muin ud-Din Chishti since his death in 1236, in order to seek the benefactions of his blessings (Excerpt 3.4). The Mughal Emperor Akbar himself went on foot from Agra to Ajmer to pay his homage to the Khwaja, in fulfilment of his vow to God on the birth of his son Jahangir in 1569.[74]

Select excerpts

3.1 Nizam ul-Mulk's advice to monarchs on using spies for the good of the country

Spies must constantly go out to the limits of the kingdom in the guise of merchants, travelers, Sufis, pedlars and mendicants, and bring back reports of everything they hear . . . In the past it has often happened that governors, assignees, officers and army commanders have planned rebellion and resistance, and plotted mischief against the king; but spies forestalled them and informed the king, who was thus enabled to set out immediately with all speed and, coming upon them unawares, to strike them down and frustrate their plans; and if any foreign king or army was preparing to attack the country, the spies informed the king . . .

(Darke 1960: 74–5)

3.2 al-Biruni's independence of mind and critical judgement

al-Biruni had great admiration for Indian mathematics and astronomy. It was only after he had completed a thorough examination of Indian life and learning that he felt compelled to make some highly critical remarks about Indian society and scholarship. They reveal to us the decline of intellectual life in the country since the days of the Imperial Guptas.

> We can only say, folly is an illness for which there is no medicine, and the Hindus believe that there is no country but theirs, no nation like theirs, no kings like theirs, no science like theirs . . . Their haughtiness is such that, if you tell them of any science or scholar in Khurasan and Persis, they will think you to be both an ignoramus and a liar. If they traveled and mixed with other nations, they would soon change their minds, for their ancestors were not as narrow-minded as the present generation is . . . Even the so-called scientific theorems of the Hindus are in a state of utter confusion, devoid of any logical order, and in the last instance always mixed up with the silly notions of the crowd, eg immense numbers, enormous spaces of time, and all kinds of religious dogmas . . . I can only compare their mathematical and astronomical literature, as far as I know it, to a mixture of pearl shells and sour dates, or of pearls and dung, or of costly crystals and common pebbles. Both kinds of things are equal in their eyes, since they cannot raise themselves to the methods of a strictly scientific deduction.
>
> (Sachau 1888: Vol. 1, 18–25)

3.3 Poetic moods of Masud Saad Salman in his Habsiyat *prison poems*

Although a renowned poet at the Ghaznavid courts in Ghazna and Lahore, Masud Saad Salman had, at one stage in his life, fallen foul of the political establishment and, consequently, been imprisoned. His prison poems, written both while serving the sentence and after his freedom, evocatively express his different moods.

His grief:

> In the prison of Maranj with these chains,
> Salih (his dead son), how can I be alone without you?
> Sometimes I cry lapfuls of blood over your death,
> Sometimes I tear my clothes to shreds in pain.

His awareness of the hopelessness of escape:

> The affairs of the world are limited to
> The prison and fetters of this weak body!

> Even in chains and prison I am not secure
> Until they have ten watchmen around me.
> All ten are seated at the door and on the roof of my cell,
> Saying to each other incessantly,
> 'Rise and see lest by magic
> He fly off into the sky through a crack in the window.
> Be watchful for he is a trickster
> And will make a bridge from sunlight and a ladder from the wind'. . .

His pride:

> I am he than whom in Arabia and Persia
> There is no one more smooth-tongued.
> If there is a difficulty in poetry or prose
> The world asks for me as an interpreter.
> In both these languages, in both arenas
> My victory has reached the heavens. . .
> One day I will emerge from prison and chains
> Like a pearl from the sea or gold from a mine.
>
> (Sharma 2000: 90–1, 96)

3.4 Urs *celebration in South Asia*

In the Sufi tradition, the term '*urs*' means the union of a saint with God and signifies the day of death of the saint. *Urs* celebrations are commonly held on the anniversary of the death of Sufi saints, *pir*s, of the past. One best known annual celebration takes place at Ajmer, India, on the anniversary of the death date of Khwaja Muin ud-Din Chishti.

> In South Asia, dervishes frequently take part in processions coming to the shrine from very distant regions in order to celebrate the urs. Sometimes they carry replicas of the shrine and shrouds as votive offerings. On their way they stop at places linked to the life of the saint they venerate. During the pilgrimage dervishes are given food and shelter for the night by the local population. Beggars often accompany the procession of dervishes and devotees from one urs to another, making use of the free communal distribution of food . . . and benefiting from the generosity of the pilgrims . . . In Pakistan, wrestling contests, sword dances, tent-pegging and similar events are held, and the dervishes dance in ecstasy.
>
> (Frembgen 2008: 39)

4

THE DELHI SULTANATE AT ITS ZENITH (1206–1351)

The Ghurid conquests between 1186 and 1206 eventually led to the establishment of what is called the Delhi Sultanate. This was essentially a regional North Indian power, although during its zenith its territorial boundaries extended southwards too. The fact that it was regional should not mislead us into thinking that other Indian states in the thirteenth and fourteenth centuries had the capacity to challenge its might. The sultanate lasted from 1206 until the advent of the Mughals in 1526, but from 1351 onwards its power markedly declined.

Before 1206, apart from Sind, Muslim authority in India was confined to Punjab and the frontier lands between present day Afghanistan and Pakistan. The Ghaznavids and Ghurids were dynasties essentially shaped and influenced by the political conditions of the eastern Iranian, Transoxanian and Afghan regions. While the Turkish and Persian influences that they had brought into India continued after 1206, from then on the sultanate matured and developed essentially within the Indian heartland; and Indian cultural influences succeeded in making it an Indian polity, distinct from the Iranian or Central Asian. It was simultaneously a Muslim authority and, in that respect, it was alien to the vast majority of Hindu Indians. Over time, however, the sultans of Delhi charted out a balanced strategy of survival and domination in an alien religio-political landscape, which helps explain the relative longevity of the system they established.

Sultans and nobles

The sultanate was an authoritarian monarchy: there was no question of democratic politics. On the other hand, the possession of the throne did not entitle a sultan to rule with unchecked powers. Although wielding great authority and sizeable patrimony, each sultan had to be conscious of the sensibility of military and religious grandees around him. Whenever a sultan trampled on that sensibility, his fate was sealed, which happened more often than not.

A snapshot of the great sultans and their dynasties

Seventeen sultans in all ruled between 1206 and 1351 (see Dynastic lists, p. 257–62). They are conventionally grouped under three dynasties: Slave, Khalji and Tughluq. The term 'Slave dynasty' reminds us of the fact that some of the Turks who started their careers as slaves became powerful monarchs (Chapter 3). Yet only three of the eleven so-called slave-kings had originally been Turkish/Mamluk slaves. The dynasty began with the assumption of power in India by Qutb ud-Din Aybeg on the death of his master, Muhammad Ghuri, in 1206. Qutb ud-Din may therefore be considered as its first sultan. During his short reign, he strove to ensure that the Ghurid victories were translated into a meaningful Northern Indian empire.[1] His death in 1210 resulted in the throne first passing to his son and then swiftly to his own slave-commander, Shams ud-Din Iltutmish, who ruled between 1211 and 1236. While it was at Lahore that Qutb ud-Din was declared the sultan, Iltutmish himself was crowned in Delhi; and, therefore, strictly speaking, the Delhi Sultanate began in 1211 rather than in 1206. During his twenty-five years' reign Iltutmish centralized power at his Delhi court, thus diminishing the influence of such outlying sultanates as in Sind and Bengal.[2] He checked the early attacks of the Mongols; and his authority was noticed even by the caliph in Baghdad.[3]

With the death of Iltutmish in 1236, there began three decades of what has been described as a period of 'disintegrating Shamsi dispensation'[4] during which his four children and one grandchild ruled. Since none of them had been slaves, it is not surprising that historians sometimes style Iltutmish's family the Shamsi dynasty, after his first name Shams ud-Din. Intrigues, conspiracies and jealousies among the court nobles characterized the reigns of the five Shamsi rulers between 1236 and 1266. The most remarkable of the five was Iltutmish's daughter, Raziya, who reigned from 1236 to 1240. Brave, and skilled at horsemanship, she was unafraid to lead an army into battle; but her enigmatic and strong-willed personality was not to the liking of court nobles who brought about her downfall.[5]

The other Shamsis continued to reign until 1266, but in the reign of Nasir ud-Din Muhammad Shah we witness the rise of his regent, Ghiyas ud-Din Balban, who had first begun his career under Iltutmish as a slave-soldier. Nasir ud-Din, a sultan of gentle disposition, and an aesthete more interested in religion and calligraphy, left all major decisions in the hands of Balban, who consolidated the sultanate by suppressing rebellions and checking the ever-increasing Mongol threat. Although his power was resented by non-Turkish elements within the court nobility, Balban himself wrested the throne after Nasir ud-Din's death in 1266. As sultan he ruled with an iron fist for the next twenty-one years,[6] when the Slave dynasty reached its apogee. A stickler for etiquette, pomp and extreme dignity, he did not allow even the mightiest of courtiers room for intrigue against him.[7] All his contemporaries remained in awe of him and, like Iltutmish, he could be considered as one of the powerful Delhi sultans of the thirteenth century. His descendants proved unfit, and the Slave dynasty ended in 1290 with the coming to power of Jalal ud-Din Khalji.

The Khaljis were a Turkic people who had long been settled in Afghanistan. Although the older Turkish nobility from Shamsi times at first looked upon them with suspicion, Jalal ud-Din was able to reconcile the various parties at the court.[8] His six-year reign ended with the betrayal by the one he trusted most, his own nephew Ala ud-Din, who murdered him in 1296, seized the throne and then got all his cousins killed or blinded.[9] This treachery and cruelty apart, Ala ud-Din Khalji proved to be a dynamic sultan in his own right and in comparison with peers of his times.[10] His reign saw a great expansion of the sultanate's frontiers; his economic policies and administrative edicts, although harsh and arbitrary by modern standards, helped to sustain his enlarged sultanate. Along with Iltutmish and Balban, Ala ud-Din Khalji is our third noteworthy sultan. The next two Khaljis turned out to be great disappointments; and in 1320 the throne was seized by Ghazi Malik, one of Ala ud-Din's commanders, known as Ghiyas ud-Din Tughluq.

The Tughluqs were a Turkic family of the Qarauna clan who inhabited the lands between Sind and Turkestan and who had migrated to India in the reign of Ala ud-Din Khalji. Ghiyas ud-Din Tughluq's four-year reign was followed by the quarter-century reign of his son, Muhammad Shah Tughluq. A charismatic and mercurial figure of abundant generosity on the one hand and, on the other, of a contradictory rashness and cruelty, Muhammad Shah Tughluq was a visionary whose boldness and ruthlessness in action earned him both respect and fear throughout his kingdom.[11] He was plagued with many rebellions, and a number of provinces broke away from his control.[12] Be that as it may, Muhammad Shah Tughluq has to be the fourth of our great sultans. After him, the Tughluq dynasty continued until 1413.

The authority of the sultans

From the foregoing brief review of the sultanate's dynastic history, it can be concluded that there was no systematic policy regarding the succession to the throne. The sultanate was not a modern day type of constitutional monarchy. The hereditary principle of primogeniture did not operate meaningfully, and a majority of the sultans were not first born. Only three of the seventeen sultans were the eldest children of their predecessors. Unlike in a constitutional monarchy, the possibility of a lifelong occupancy of the throne was minimal: only five of the seventeen sultans retained the throne until their natural death. The absence of a genuine hereditary principle and the relatively short reigns of most sultans indicate the turbulent nature of the political system. In an age when violent disputes were only settled by more extreme violence, it was the force of personality that made or deposed a sultan. Without exercising a ruthless streak of violence and cruelty, no sultan of Delhi could survive; for the sake of power nearly all of them executed or blinded some of the nearest and dearest in their families.

Although the sultans adhered to the Persian tradition of the divine right of monarchs to rule, there were certain critical obligations that they could not overlook. The most cogent listing of these obligations can be found in the *wasaya*, or precepts, of Sultan Ghiyas ud-Din Balban, in his advice to his sons on the duties of kingship.[13] Good kings, in his eyes, feared only God and none else; they had to rise above and to keep distance from the commonalty; they must be impartial in their judgements; they had not to flinch from applying the severest punishments upon those who defied orders; and they had to practise piety and moderation in daily life. Applying Balban's rigorous tests, the seventeen sultans can be typified by three kinds of personality. First, the four most impressive sultans – Iltutmish, Balban, Ala ud-Din Khalji and Muhammad Tughluq – behaved with decisive ruthlessness. Second, there were sultans like Raziya who could not rise above court intrigues and unwittingly allowed themselves to be manipulated by the nobility, ultimately resulting in their loss of power.[14] Third, there were the ephemeral sultans who were either too weak or unskilled and who lacked good impartial advice.

The personality of the monarch undoubtedly determined the level of recognition and support that he could expect from the nobles and courtiers. Additionally, each sultan attempted to bolster his authority through an oath of allegiance (the *bayat*), to be taken by a wide circle of people, including the nobles at the court and the leading citizens of Delhi.[15] The less confident a monarch felt about himself, the greater the urgency with which the pledge was administered. On rare occasions, the sultans followed the Ghaznavid practice of taking an oath of allegiance to the Abbasid caliph in Baghdad, in return for the recognition of their authority in India. It happened first in 1229, when Sultan Iltutmish received the caliph's ambassador who conferred on him such exalted titles as *Yamin Khalifat Allah* (Right Hand of God's deputy) and *Nasir Amir al-Muminin* (Auxiliary of the Commander of the Faithful).[16] A hundred years later, Sultan Muhammad Tughluq was also delighted to receive a deputation from the Abbasid caliph in exile in Cairo.

The nobility

The ablest, the bravest and the most devious of the Mamluk/Turkish slave-soldiers, whose ancestors had been in the vanguard of India's invasions during the Ghaznavid and Ghurid times, constituted the central core of the sultanate's nobility.[17] They were the elite corps, called the *Bandagan-i Khass*; and their heyday was in the thirteenth century, when three of the sultans – Qutb ud-Din Aybeg, Shams ud-Din Iltutmish and Ghiyas ud-Din Balban – had themselves served their apprenticeships as slave-soldiers. Amidst them was an inner circle of nobles, known as the *Chihilgani*, or the Forty.[18] Whether there were just forty of them or a certain number who themselves owned forty slaves each is not entirely clear; but they considered themselves to be above all other nobles and, when threatened with the loss of power, were prepared to kill their rivals. The four monarchs after Iltutmish fell victim to the Forty because they showed partiality towards non-Turkish nobles.[19] What in the end dented the strength of the Forty was the fact that in India, unlike Egypt, the Mamluks did not possess a monopoly of offices. There were many free Turks and Tajiks who had also gained, in the course of time, honoured places for themselves at the Delhi court.[20] Sultan Balban introduced new blood at the court by ennobling Mongols, Hindus and Indian Muslims.[21] Sultan Ala ud-Din Khalji further diversified the court by replacing some of the older nobility with members of his Khalji clan.[22] In the early fourteenth century, owing to a flourishing trade in Turkish slaves in the markets of Central Asia and Khurasan, large numbers of slaves were bought by Sultan Muhammad Tughluq; and the Turks again predominated at the court.[23] But by the middle of the fourteenth century, however, the Afghans and native Indians began presenting a strong challenge to the Turks.

Just as there was no fixed principle of hereditary kingship, so it was with the nobility. While members of an individual family frequently retained power during the reigns of successive sultans, the idea of a hereditary aristocratic title was considered unacceptable.[24] Merit was the key criterion in the recruitment of nobles. Only when a slave-soldier had proved his mettle on the battlefield by his fighting and leadership skills did he have a chance of moving upward. The process began normally in the imperial household, where a new entrant might be assigned any one of a number of posts, such as the supervisor of royal kitchens or the keeper of the royal wardrobe. Moving up the ladder, there were such higher posts as those of the master of ceremonies, the commander of the sultan's bodyguards, or the head of the royal household. Yet further up the scale, there were even more lucrative posts to be filled in the great departments of state, such as the head of the armed forces, the head of Chancellery, the Chief Justice or the principal news writer of the kingdom. Over and above all the offices was that of the *wazir*, the chief adviser to the sultan.[25] Even he could not divorce himself from military activity.[26] A key department controlled by the *wazir* was the Treasury (*diwan-i-wazarat*). The vast revenues that accrued from land, along with a variety of other taxes, dues and tributes, and the disbursements of these monies, were all handled at the Treasury where an army of auditors and clerks checked the accuracy of figures. Year by year the sultanate's wealth continued to increase, and by holding the purse strings of the kingdom the *wazir* secured a strategic position at the court.[27] *Wazirs* could raise private militias, and a number of them had the potential to lead rebellions against the sultans. Conversely, they too could be at the mercy of powerful and capricious masters.

Of all the perquisites made available to the nobles, none was more coveted than the holding of an *iqta*, a portion of land of varying size, assigned by the sultan. From the early days of Islam, an *iqta* concept has existed as a form of reward for services to the state.[28] In the Delhi Sultanate all land was grouped into two classes: the *khalisa* or the royal demesne and the *iqta* lands. From the royal demesne, the sultan's officers collected revenue directly from the people in the form of land tax, poll tax, levies or imposts for meeting the expenses of the royal household. For the *iqta* lands, there were two types of assignments: first, the lesser lords, granted very small parcels of land, could retain the revenue from their grant of land in lieu of responsibility for providing up to three horses with all the supplementary equipment and subsistence for the sultan's military needs.[29] The more important grandees and provincial governors were, in recognition of their honour and status, bequeathed huge grants of land. Depending on the size of the land granted, a nobleman might be asked to raise over 500 horses or more for the army. Larger *iqta* holders had also a duty to be proactive in all matters affecting stability and peace in their locality, and were obliged to maintain a balance between their sojourn at the royal court and their duties in the provinces. Quite often, in the absence of strict monitoring, the *iqta* rights of the lords quietly passed on to their families when they died, even though the successors offered no service to the state. Since an *iqta* was a privilege for an individual during his life and had no hereditary connotation, this was avowedly illegal. This rule was occasionally lifted; and, from the middle of the fourteenth century, the Tughluqid monarchs rashly granted lands in perpetuity to many of the nobility.

Territorial expansion and the Mongol threat

Throughout the thirteenth century, the sultanate effectively remained confined to a fairly narrow territory extending from the Afghan frontier to Bihar and Bengal. This was the inheritance bequeathed to Iltutmish by Muhammad Ghuri and Qutb ud-Din Aybeg. The local Muslim potentates of Punjab and Sind retained certain autonomy, but the many expeditions sent from Delhi to curb their ambitions indicate that they were, to all intents and purposes, considered as part of the sultanate. In Rajasthan a secure town was Ajmer, where Muin ud-Din Chishti had established his Sufi school (Chapter 3).[30] To the east, in the Doab, the sultanate controlled large towns like Delhi, Mathura, Badaun and their immediate surrounding countryside. Two great Hindu centres on the Ganges, Kanauj and Benares, were also in Muslim hands. South of the River Yamuna, the fort of Gwalior was taken, lost and retaken several times, and finally incorporated into the sultanate.[31] Further east, Bihar and Bengal were held by governors appointed by the sultans. The rest of the subcontinent remained in the hands of numerous Hindu rulers. The Delhi sultans of the thirteenth century strove greatly to expand their frontiers, but the magnitude of the task proved impossible to achieve for a very long time. Terms such as *mleccha* or *Turushka* in Hindu inscriptions of this period refer to Turks or Mongols,[32] hinting at the military impact of the sultanate upon Hindu kingdoms. Scanty evidence of the possible expansion of the sultanate's territory may also be gathered from studying the historical records of the distribution of *iqta* lands which, in a majority of cases, belonged to the Muslim aristocracy.

The sultanate faced a tenacious enemy not just in the official Hindu armies but in the many irregulars and marauders. These latter groups of people, quite often nomadic, found sanctuary in many inaccessible, heavily forested, hilly or remote areas, which the sultanate's cavalry had difficulty reaching or penetrating.[33] The authority of the sultanate was based on large forts and garrison towns, inhabited by both Muslims and Hindus, while the countryside remained predominantly Hindu. The major urban centres, like Peshawar, Lahore, Multan, Delhi or Lakhnauti, were heavily fortified and their routes to food supply sources in the immediate countryside were well patrolled. Yet, not too far from the Muslim boundaries, there were other gigantic forts belonging to the Hindu kings, which remained impervious to attacks by the sultanate armies. The forts of Ranthambore, the 'key to the south', and Chitor in Rajasthan, were besieged by the thirteenth-century sultans time and again, but they always reverted to their Hindu Chauhana and Paramara defenders.

After a hiatus of some hundred years, the situation changed dramatically with the rise to power of the Khalji and Tughluq dynasties. The expansion of the sultanate into an almost pan-Indian empire began essentially with Sultan Ala ud-Din Khalji and resumed under Sultan Muhammad Tughluq. Economic plunder and territorial usurpation were the prime motives.[34] Hindu India to the sultans was an endless source of gold, foodstuffs, cattle, elephants and slaves. The acquisition of more land meant greater opportunities for disbursing favours through *iqta* lands and raising profitable revenues. Successful campaigns against Hindu kingdoms owed to military mobilizations on a scale hitherto unachieved, with the deployment of swiftly moving cavalries, siege engines and massive earthworks.[35] Under Ala ud-Din Khalji, the province of Gujarat, along with the great forts of Rajasthan and Malwa, surrendered. His slave-commander, Malik Kafur, destroyed the rich Deccan city of Deogir and led the expansionist trail through the south.

Under Muhammad Tughluq the Chandella and Parmara dynasties came to an end, and practically the whole of the northern Deccan was brought under the sway of the sultanate. Muhammad Tughluq permitted the Hindu family of Sangamma a measure of autonomy in the lands south of the River Krishna (Map 4.1).

Within a period of two or three decades, under the Khaljis and the early Tughluqids, the imperial ambition of the sultanate had come to fruition.

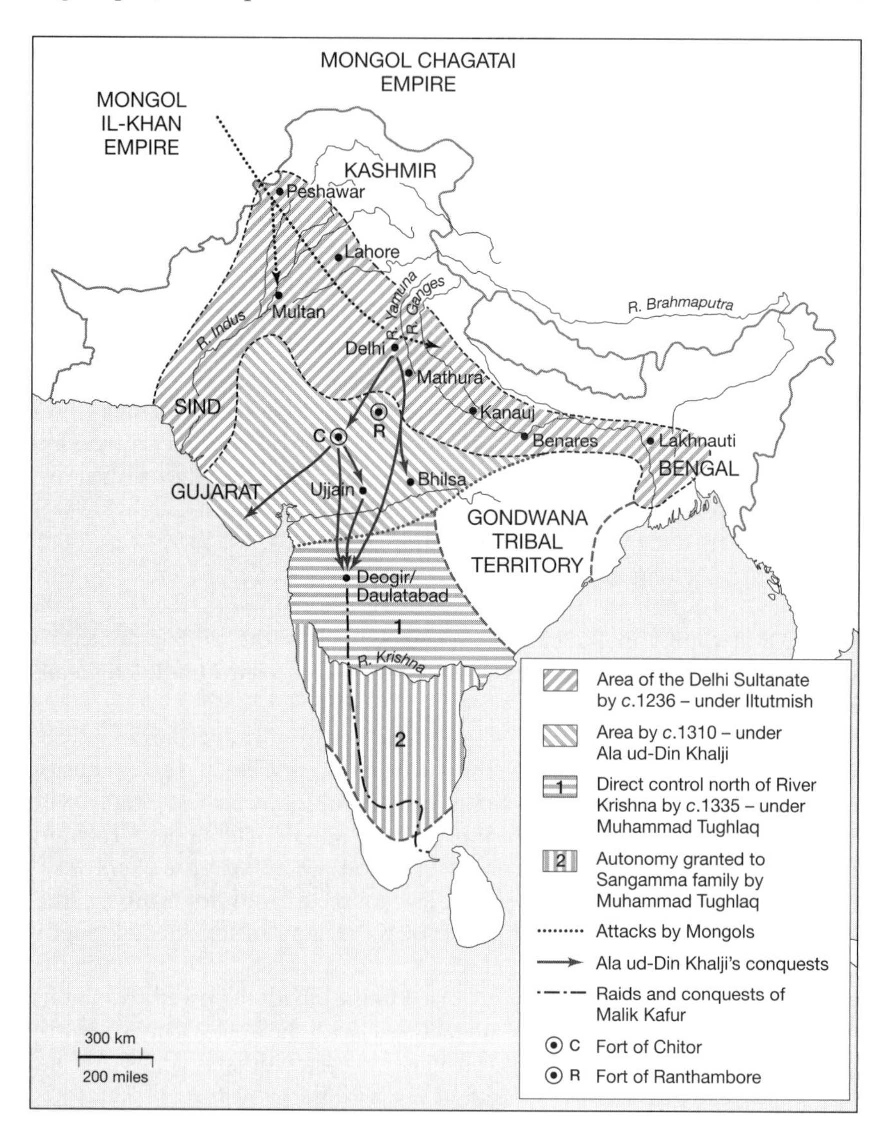

Map 4.1 The southern expansion of the Delhi Sultanate

This project, however, did not endure owing to the lack of an impersonal bureaucratic system which could have welded heterogeneous cultural regions into a single polity. The ambitions of individual sultans were not substitutes for a long term strategic plan for political unity. One particular episode from the reign of Sultan Muhammad Tughluq is instructive in the way it illustrates the impetuosity of the sultan and the inadequacy of his planning. When nearly the whole of the subcontinent had been conquered, Muhammad decided that Delhi was too far north to be a true capital for the whole country; and in 1327 he passed a decree making Deogir (Daulatabad) in the Deccan his capital. While Delhi was to be turned into a major military fortification zone, partly as a defence against the Mongols and partly to prepare for the sultan's grandiose project of invading Khurasan, all the non-military personnel, particularly the Muslim aristocratic elite, were ordered to move to the new capital. The task of depopulating Delhi proved to be so enormous that, within a few years, it had to be abandoned. The over-ambitious territorial plans of the sultanate only ended up in revolts not only in the Deccan but in other provinces too.

The challenge from the Mongols

One of the reasons why the kingdoms beyond the Delhi Sultanate could not be confronted vigorously enough in the thirteenth century lay in the external threat from the Mongols at that time. The Mongols shared Eurasia with the Turks: they inhabited the eastern Eurasian steppe lands, while the Turks held the west.[36] Although the Turks had been migrating out of their traditional lands long before the thirteenth century, the Mongol stirrings had only just begun. In the very year 1206, when Qutb ud-Din Aybeg established the sultanate, the warlord Genghis Khan was chosen as the Mongol leader or Great Khan. After first making raids into China he turned his attention towards the Muslim world in the south-west. By 1221, he had defeated the kingdom of the Khwarazm Shah in northern Iran.[37] Believing passionately in the Mongol right to rule the world, he and his successors took the sword into the distant corners of the Eurasian land mass. Striking fear in the hearts of many nations and peoples, the Mongol armies warred incessantly throughout the thirteenth and the early fourteenth centuries.[38] The lands of Russia and Ukraine as far west as Moscow and Kiev, and Persia and China, all became part of the Mongol empire. However, family dissensions among Genghis Khan's successors ultimately led to the division of the empire into four parts: the Golden Horde of Russia, the Il Khanate of Persia, the Chagatai Khanate of Central Asia and the Great Khanate of China. The unitary nature of the empire thus came under great strain.

Apart from their militaristic ideology, there were other factors that drew the Mongols towards India. One was the *realpolitik* prevailing in the Indus basin and the north-western frontier areas of India where rival Muslim potentates were prepared to use the Mongols for their own ends in their internecine struggles as well as in their defiance towards the Delhi Sultanate. In turn, the Mongols encroached into India either as friends or foes of the local princes, while traducing and showing scant respect for the sovereignty of the Delhi Sultanate.[39] There was also the economic motive behind the Mongol forays. Basically a nomadic people, they were always seeking fresh pasturage; and parts of the Punjab provided good pasture lands for their animals. Additionally, in some border areas, there were fine breeds of horses available for their war machine. India was also the place from where they could steal much gold and silver and, above all, capture numerous slaves. The temptations for the Mongols to loot from India were ever present, and they rapidly increased as the wealth of the sultanate soared after the territorial expansion into the south by Sultan Ala ud-Din Khalji.[40]

For nearly a century, from the 1220s to the 1320s, the Mongols attacked parts of India for varied reasons and in different circumstances; but the end result for them in India was negative. The last wave of attacks, led by the forces of the Chagatai Khanate, came in the late thirteenth and early fourteenth centuries; but Ala ud-Din Khalji had prepared his defences well.[41] He and his generals built gigantic new forts and kept the old ones in repair all along the north-west frontier; they placed entire populations on a war footing; and they enforced a wartime economy on the citizens of Delhi. Whatever hardships that may have caused, it would have been generally far worse for India's people had the Mongols been the victors. Terrible vengeance was wreaked upon the Mongols by Ala ud-Din Khalji through executions and the trampling of their warlords under elephants' feet. The Mongols had earned a reputation for great ferocity and savagery through their lust for a Mongol world empire: but it was in India that the tables were reversed.

Peasants, craftsmen, slaves and traders

The economic history of the sultanate is basically a story of the enrichment and enhancement of the military aristocracy at the expense of their subordinate classes. As in most pre-modern economies, it was the wealth accrued from the rural economy that mainly sustained the power of the ruling class.

Three types of rural taxes were levied: on cultivation (*kharaj*), on milk cattle (*charai*) and on houses (*ghari*). It was a crushing burden for the peasants. The cultivation tax itself was set at 50 per cent of the produce of the land, and a large number of peasants had often to sell everything they owned before they could pay the tax. Ala ud-Din Khalji enforced a tyrannical tax regime, but his reputation for not taking lightly any dissent meant that there was disgruntlement but no rebellion.[42] The situation changed drastically when Sultan Muhammad Tughluq attempted aggressively to enforce the collection of all taxes throughout his much larger-sized kingdom. A great peasants' rebellion broke out in the countryside in the early 1330s, with the consequence that the rural receipts dramatically diminished. A great famine, lasting seven years, from 1334 to 1341, caused the ruin of agriculture.[43] Although famine relief was provided (Excerpt 4.1), an insurgency raged in the countryside, according to an account by the Moroccan traveller, Ibn Battuta, who resided at the Tughluqid court during these years.[44]

The revenues were traditionally collected by local headmen on behalf of the superior landlords who, in turn, were expected to remit a portion to the king. From the late thirteenth century onwards, the sultans were creating a new rural aristocracy. The members of this class, the *chaudhuri*s, eventually became the right-holders of lands or the *zamindar*s; they became the agents of the sultanate for the collection of revenue from peasants.[45] Most of the *chaudhuri-zamindar*s were Hindus, but Muslim aristocrats with *iqta* rights also had their own claims over land. Yet another category of people with a stake in the land consisted of those who were granted the revenues of lands and villages for their lifetime or in perpetuity. A recipient could be an individual or a religious-educational institution, like a mosque, a *madrasa* or a Sufi *khanqah*. Such grants of land, known as *madad-i maash*, were valued at approximately 5 to 6 per cent of the total revenue of the sultanate.[46] All in all, the peasants suffered from the various groups of people above them.

Rural produce and high taxation induced the peasants to sell as much as possible to urban dwellers in places like Delhi, Lahore, Multan, Deogir, Anhilwara and Lakhnauti, where craftsmen and merchants abounded. Much of the trade was local, but the towns served as entrepots and clearing centres for long distance trade in such commodities as horses, slaves, indigo, silk and clarified butter. While political power in the sultanate was in Muslim hands, control of trade stayed with Hindu or Jain merchants and entrepreneurs. Sometimes known as Multanis, they provided cash and ready loans to the Muslim aristocracy.[47] It was during this period that brokerage became prominent with the rise of an intermediary class, that of the *dallal* or broker. In and around the towns, there operated the key non-agricultural sectors of production. These included, primarily, manufactures of metals and minerals, such as iron ore, copper, gold and silver, diamonds and pearl jewellery. Many crafts like cotton carding, silk weaving and paper manufacturing, which were well developed in the Islamic lands west of the subcontinent, were introduced into India by waves of Muslim immigrants from Iran and Central Asia.[48]

The condition of urban labour was not much better than that of slaves. Slavery was not unknown in ancient India, but it was in the heyday of the sultanate that it became an institutionalized part of Indian labour in general. The sultanate continued the Ghaznavid and Ghurid tradition of raiding and plundering, first, in areas outside its boundaries, then in rebellious areas within its boundaries and, lastly, any region that defied its authority.[49] Huge numbers of slaves, mostly Hindu, were brought into towns and sold off to the highest bidders. Delhi had a great slave market of its own. After the middle of the fourteenth century, however, slavery declined dramatically; and by the fifteenth century the slave markets had gone. The reason was chiefly economic: the rise of free and cheap labour.[50] It was free in the sense that many Hindu slaves converted to Islam and gained their liberty; and it was cheap because the numbers of people seeking employment keenly competed with each other (as they still do in South Asian economies today). Another reason was the increasing Indianization of the sultanate during its declining years when it was no longer possible for Muslim grandees to continue their disdainful treatment of Hindus.

The economic activity reviewed so far suggests a great deal of self-sufficiency on the part of the sultanate. Most of the trade, inland and overseas, was land-based; and for long the Delhi sultans seemed to show little interest in maritime trade. In Bengal, however, a flourishing oceanic trade with China and South-East Asia was conducted by Arab and Persian traders. Some of the customs revenue reached the royal treasury. The situation changed dramatically at the beginning of the fourteenth century when Gujarat was partially conquered by Ala ud-Din Khalji. The Muslim traders of this province had for centuries engaged themselves in the maritime trading activities reaching the Middle East and the Mediterranean.[51] The economic size and wealth of the sultanate meant greater opportunities for Gujarati entrepreneurs for further trading opportunities in the deeper hinterland. The Gujarati ports could export products secured from far and wide, thus benefiting the sultanate with increased customs revenues. A more consequential development was the importation of much bullion, since foreign goods were no match to Indian goods in value. The economic power of the sultanate, as witnessed in its large scale currency circulation, tax structures, growing populations and markets, helped to absorb large inflows of precious metals.[52]

That an individual sultan's eccentric economic policy could clearly destabilize the financial system of even such a powerful state as the Delhi Sultanate is evidenced by what happened under the temperamental sultan, Muhammad Tughluq. Facing a financial emergency, caused by his over-ambitious military plans and the expensive building of the new capital at Deogir (Daulatabad), Muhammad devised a bizarre financial scheme of issuing a type of 'fictitious' currency, which was his token currency.[53] The token currency normally consists of currency notes and coins with a value printed on them but which are intrinsically worthless. The practice is basically meant for our convenience. Such coinage may show a face value much higher than the cost of its metal content, but it is in the final resort backed up by bullion reserves in the vaults of the central bank. Muhammad Tughluq was perhaps aware of a form of token currency in use in the Il-Khanate kingdom of Persia, but his economic illiteracy made him sign a decree stating that the copper and brass tokens had the same value as gold and silver coins. Since no qualification or clarification was issued, people literally took Muhammad at his word and, with the worthless tokens, began to purchase gold and silver coins for which they were legal tender. Soon, the revenue began to be paid in tokens, and counterfeiting became rife. Since the assayers were mostly from the Hindu goldsmith caste, it is clear what the great historian Barani meant when he wrote that every Hindu's house had become a mint.[54] Within eighteen months, Muhammad Tughluq had to withdraw the tokens. In his desperation to end the ill-conceived experiment, he ordered his officers to buy up the tokens with gold and silver! A similar experiment with token paper currency, tried out by the Ming emperor of China in the late fourteenth century, also ended in failure.[55]

Guardians of Islam and attitudes to Hindus

Believing themselves to be good Muslims, the sultans of Delhi naturally wished to give pride of place to Islam within their kingdom. They were patrons of learned and pious co-religionists, some of whom were *ulama*, who learnt the Quran by rote, studied the *hadith*s of Prophet Muhammad and specialized in exegetical literature.[56] Most of these experts had completed their training at home rather than in any mosque or *madrasa* but, once fully qualified, they expected patronage from the state. As the sultans grew in wealth and power they indeed supported Islamic institutions dominated by the *ulama*, some of whom even engaged in worldly activities like commerce, keeping them close to the centre of power.[57] The vast majority of the *ulama* were Sunni, strongly opposed to the Shia; and they expected the sultans to be harsh towards all whom they considered heretics within Islam.[58] The sultans took their advice when necessary but, quite often, rejected it too. They, however, expected the full endorsement of the *ulama* in their efforts to maintain not only faith solidarity among their Muslim subjects but also to encourage loyalty to the state.

Islam as a religion received a major boost in India in the early thirteenth century with the arrival of heterogeneous groups of people fleeing Afghanistan, Khurasan and Central Asia, the lands ravaged by the Mongols. Sultan Iltutmish not only welcomed but encouraged the new Muslim migrants.[59] This immigration continued sporadically throughout the century; and there can be no doubt that new blood greatly strengthened the relatively small

Muslim community of India. Among the immigrants there were many Sufis; and their work and influence nourished Islam in India.[60] They belonged to different orders or *silsilahs*, of which the two most influential in medieval India were those of Chishti and Suhrawardi.[61] The former became well established through the work of Muin ud-Din Chishti (Chapter 3). A typical Sufi centre of authority, called the *khanqah*, which was the resident seat of a *pir*, combined the functions of mosque, *madrasa* and recreation centre. Most of the *pirs* were *Shariah*-minded persons of great learning, but they differed from the traditional *ulama* in that they combined traditional theology with mysticism and mystic insights.[62] The most famous Chishti *pir* of the second half of the thirteenth and the early fourteenth centuries was Nizam ud-Din Auliya (1238–1325), a towering figure in Delhi during the reigns of seven sultans: Balban, Muizz ud-Din Kaiqubad, Shams ud-Din Kayumarz, Jalal ud-Din Khalji, Ala ud-Din Khalji, Mubarak Shah and Ghiyas ud-Din Tughluq.[63] He exhorted his followers to keep a distance from the royal court and the nobility but to remain close to people at grass-roots level (Excerpt 4.2). Some Sufis carried this advice to its extreme, and in their excessive zeal for other-worldliness sometimes collided against the sultans' worldly authority. Despite all the differences among Muslim groups – Sunni, Shia, Sufi etc. – the steady rise of Islam continued during this period.

The treatment of Hindus

Since the second half of the nineteenth century, a populist anti-Islamic sentiment has taken root amidst a large number of Hindus. It is often fuelled by writers and propagandists using evidence from either the horses' mouths of medieval Muslim historians of India who were highly biased against Hindus, or from imperial Victorian historians who asserted the benefits of British rule, or from foreign historians who often uncritically relied on medieval accounts (Excerpt 4.3). It is therefore necessary to try to understand the position of Hindus in the Delhi Sultanate in a more measured way.

There can be very little doubt that some of the general prejudice against the Hindus arose out of the several discriminatory provisions in Islamic *Shariah* law.[64] A non-Muslim's word was often not accepted against a Muslim; the murder of a non-Muslim was not as heinous a crime as that of a Muslim; non-Muslims could be forbidden to wear clothes fashionable among Muslims or to ride a horse; and so on. The vociferous protagonists of such discriminations were mostly the ill-educated and fanatical section of the *ulama* who urged the sultans of Delhi to harass and oppress the Hindus. Their attitudes and behaviour were an extension of Islamic triumphalism of the period when North Indian cities and provinces were succumbing to Turkish armies. They were also joined by those sycophantic writers, historians and courtiers whose religious prejudices overruled their rational thinking. Since every sultan had to be mindful of the potential threat from the Islamic clergy and elite, each one had to placate them periodically by either drawing up harsh policies towards the Hindus or at least making some symbolic anti-Hindu gestures.[65] As hard-headed realists, however, the sultans realized the limitation of their power vis-à-vis the Hindu population. It was the Hindu financiers who bank-rolled them for their wars; without Hindu labourers most of the urban construction works could easily be jeopardized; skilled Hindu craftsmen were needed even for Islamic architectural designs and motifs; without the educated upper-caste Hindu civil servants the administration of the state could not function effectively; Hindu assayers were critical to the successful running of the mint; and thousands of Hindus were enlisted in the sultans' armies.[66] For these reasons the sultans hearkened to the counsel of their Islamic advisers but without necessarily acting on their advice.[67] Indeed, some sultans went out of their way to impress the Hindus by adopting such Hindu royal customs as riding on elephants, having their horoscopes prepared by Hindu astrologers,[68] imprinting the Hindu deity of Lakshmi on the reverse of their gold coins and honouring Hindu poets, musicians and *Brahman* intelligentsia at the royal court. The loyalty of the vast mass of the Hindu population within the Delhi Sultanate was secured by these means.

Three particular questions regarding the treatment of the Hindus require our further attention. The first was the imposition of the *jizya* poll tax on non-Muslims; the second was the extent to which conversion to Islam was forcibly carried out; and the third was the issue of Hindu temples.

From the time of Prophet Muhammad onwards, Muslim preachers and potentates had distinguished among non-Muslims, the *dhimmi* or the People of the Book, and the *kafir*s who did not possess any heritage of written sacred literature. None of the Delhi sultans ever officially proclaimed the Hindus as *kafir*s, whatever they might have thought about them in private. The life of the *dhimmi* under Muslim governance was not easy, but there were at least clear rules that obligated both rulers and the ruled. The main contract stipulated that the *dhimmi* could continue to follow their religion and to expect and receive the protection from the Islamic ruler as long as they paid a poll tax, the *jizya*.[69] In a sense, this was a tax in lieu of military service which, it was presumed, a non-Muslim could not be expected to offer. Since, however, a sizeable number of Hindus enlisted in the armies of the sultans, the *jizya* in India ceased to be a universal tax on non-Muslims. There were always several exceptions which had to be accommodated. The collection of the *jizya* was also varied. Although intended as a poll tax on each individual non-Muslim, quite often it came to be collected from individual villages in the countryside.[70] The *jizya* yielded substantial sums for the sultans' treasuries – a compelling factor to be taken into account by any sultan who ever wished to embark on a fanatical policy of converting the Hindus.[71]

Since Islam, like Christianity, is a missionary religion, conversion of the heathen or the *kafir* has always been both overtly and implicitly accepted as a chief *raison d'être* behind its wars of expansion. In India, the payment of the *jizya* certainly pressurized many a poor or lowly Hindu to convert to Islam; but there is very little evidence of forced conversion within the heartland of the sultanate during our period. Most of the conversions had taken place in the frontier lands or in regions like Sind, Bengal, Punjab or Malabar.[72] It was there that many of the Muslim converts had been drawn from indigenous groups which had not been integrated in the Hindu social order or the caste system. These groups included forest tribes or peasants, as in Bengal, or the pastoral and agrarian peoples like the Khokars and Jats of Punjab. Quite often, in places like Sind, Gujarat and Malabar, conversion went hand in hand with intermarriage among the trading classes.[73] But while any of the Delhi sultans would have been pleased to offer a robe of honour and a gold ornament 'to a Hindu who presented himself for conversion',[74] none of them was vicious enough or stupid enough to attempt their forcible conversion to Islam.

The record of the sultans is more mixed over the question of Hindu temples. The mindless iconoclasm initiated by Mahmud of Ghazna influenced some of the Delhi rulers. Qutb ud-Din Aybeg is supposed to have destroyed nearly 1,000 temples and raised mosques on their foundations.[75] Iltutmish, despite his otherwise relatively liberal outlook, razed temples at Ujjain and Bhilsa.[76] Much greater destruction was wrought by Ala ud-Din Khalji at places like Chitor and Bhilsa, while his deputy Malik Kafur carried out an orgy of desecration in his southern campaigns on his master's behalf.[77] The figures are, however, quite often exaggerated. None of the sultans ever destroyed Hindu temples *en masse*; the destruction was limited to only those temples that were seen in their eyes, and in the eyes of the Hindu devotees themselves, as the potent symbols of any Hindu ruler who challenged the authority of the sultans. On that basis, according to a distinguished historian, only eighty Hindu temples were desecrated by Muslim rulers between 1193 and 1729.[78] Occasionally, the sultans also provided funds to repair the temples.[79]

The Sultanate architecture

The mosque and the tomb were the two eminent Islamic buildings that were new to India. Resting on a plinth, the mosque (*masjid*) normally has an extensive courtyard (*sihn*) with a tank in the centre for ablutions; and is surrounded by pillared cloisters (*liwan*), on three sides, with several entrances. Since a Muslim has to face Mecca while praying, all Indian mosques have domed prayer halls at the western end of the courtyard. In the back wall of the prayer hall is a niche or a recess (*mihrab*) indicating the direction for prayer (*qibla*), while the pulpit (*minbar*) is generally to its right.[80] The Islamic tomb is normally a square building (*hujra*), with one chamber standing on a raised platform which is covered by a dome (*gumbad*). In the centre of the square building stands the cenotaph (*zarih*), a rectangular structure placed above the grave (*qabr*) in the chamber below (*maqbara*). The niche (*mihrab*) is in the western wall of the building.[81]

The Hindu masons, who were the mainstay of the labour force, had to learn the techniques of constructing four important architectural features of Islamic architecture: the pointed arch with a keystone at its centre, the dome which required transforming the rectangular or square shape of a room in order to accommodate its round base, the tower (*minar*) and the geometrical patterns and calligraphy as decorative motifs (Plate 1). These were architecturally complicated skills, even for masons of India with a long tradition of building beautiful temples; but they were mastered over time with help from specialist immigrant Muslim architects.[82]

Indo-Islamic architecture can be divided into three stylistic periods: the sultanate, the regional and the Mughal. Each period saw the evolution of new and developing styles in the building of cities, walls, mosques, tombs, palaces, forts and many other types of buildings. The monuments built by the Delhi sultans until the mid-fourteenth century will be dealt with in this section, while those of the later sultanate, the regional and the Mughal eras will be examined in the next two chapters.

It is in Delhi that we can best witness the ruins and monuments from the sultanate era. In 1193, on the site of the old Hindu citadel of Qila Rai Pithora, Qutb ud-Din Aybeg founded the first of the six cities of Delhi under Islamic rulers (Plate 2). He commissioned the building of a great mosque, the Quwwat ul-Islam (The might of Islam), and a monumental tower, the Qutb Minar, in celebration of Muhammad Ghuri's great victory over Prithviraj Chauhan (Chapter 3). Small mosques had been built earlier by Muslims in Sind, Punjab and Malabar, but the Quwwat ul-Islam was the first mosque in India conceived on a grand scale. A few years into its construction, Qutb ud-Din ordered the building of an arched façade, with five openings, in front of the sanctuary, serving as a screen. Most of the material for the cloister pillars, walls, roofs and paving stones were pillaged from the twenty-seven Hindu and Jain temples that had once stood on the Qila Rai Pithora.[83] The original sculptures 'were mutilated or were so set in walls that the unworked sides of stones were all that could be seen'.[84] The Hindu floral motifs and ornaments remain as mute evidence.

Qutb ud-Din died in 1210 without seeing the completed mosque or the minaret. It was under his successor, Iltutmish, that the project came to be extended and further developed architecturally. He doubled the size of the mosque complex, by widening the courtyard and extending the screen of arches on each side. The replacement of florid Hindu motifs by arabesque patterns and Quranic inscriptions was now gathering apace and authentic Islamic influences and styles from the Middle East were making themselves more pronounced.[85] The Qutb Minar, whose upper levels were finally completed in 1368, is considered as one of the finest monuments of this period; and so is Iltutmish's tomb.[86] Further architectural developments occurred under Ala ud-Din Khalji who built the second city of Delhi at Siri.[87] His majestic gateway at the entrance to the Qutb complex extensions, the Alai Darwaza, provides us with the first indication of the future glory of later Mughal architecture a few centuries hence.[88]

Islamic architecture became more assertive under the Tughluqs, particularly with the standardization of Islamic techniques and materials within a highly organized structure that received substantial patronage from the sultans. The first of the Tughluqs, Ghiyas ud-Din Tughluq, ruled for only five years but began an ambitious scheme to create a formidable new fortress at Tughluqabad, the third city of Delhi.[89] His successor, Muhammad Tughluq, began the building of the fourth city, the Jahanpanah (the World's Refuge), whose ruins are still extant.[90] However, his true fame, or rather notoriety, lies in the fact that he ordered the transfer of the capital from Delhi to Daulatabad in central India and the transportation of all the skilled Muslim artisans of Delhi.[91] This move, along with the depletion of the treasury, meant that the fine stonework of the Delhi architecture of earlier reigns was replaced by cheaper material, more of a plaster over a rubble core. On the other hand, Delhi's loss proved to be a gain for the Daulatabad and the Deccan region where the seeds of a future great tradition of Indo-Islamic architecture were planted.

Narrating and composing in the Persian tradition

The dominance of Persian was marked in the Islamic literature of this period, be it historical, religious or poetic. This dominance was the result of migration and settlement of Persian scholars throughout the thirteenth century when the Mongol pressure on Persia became intense.[92] From their works, we can learn much about the political, social and cultural spirit of the age. As they sought protection and patronage of the royal court, they were sometimes prone to flattering their patrons obsequiously; therefore, their historical texts should be treated with circumspection.[93] Three types of historical texts were written: the *manaquib* which used florid and sentimental language to describe episodic themes, the *tarikh* or general history, and the *adab* or didactic texts about codes of conduct. The last two yield us relatively greater historical information, but they still lacked the standards of objectivity and detachment that we look for in good modern historical works.[94] Their authors were Islamo-centric in the extreme, and history for them was essentially the story of God's plan to bring about the victory of Islam everywhere. Notwithstanding this limitation, the works of four historians of India, writing in Persian during this period, merit attention. Two of them – Fakhr-i Mudabbir and Juzjani are the historians from the thirteenth century; while Isami and Barani belong to the first half of the fourteenth.

Fakhr-i Mudabbir, a first generation immigrant arriving in India in Qutb ud-Din Aybeg's time, was the author of a didactic military text, *Adab al-Harb wal-Shajaa*, which deals with instruments of warfare and Islamic principles and practice for the sultans to follow. His detailed analysis of *jizya* also throws light on this complex and controversial levy.[95] The second, and more weighty a historian, was Minhaj al-Din bin Siraj al-Din Juzjani, a native of Ghazna, who first came to India during the reign of Iltutmish. His magisterial work of history, named after sultan Nasir ud-Din Muhammad, the *Tabaqat-i Nasiri*, was completed in 1260. The text, comprising of 23 *tabaqats* or sections, is a general history of the world, with sections 19 and 20 dealing with the history of the sultanate from 1206 to 1260. A great deal of personal information about the sultans and nobles at the court, along with economic data concerning the *iqtas*, can be elicited from the text, despite the confusing and contradictory presentations of events and episodes.[96]

Juzjani did not go beyond 1260 and for that we have to rely upon a third historian, Isami, who lived in the first half of the fourteenth century and who wrote an epic chronicle, the *Futuh al-Salatin*. This text deals in detail with the events from 1260 to 1349, particularly his family's forced transfer from Delhi to Daulatabad by Sultan Muhammad Tughluq. Isami possessed great ability to marshal evidence and write expressively.[97] His contemporary, Zia ud-Din Barani (1285–1357), perhaps the most outstanding scholarly historian of the fourteenth century, is the fourth great authority on the sultanate. His faults lay in his religious and class prejudices; for him divine providence as ordained by Islam, rather than human affairs, was the key to understanding history; and he had a low opinion of both Hindus and Muslim converts from lower-caste Hindus. His book, *Tarikh-i Firuz Shahi* is, however, one of the grandest histories of medieval India, containing an elaborate philosophy and rationale of history, with eloquent and dramatic narrations of famous episodes from the history of the sultanate. Whatever reservations we may have about his personal views, Barani remains the seminal primary source for this period.[98]

Both Arabic and Persian were the mediums in which much of the religious literature of the period was expressed. The texts, most of which were not particularly original, covered the four great branches of Quranic studies: the traditions of the Prophet (*hadiths*), jurisprudence (*fiqh*), exegesis (*tafsir*) and mysticism (*tasawwuf*).[99] The systematic study of Islamic law in India began with the arrival of Maulana Burhan ud-Din from Balkh in the reign of Sultan Balban. He brought with him the *Hidaya*, the classic legal textbook, written in Arabic. Balban ordered its translation into Persian and he and his court sat at the feet of Burhan ud-Din after Friday prayers, learning the subtleties of Islamic law.[100] While jurisprudence and exegesis remained specialist fields in the hands of jurists and scholars, the mystics led the way in *hadith* studies and Sufi devotional writing. The most important work on the *hadith*, *Mashariq al-Anwar*, was a collection of 2,253 sayings of the Prophet, written by al-Hasan as-Sagani al-Lahori, a mystic of the early thirteenth century.[101] It remained the guidebook for ordinary Muslims of India for subsequent generations, partly owing to the teachings of the Sufi saint Nizam ud-Din Auliya, a central religious personality of the sultanate in this period.[102] The collection of sayings of Nizam ud-Din and other Chishtiya spiritual masters gave rise to a new class of Islamic religious literature in India, known as the *malfuzat*.[103] The most famous *malfuzat*, the *fawaid al fuad* (Things profitable to the Heart), a record of conversations by Nizam ud-Din between 1307 and 1322, was written by Hasan Sigzi Dihlawi, an early fourteenth-century poet.[104]

There were a number of poets writing in Persian during this period, but the one figure who stands head and shoulders over all is Amir Khusrau, a disciple of Nizam ud-Din Auliya. Their intimate closeness is symbolized by their adjacent tombs in Delhi.[105] Born in 1253 to Indo-Turkish parents, Amir Khusrau was a literary child prodigy, excelling from the earliest age in the art of poetry and poetic writing. His genius at writing and composing was appreciated at many royal courts in India; while adroitly displaying his loyal allegiance at each one, he always prepared to move on to the next.[106] He was briefly imprisoned by the Mongols in 1284, but finally settled in Delhi where he died in 1325. He was familiar with all the different styles of Persian poetry, such as the *qasida*, *gazal*, *tarana* and *khyal*, but his true talents are to be witnessed in his long epic compositions, the *masnavi*s.[107] His skills of word-play and riddles, along with the use of terms from different languages like Arabic, Persian, Turkish and Hindi, have made his poems a joy for both the cultured and the unlettered people of South Asia unto this day (Excerpt 4.4). His *masnavi*s, arranged in five *divan*s or sections, with each containing a collection of poems from different phases of his life, are not only highly dramatic in effect but provide much useful knowledge of the culture of his times.[108] A number of *masnavi*s deal with royal episodes celebrating victories, marriages or meetings. From his well known composition, *Nuh Sipihr* ('Nine Spheres'), one becomes acquainted with his perceptions of India, his love and enthusiasm for his native country, its customs, flora and festivals, its legends, chess and music.[109] India, to Khusrau, was an earthly paradise.[110] Historians of Indian classical music credit Khusrau with the development of the instruments of *tabla* and *sitar* and, for Khusrau, 'this music has a peculiar charm not only for human beings, but for animals also'.[111]

Select excerpts

4.1 Distribution of food during the famine

The Delhi Sultanate lacked administrative machinery to oversee famine relief throughout the kingdom but, in the interest of its own security, made sure that the capital's populace was not unduly affected. This has been a recurring phenomenon throughout history. The metropolis always wins over the periphery.

> When the severe drought reigned over the lands of India and Sind and prices rose to such a height that the mann (about 15.5 kgs) of wheat reached six dinars, the Sultan (Muhammad bin Tughluq) ordered that the whole population of Dihli should be given six months' supplies from the (royal) granary, at the rate of one a half ratls (lbs) per day per person, small or great, free or slave. The jurists and qadis went out to compile the registers with the (names of) inhabitants of various quarters; they would then present them (to the authorities) and each person would be given enough to provide him with food for six months.
>
> (Gibb 1971: 695)

4.2 Political wisdom of Nizam ud-Din Auliya

Acting on the advice of his spiritual mentor, Shaykh Farid Ganj-i Shakar, Nizam ud-Din Auliya asked all his followers to avoid contacts with kings and princes or seeking favours from them. Some in the political establishment were suspicious of the saint's motives; they were worried that his popularity with the ordinary populace in Delhi could pose a political danger. His consistent stand on the matter, however, persuaded the sultans of his sincerity.

> Sultan Jalal ud-Din Khalji offered him (Nizam ud-Din Auliya) some villages, which he declined to accept. The Sultan then sought an interview with the Shaykh but was turned down. He then thought of a surprise visit. 'My house has two doors,' warned the Shaykh; 'if the Sultan enters by one I will make my exit by the other'.
>
> (Lawrence 1992: 35)

4.3 Negative stereotyping based on uncritical medieval Muslim histories

A typical example of stereotyping from a book written by the distinguished American historian Will Durant is as follows:

> The Mohammedan conquest of India is probably the bloodiest story in history. The Islamic historians and scholars have recorded with great glee and pride the slaughter of Hindus, forced conversions, abduction of Hindu women and children to slave markets and destruction of temples carried out by the warriors of Islam during 800 AD to 1700 AD. Millions of Hindus were converted to Islam by sword during this period.
>
> (Durant 1935: 459)

The assertions contained in the above statement are a god-send to those Hindus who wish to inflame religious passions and hatreds against the Muslims in India today.

4.4 The poetic genius of Amir Khusrau

Amir Khusrau was not a stuffy poet hide-bound by conventions. His poems are in the older Persian tradition of defiance of taboos that society and Islamic religious authorities imposed on writers and poets. The following two odes are good examples of Khusrau's lightness of touch.

On life

> The times are good, so let's all drink
> From having fun we shouldn't shrink
> Let's all in wine put our trust
> For in the end, we'll all be dust
> Listen to the bell and do not fuss
> For one day, it will toll for us
> And let's enjoy our girls and wine
> Let's go out and drink and dine
> Half dead we are and half alive
> Let's with wine our hearts revive
> Being in love, we cannot sleep
> We're like dogs and watch we keep.
> In boasting, Khusro, we know you wallow
> But the drum you beat is so very hollow.

On wicked women

If she is yours because you are greedy
Better it is to be lonely and needy
And if she is yours because of money
She is no honey, and it is not funny.
And if she shows no mercy or pity
Delay you not and leave the city.

(Shaida 2008: www.booksurge.com)

5

NEW CENTRES OF MUSLIM POWER AND CULTURE (1351–1556)

The two centuries covered in this chapter witnessed a dynamic period of change in Muslim power and culture in India. Imperial politics gave way to regional politics, and Delhi's political authority mattered less and less after the middle of the fourteenth century. Muslim polities and structures flourished in other Indian centres of power (Map 5.1). This centrifugal tendency continued almost unchecked until 1526 when Babur and, after him, his Mughal successors began a process of restoring power to the centre. Through all the political upheavals of this period, however, Islam continued to consolidate itself quietly and almost imperceptibly, while new and bold attempts at Indo-Islamic fusion in forms and styles were beginning to foreshadow the dawn of a unique cultural synthesis under the Mughals.

Fluctuations of imperial power and fortunes

The fading of the Delhi Sultanate

After 1351 the Delhi Sultanate was in crisis. Its decline can be explained by three main factors: the financial profligacy of Firuz Shah Tughluq, the first of the Delhi sultans of this period (see Dynastic lists), the rivalries among his successors, and the sack of Delhi by Timur the Great in 1398. Firuz Shah's predecessor, Muhammad Shah Tughluq, had left the treasury nearly empty by his impetuous schemes of expansion (Chapter 4). While Firuz Shah secured a degree of prosperity for the general populace, he showered uncalled for generosity upon the most powerful and wealthy in the kingdom. The army commanders were remunerated by assignments of land rather than by fixed salaries; and no system of checks existed to prevent those who had been assigned lands from greatly exploiting their peasants. This contributed to increasing inequality and corruption during Firuz's reign.[1] Firuz also granted away the *iqta* lands in perpetuity on an hereditary basis, a practice that was never permitted during the heyday of the sultanate. Such assignments significantly degraded military capacity, with the result that Firuz could not reclaim the provinces that had seceded.[2] With Firuz also becoming an anti-Hindu bigot in his advancing years and losing popularity among his majority subjects, his reign ended in political chaos.[3]

A succession struggle weakened the sultanate further. During the next decade, the crown of Delhi passed from one of Firuz Shah's great-grandsons,

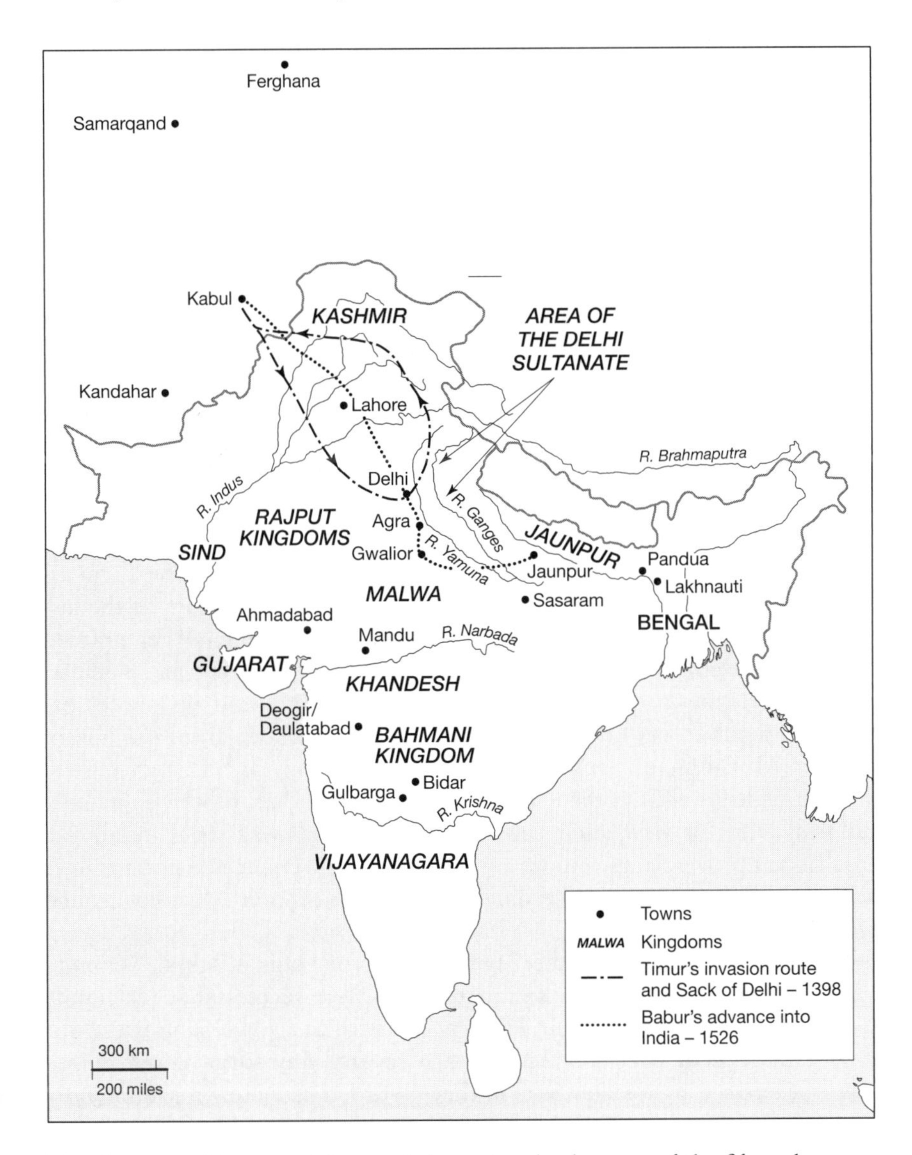

Map 5.1 From Timur to Babur, and the regional sultanates of the fifteenth century

Ghiyas ud-Din Tughluq Shah II to a grandson, Abu Bakr Shah, then to his third son, Nasir ud-Din Muhammad Shah, followed by the latter's son, Ala ud-Din Sikander Shah, resulting in a further civil war between the followers of another great-grandson, Nasir ud-Din Nusrat Shah and those of a further grandson, Nasir ud-Din Mahmud Shah who happened to be the last of the Tughluqids (see Dynastic lists). It was no wonder that, amidst this strife, the invader Timur was able to fatally wound the Delhi Sultanate.

Timur Lang/Tamerlene (1336–1405) was a Turk, not a Mongol; with military prowess and deceit he controlled much of the Mongol-Chagadayid realm by 1370.[4] After destroying nearly all Persia, Georgia and the Tatar Empire he vented his wrath on the Delhi Sultanate. In December 1398 the city of Delhi was comprehensively sacked by his troops who massacred its citizens and plundered their wealth. It might be that he lost control of his troops and that they inflicted damage 'which he did not intend'.[5] The end result, nevertheless, was horrendous. Except for a small section of the Muslim intellectual and religious elite that could be manipulated by him, very few were spared. Hindus and Muslims suffered in equal measure, and their resistance was in vain.[6] Timur took away with him not only their material assets but also Delhi's finest craftsmen and builders with whose skills he was to make Samarqand the queen of Central Asian cities (Excerpt 5.1). For two centuries, the Delhi rulers had denied a single victory to the Mongols; and yet within a period of a few days Timur had devastated the Sultanate. The Tughluq dynasty carried on ingloriously until 1413.

Two lesser dynasties – Sayyids and the Lodis – were to follow the Tughluqs over a period of more than a hundred years. Within two years of the last Tughluqid's death, the Amir of Multan, Khizr Khan, who had been granted a hereditary right over his *iqta* by Firuz Shah Tughluq and who had a helpful patron in Timur's son, became the new sultan. He and his successors claimed, without any credible evidence, to be descended from the line of Prophet Muhammad's family, calling themselves Sayyids.[7] This gave them a spurious legitimacy in the eyes of Muslim masses; but, otherwise, from a political point of view, their rule of 37 years in all was spent in fighting recalcitrant provincial governors and nobles. As two eminent historians have said, their 'political vision was confined to a radius of some 200 miles around Delhi'.[8]

Between 1451 and 1526 the Afghan family of Lodis – Bahlul, Sikander and Ibrahim – controlled the sultanate. They were successful at reclaiming lost provinces;[9] however, their problem lay with their own supporters, the dominant Afghan nobility. The Afghans resisted any form of centralized power exercised by a sultan whom they regarded as merely a *primus inter pares*.[10] The dynamic personality of Sikandar Lodi was able to bring them to heel, but when his successor, Ibrahim, fell foul of them, they conspired with his uncle and invited Babur, the Timurid ruler of Kabul, to invade India.

Ibrahim was defeated and slain at the Battle of Panipat in 1526, and the Delhi Sultanate ceased to exist.

The first two Mughals and the Surs

Babur, an Uzbek warlord, was the founder of the Mughal dynasty and empire. The term 'Mughal', used by the Persians to refer to the Mongols, is now universally associated with Babur's dynasty; but it is worth remembering that the Mughal emperors called themselves Timurids. Babur came from a complex family background that may be described as Turco-Mongol. On his father's side, he was a fifth generation descendant of Timur Lang, which made him a Turk. From his mother's side, he was a descendant of the Chagadayid successors of Genghis Khan, bringing the Mongol element into prominence. Babur's personality was therefore shaped by his Turco-Mongol family and cultural inheritance.[11] Emotionally, however, he identified himself with Timur and the Timurids; and many of his empathies lay with his Turkish ancestors.[12] Like many Timurids, he also greatly admired Persian culture.[13] His ambition was to be the new, perhaps a gentler, Timur; but this was not to be easy on account of many enemies from within his extended family and other Uzbek potentates. Born in 1483, he had inherited at the age of eleven the small principality of Ferghana, near the Timurid metropolis of Samarqand in Uzbekistan. This was never to be enough for a man of such restless and prodigious energies, whose dream was to be the ruler of Samarqand itself. But fate had decreed otherwise. Thrice, between 1494 and 1512, he came to possess Samarqand, sometimes with the help of the Persians, only to lose it to his enemies for one reason or another. His conquest of Kabul in 1504 was like a second prize for him; but, having been thwarted decisively in Uzbekistan, he decided to build on what he had. From Afghanistan he eventually turned to India in 1519.[14]

Ibrahim Lodi's rule in Delhi had become highly irksome to his Afghan supporters, and it was they who invited Babur to interfere for their own reasons. He considered India to be a useful prize, partly because its great wealth could enhance his meagre Afghan resources and partly because it could offer him protection and refuge in case of an attack from the Uzbeks. His victory at Panipat on 21 April 1526, was due to two specific military tactics: the use of heavy artillery, the first of its kind to be deployed on an Indian battlefield, and coordination between artillery and cavalry in cornering and confusing the Lodi forces.[15] Once established in India, Babur distanced himself from Timur's legacy, as the latter's memory was not cherished by either Hindu or Muslim elites.[16] Instead, he came to portray himself as a *ghazi*, the holy warrior of Islam, in the image of his new hero: Mahmud of Ghazna.[17] Unlike Mahmud, however, Babur decided to make India his home; this meant that roles were to be reversed between Kabul and Delhi. Kabul ceased to be the metropolis and became a periphery. The Afghan nobles were none too happy about that. Babur further consolidated his hold on the centre of power in India by three decisive victories against the rulers of Malwa, Mewar and Bengal. He sat on the throne of Delhi as *Badshah*, the emperor, over and above the greatest and strongest of Indian princes (Plate 4.1).[18] He ruled India for just four years; but as the founder of one of the world's most famous dynasties he has a permanent place in the pages of history.

Babur's son, Humayun, came to the throne in 1530 and lost it within ten years. Three powerful forces were ranged against him: his own brothers; a powerful sultan of Gujarat, Bahadur Shah; and Sher Khan, an Afghan soldier from the Sur tribe in the service of Babur, who had risen to become the Prince of Bihar. Humayun fought all three vigorously, but failed to follow up his victories with decisive control and retribution. In the end, it was Sher Khan who got the better of him and who in 1540 forced him into exile from Delhi.[19] Seizing the Mughal throne Sher Khan was crowned as Sher Shah Suri. His Sur dynasty was short-lived, although during his five years as ruler he brought most of North India under central control, carried out remarkable administrative reforms and provided patronage for arts and public works. After his death in 1545, the dynasty lingered on for a decade under his successors. In the meantime, Humayun fled to Iran in 1544 after encountering many difficulties and misfortunes within India. Within a year, with help from the Iranian ruler, Shah Tahmasp (Excerpt 5.2), he captured Kabul and Kandahar from his brothers; he handed over Kandahar to the Persian king for his help but kept Kabul for himself. From there, during the next ten years, he organized himself and watched the steady decline of the Sur dynasty. With further Persian help, he struck finally in 1555 and re-conquered Delhi.[20] Although he was dead within six months, after falling from the stairs of his library in 1556, Humayun had redeemed himself by restoring the Mughal state.

Administration under Sher Shah Suri

The restoration of the Mughal state by Humayun reversed the process of fragmentation and provincialism and foreshadowed the pan-Indian imperial project of Akbar, his son and successor. It is, however, worth acknowledging that Akbar had a forerunner in Sher Shah who had a genius for administration that few rulers could match. During his five years' reign (1540–5), with iron discipline imposed upon his army, he recovered most of the lands lost since the later Tughluqid times. His writ ran from the north-west frontier and Baluchistan up to Bengal. Ruling from Agra (rather than Delhi), he imposed administrative units called *sarkars* (meaning 'government' or 'authority' in Persian) throughout his kingdom.[21] Over each *sarkar* a trusted official acted as his military agent. The later *subahs* of Akbar's reign and the districts under the British were more or less based on the model of Sher Shah's *sarkars*. Within each *sarkar* there were local territories, called *parganas*, where civil officials were employed to keep the peace and collect revenues. Revenue administration was guided by two principles: equity and humanity. Revenue demand by the state was based not on arbitrary assessments but on the method of measurement.[22] This was followed by a classification of land into good, middling and bad. Revenue demand from the peasants was for a third of the produce in kind or cash within each category of land; and specific documents, known as *kabuliyat* (deed of agreement) and the *patta* (title deed), were exchanged between the peasant and the revenue officer. The peasants had the right to question the basis on which the measurement was carried out and the land thus classified; but once agreed, had to meet the state's revenue demand without question. Sher Shah's success in raising large land revenues without crushing the spirit of the peasants might be explained by the fact that he had gained practical experience of farming and management of peasants before becoming a king.[23]

Other achievements of Sher Shah may also entitle him to be called a harbinger of a Mughal administration. An incipient police system, aided by harsh but clear guidelines on matters of law and order, along with much espionage, was established.[24] An extensive road network, with *sarais* (rest houses for boarding and lodging) constructed every five miles, was built and made safe for all travellers.[25] Irksome internal imposts and duties were abolished and trade thereby encouraged.[26] A new standardized silver-based monetary unit called the *rupiyah*, the ancestor of modern Indian rupee, was introduced and remained in circulation until 1835.[27] Sher Shah's concern for speedy and fair justice for his people and the absence of religious prejudice against Hindus or other non-Muslims are also indicators of a sovereign who was ahead of his times and who thought in secular terms.[28] Greater the pity therefore that he ruled only for five years and that his successors proved singularly unable to carry on with his good work.

The rise of regional sultanates

Until the coming of Babur in 1526, three sets of political structures existed within South Asia. The first was the declining Delhi Sultanate. Second, there were a number of Hindu kingdoms, the most powerful being Vijayanagara in the south and some of the Rajput principalities. Third, there were eight Muslim regional sultanates, which we review in this section. Six of them arose out of the varied rebellions that the Tughluqid sultans of Delhi faced from the late 1330s onwards. The other two – Sind and Kashmir – had developed independently.

Sind and Kashmir

Muslim rule in Sind antedated the Delhi Sultanate by many centuries; but, although its rulers could not isolate themselves from Afghan invaders or the imperial Delhi rulers, their politics remained parochial. The long established Sumra dynasty gave way to an obscure local dynasty of the Sammas in the fourteenth century; and the latter were in turn ousted by the Afghan Arghuns in the fifteenth. Notwithstanding their dynastic conflicts, Sind retained its independence and autonomy until Emperor Akbar annexed it in 1593. Islam had become the dominant faith in the region, but the model of tolerance established in the early eighth century by Muhammad bin Qasim, along with the influence of Sufi saints, and the continued presence of Hindus and heterogeneous Shia groups,[29] all helped to create a society in Sind largely free from religious tensions.[30]

Like Sind, the other Muslim kingdom, without strong ties with Delhi, was Kashmir. However, unlike in Sind, Islam gained a strong foothold in Kashmir only as late as the fourteenth century. Until then, the region was ruled by Hindu or Buddhist dynasties. Muslim rule began with Shah Mirza Shams ud-Din in 1339, and his line continued until 1561. A local family of the Chakks then ruled for twenty-seven years, before Kashmir was finally absorbed into the Mughal Empire by Akbar in 1588. From among the many Muslim kings during this period, the names of two stand out: Sikander (r. 1389–1413) and Zainul Abidin (r. 1420–70). Medieval Kashmir was transformed under their rule. There were periods of religious tension, but attempts were also made to reconcile Hindus and Muslims. Arts and crafts were greatly patronized, encouraging many skilled immigrants from Persia and Central Asia to settle in Kashmir.[31]

Bengal and Jaunpur

Of the six renegade sultanates that had seceded from the control of Delhi the most important was Bengal. Owing to its great wealth from natural resources and trade, the Delhi Sultanate had never tolerated any dissent or attempt at secession of this province, but in 1339 Shams ud-Din Ilyas Shah broke away from imperial control and transferred his capital from Lakhnauti to Pandua. His Ilyas Shahi dynasty ruled Bengal during two separate periods: 1342 to 1415, and from 1433 to 1486; and under its rulers a composite Bengali identity, encompassing both Hindu and Muslim symbols and motifs, came to be reflected in coins and architecture.[32] Although a powerful Hindu landlord, Raja Ganesh, carried out a *coup d'état* in 1415 and managed to put his son on the throne, the latter converted to Islam and then continued the Ilyas Shahi tradition of compromise and fusion.[33] Similarly, when another dynasty, that of the Husain Shahs, acquired power in 1493, it, too, followed in the path of the Ilyas Shahis. Under the first two Husain Shahs, Ala ud-

Din (r. 1493–1519) and Nasir ud-Din Nusrat (r. 1519–32) Bengal not only prospered but strengthened its cultural distinctiveness.[34] Through the ups and downs of Bengali dynasties in this period, one factor remained constant: the continuing process of indigenization incorporating Hindu and Islamic representations in politics, art, literature and coinage.

To the west of Bengal lay the sultanate of Jaunpur, a fertile area of the mid Gangetic plains, always considered by the Delhi sultans as their patrimony; but, after witnessing the Tughluqid chaos that followed after Firuz Shah's death in 1388, Malik Sarwar, the governor, seceded in 1394 and established his own Sharqi dynasty. The term 'Sharqi' was derived from his title, *Malik al-Sharq*, or Lord of the East, originally awarded by Firuz Shah.[35] The dynasty prospered; and its army proved to be more than a match for the imperial forces. By 1483, however, the Jaunpur kingdom had ceased to exist, after suffering great defeats at the hands of the Lodi sultans of Delhi.[36] During its heyday, under Ibrahim Shah Sharqi (r. 1402–40) and Mahmud Sharqi (r. 1440–57), Jaunpur became both a centre of Islamic culture and Hindustani learning and music; and it attracted the finest scholars, devout Sufis and the most skilled artisans from other parts of India, Iran and Central Asia.[37]

Gujarat

A little later than Jaunpur, the Gujarat Sultanate was established in 1407. Gujarati prosperity was born out of trade and maritime contacts with the Middle East. The Arabs and the Persians were not just frequent visitors; a number of them had settled too. A local Gujarati Muslim population grew through conversions.[38] Although from the year 1000 onwards there were attacks by the Ghaznavids and Ghurids, the indigenous control in Gujarat remained firm until 1296 when it was absorbed in the Delhi Sultanate after Ala ud-Din Kalji's conquests (Chapter 4). From 1296 to 1407, Gujarat was ruled from Delhi. In the chaos of the late fourteenth and early fifteenth centuries, however, in the dying days of the Tughluqid monarchy, the renegade governor Zafar Khan usurped power and crowned himself as Muzaffar Shah. Under the three sultans who followed him – Ahmed Shah I (r. 1411–42), Mahmud I Beghara (r. 1459–1511) and Bahadur Shah (r. 1526–35, 1536–7) – the Gujarat Sultanate became a substantial regional power. The new city of Ahmadabad testified to its rising commercial prosperity that was shared by both Hindus and Muslims.

Malwa and Khandesh

The kingdom of Malwa, strategically placed as a buffer between Delhi, the Rajput states, Gujarat and the Deccan, had belonged to the Delhi Sultanate for nearly a century since its invasion by Ala ud-Din Khalji in 1305.

A Mamluk chieftain, Dilawer Khan Ghuri, originally a tributary of Firuz Shah Tughluq, declared independence in 1401, but in turn lost power to his feudal chieftain, Mahmud Khalji, in 1435. From the capital Mandu, Mahmud Khalji (r. 1435–69) and his successor Ghiyas ud-Din Khalji (r. 1469–1501) ruled over a prosperous Malwa where art and architecture blossomed.[39] Internal dissensions in the early sixteenth century, however, compromised Malwa's independence.

Squeezed between Malwa and the Deccan was the sliver of a land along the banks of the Tapti River, known as Khandesh. In 1370, the Delhi sultan Firuz Shah Tughluq granted extensive lands to a fugitive, Malik Raja, who had been exiled from the southern Bahmani kingdom; but with Firuz Shah becoming increasingly ineffective in his dealings with rebellious provinces, Malik Raja seized the opportunity to declare independence from Delhi in 1382. Claiming unproven descent from the second Caliph, Umar, Malik Raja established his Faruqi dynasty, which lasted until 1601 when Emperor Akbar annexed Khandesh. Despite its relative longevity, the kingdom remained in the shadow of the Gujarat sultanate.

North-central Deccan

The last of our eight regional sultanates was the Bahmani kingdom of north and central Deccan. Muslim power was first projected in the Deccan by the victories of Malik Kafur, the general of the Delhi sultan Ala ud-din Khalji; but while Ala ud-Din was content with the loot from the south, it was Muhammad Shah Tughluq who developed an imperial policy towards the south. The key to this policy was his decision to transfer the capital from Delhi to Daulatabad (Deogir) in 1326–7 and to order at least a tenth of the Muslims of Delhi from all classes to make new homes 600 miles away in the new city. The settlement of so many Muslims in Daulatabad was to help gradually diffuse Islamic culture in the south. The other important decision that Muhammad Tughluq took was to establish two different imperial policies on the two sides of the River Krishna. In the area of the Deccan north of that river he adopted a more intrusive, colonial type of Tughluqid authority; while in the lands south of the river he was prepared to accept the local control by Hindu rulers of the Sangamma family who accepted his suzerainty (Chapter 4).[40] By the 1330s, the Sangammas had broken from the Tughluqids and established their Hindu kingdom of Vijayanagar. Rebellion also followed in the Daulatabad-controlled northern area where Tughluqid rule became highly unpopular owing to Muhammad Tughluq's authoritarian instincts; and by 1347 the Tughluqid governor Zafar Khan threw off Delhi's authority and, with the support of the influential Chishti Sufi sheikhs, was crowned as Sultan Ala ud-Din Bahman Shah. To ensure his safety and complete independence from Delhi, Bahman Shah ordered a new capital to be built further south at Gulbarga.[41] Thus began the Bahmani dynasty and the Bahmani kingdom that lasted approximately until 1518.

The anti-Tughluqid sentiment of the Bahmani sultans also expressed itself in extending a warm welcome to Persians, Arabs and Central Asians from outside North India. The sea ports of the Konkan coast of India, controlled by the Bahmanis, facilitated extensive trade with the countries of the Middle East where Persian culture was much admired. The Persians were the most sought after foreigners, and large numbers of them came to Bahmani lands as merchants, travellers, and settlers. With their arrival, Shiism gained great favours at the Bahmani court. One of the Bahmani sultans, Ahmed Shah I (r. 1422–36), even decided to transfer the capital once again, this time to Bidar, north-east of Gulbarga, to celebrate his kingdom's pride in associating with the Persian world.[42] It was not long, however, before serious frictions arose between the indigenous Muslim elite and the Persians and other foreigners.[43] The native Muslims called themselves the Deccanis and described the foreigners as Westerners (Excerpt 5.3). The rivalry between the two groups reached its height in the 1460s and 1470s as a result of the arrival in 1453 at the Bidar court of a highly polished and seasoned Persian intellectual, merchant and diplomat called Mahmud Gawan.[44] Within five years Gawan had risen to become the *wazir*, serving the Bahmani rulers with great distinction for over twenty years. The Bahmanis were courted by no less a person than the Ottoman sultan Mehmet II, the conqueror of Constantinople. Under Gawan, Bidar became a cosmopolitan metropolis, and perhaps his greatest monument was the college (*madrasa*) that he endowed. Unfortunately, despite all his hard work, Gawan fell victim to the poisonous atmosphere generated by the conflict between the native Deccanis and the Persians; and in 1481 a treacherous plot by the former brought about his execution at the court. Within nine years of his death, the Bahmani state began to disintegrate, to be eventually carved up into five different principalities: the smaller sultanates of Bidar, Berar, Ahmednagar, Bijapur and Golconda.

The Muslim community: a snapshot

Diversity

The first reliable estimates for Muslim population in South Asia only emerge from the 1872 census and, therefore, it is impossible for us to be sure about their numbers in this period. From the various contemporary accounts, however, we can surmise with some degree of certainty the extent of Muslim settlements across the subcontinent. The oldest such settlements were in Gujarat, Kannada lands and Malabar which, as we mentioned in Chapter 2, had been the main trading destinations of early Muslim sailors and travellers. Although immigrant Arabs and Persians intermingled easily with local Indians, the proportion of Muslims remained relatively small in these areas. The greatest concentration of Muslims was on the western periphery, in the frontier regions between the River Indus and Afghanistan, Sind and Punjab. In the heartland of the Delhi Sultanate, that is in the Gangetic Doab and the lands further east, where Muslim power was well entrenched since the twelfth century, the numbers were actually more modest, except for the Delhi area. This, therefore, casts doubts on assertions about conversions being carried out by force. Further east, Bengal had a large urban population of Muslims in our period; but the vast mass of Bengali peasantry in the interior and the delta became Muslim only after the mid-sixteenth century.[45] Smaller Muslim communities were dispersed in different regional sultanates.

From the standpoint of ethnic diversity, we can classify the Muslim community into three broad categories: those from an immigrant background, those converted locally and those with a mixed background resulting from intermarriage. The first and the most prominent immigrant group consisted of Central Asian Turks, whose Mamluk ancestors started invading India in the eleventh century. Whether they occupied the highest echelons of power at the court or served as ordinary soldiers and servants, the Turks maintained a certain distance from other Muslims; and the children of mixed marriages retained pride in the Turkish half of their ancestry.[46] The second immigrant group consisted of Afghans who are quite often confused with the Turks, partly because the Turks had long been in Afghanistan before their entry into India and because both groups are quite often called Turco-Afghans. The Afghans nevertheless considered themselves separate from the Turks; and during the period of the Sayyid, Lodi and Sur dynasties (between *c.* 1410s and 1550s) they took much power from the Turks.[47] The third group consisted of Persians and Farsi-speaking ethnic groups from Khurasan. Considered generally as the most refined and cultured of the immigrants, they were generally refugees fleeing from either the Mongol aggressors or their own intolerant Shia regimes. Their skills and learning were much respected by both the imperial and regional sultans of India who welcomed them. Their presence excited jealousy among other groups of Muslim nobles, as in the Bahmani kingdom where a civil war broke out between them and the local Deccanis, but they managed to retain influential positions within the administrative structures of Muslim governments in India throughout this period and in the succeeding two centuries. The fourth immigrant group was that of the Mongols, sometimes also known as New Muslims. Despite their reputation as fierce fighters, they had been unable to overrun India, but pockets of them had managed to stay on in the north-west and in and around Delhi.[48] The fifth group consisted of those from the Arab immigrant background, mostly to be found on the western coast.[49] Lastly, there were the African Muslims, particularly in the Deccan; a few of them were able to reach the heights of military and political power (Excerpt 5.4).[50]

Even today a large group of African Indians, known as Sidis, who are the descendants of medieval African slaves, soldiers and sailors, continue to reside in Gujarat – a fact that is little known to most Indians.

Besides immigrants there were the more numerous Muslims who had been locally converted from Hinduism. Most of the conversions in this period can be attributed to a desire among some Hindus to escape from the economic and social marginalization that they could be subject to under a Muslim authority and to the efforts of some of the Sufi teachers who could inspire people by both precept and example. Only a tiny fraction of the converted could aspire to positions of authority and power, while the vast majority of them continued with the various manual trades and skills associated with the Hindu sub-castes or *jatis* to which they had earlier belonged. Some of the Muslim *jatis* of Bengal, for example, were those of weavers, butchers, loom makers, circumcisers, bow makers, paper makers, tailors or dyers. Islam endowed them with a certain sense of egalitarianism, but essentially most remained trapped within the confines of a modified and Islamized caste system.[51] Those who converted from Buddhism carried little of the caste baggage and were therefore able to become upwardly more mobile within the Muslim polity, as in Sind, Bengal or Kashmir. The intermarriages between the local Indian Muslims and the immigrant Muslims, in time, produced new generations of Indo-Muslims; and, in the succeeding centuries, they would become the dominant Muslim voice on the subcontinent.

The diversity of denominations and religious traditions should also be noted. Most Muslims from Turkish, Afghan, Arab, African and local Indian background were Sunnis; and so were the sultans and nobles of various dynasties. The majority of those from Persian backgrounds, those from the Ismaili background and their Indian converts such as the Khojas and Bohras, were Shia. The Shia were influential in the regional Bahmani Sultanate owing to the strong Persian presence there. Across the Sunni–Shia polarity there were the Sufis, who acted as missionaries of Islam in the subcontinent.[52]

Indigenization

Increasingly, during this period, the diverse Muslim community was being indigenized, owing to two major factors. The first was that wherever the Muslims lived in India they were surrounded by the Hindu humanity. As more and more Muslims were born within India itself it was inevitable that they were emotionally attached to the very soil and physical landscape of the country and psychologically familiarized with the rituals, immemorial customs and conventions of the native inhabitants from whom they had sprung. A common vocabulary of sentiment transcended many theological and doctrinal differences between the Hindu and the Muslim. This was a time when Hindu reformers – Ramananda (1400–76), Kabir (1440–1518),

Guru Nanak (1469–1539), Mirabai (*c.* 1498–1547), Chaitanya (1486–1533) and Vallabhacharya (1479–1531) – were responding to the defining doctrine of Islam, the *Tawhid*, which proclaims absolute monotheism and unity of God, by initiating what might be called syncretic movements within Hinduism.[53] The Bhakti movement and the Sikh Panth were attempts to make Hinduism less exclusive. The Sufis responded positively to Hindu reforms by a broad-minded receptivity to new ideas. They demonstrated an independent outlook by not being bound to any one particular school of Islamic jurisprudence. The varied reconciliations in thoughts and attitudes, as displayed by the Hindu reformers and the Sufis, did not necessarily bring about a striking or a long term improvement in relations between Hindus and Muslims, but they certainly helped the indigenization process.

The second crucial factor in the rooting of the Muslim community within the Indian environment was that India herself had now, for many centuries, become a major constituency of what historians describe as an Islamicate world system that linked up North Africa, East Africa, the Near East, the Middle East, Central Asia, South Asia and South-East Asia. The strength of the system depended not upon the religion of Islam but on the vast and interlocking trade networks and migratory movements of peoples across these regions. In the Indian Ocean arena, around South Asia and South-East Asia, Islam was actually a minority faith, but it was trade in varied wares and products that nonetheless made the Muslims an influential force.[54] The non-Muslims were also drawn into the Islamicate system, and they and the Muslims learnt to be interdependent upon each other. Powerful Muslim rulers might rule the sultanates, and Hindu rulers might be the lords of Rajput kingdoms or the great state of Vijayanagara in the south of India; but the rulers of neither faith could stay immune to the forces generated by the Islamicate system. What happened as a consequence was that the indigenous cultural traditions and practices became interwoven with Islamic forms in such areas as architecture, literature, dress code and even cuisine. In the coastal areas of the south, which had been part of the Islamicate system since the seventh century, Muslims had gone through the experience of indigenization over a long period of time. For the period covered in this chapter, it is from the regional sultanates that we get the greatest evidence of indigenization. In the succeeding Mughal centuries, it would affect practically the whole country.

Art and architecture: imperial and regional

Imperial architecture

Imperial sultanate architecture reached its apogee in the reign of Firuz Shah Tughluq, the first of the Delhi sultans in this period. He might not have been a sultan who projected power or prevented his empire gradually collapsing all around him, but he was certainly a great patron of architecture. A designer of buildings, it is said that he commissioned the construction of over 200 towns, 100 mosques, 30 *madrasas*, 150 bridges, along with numerous dams, reservoirs, hospitals and baths.[55] He planned the fifth city of Delhi, Firuzabad, and adorned it with splendid monuments, mosques and gardens. The ground on which may be found a few sparse remains and where Delhi's huge cricket stadium is located is known as Firuz Shah Kotla, which contained the ruins of a palace complex, a mosque and the so-called Lat Pyramid, on whose heights Firuz Shah mounted an ancient Ashokan pillar from the third century BCE.[56] In another long range of buildings, known as Hauzz i-Khass, is Firuz Shah's own tomb, an austere and plain block of grey sandstone and a work of great dignity.[57] With Delhi ruined by Timur in 1398 and with the succession of the Sayyids and the Lodis, the imaginative vigour of the architecture of Firuz Shah's reign is replaced by mostly funerary architecture. The Lodi Gardens of Delhi contain a number of square or octagonal tombs of Sayyid and Lodi rulers; they are not particularly impressive owing to their lack of height and insufficiently raised domes.[58] Perhaps the most notable of these tombs, with an Iranian-style double dome and lotus motifs decorated in Hindu style, is that of Sikandar Lodi.[59]

There is very little monumental architecture from the reigns of the first two Mughal emperors, Babur and Humayun, because both were engaged in establishing their authority or fending off enemies. The latter's Afghan rival, Sher Shah, built some impressive forts and mosques. The greatest of his monuments is his own tomb, built between 1540 and 1545, in the village of Sasaram, Bihar.[60] An immensely high five-storey structure built out of Chunar sandstone and standing on a stepped square plinth, it is magnificently set out in an artificial lake and within a walled court with pavilions at the corners. With an astounding perspective, it has been rightly described as 'a fitting climax to the long tradition of pre-Mughal Islamic architecture in India' (Plate 3.1).[61]

Architecture in regional sultanates

Impressive landmarks and monuments testify to the Indo-Islamic architectural legacy of the regional sultanates of this period. Although most of the sultanates had broken away from Delhi, the imperial architectural model developed particularly under the Tughluqs was the preferred model. As a general rule, however, and ironically enough, the best of this architecture developed in areas where there had been a long tradition of Hindu temple construction and where masons and craftsmen were readily available. The great strength of Hindu labour was its unbounded flexibility and its capacity to learn to produce new styles required by their Muslim rulers. The latter were happy to approve the new forms of Indo-Islamic architecture being created, as long as certain basic requirements of Islamic ritual and convention were met. The one exception was the Bahmani kingdom where, despite much multiculturalism officially encouraged by the sultans, Muslim architecture paid more attention to Persian models than to the indigenous Hindu models that abounded in their land. Regional architecture also flourished in areas where there was sufficient wealth accrued through trade. The Gujarat Sultanate provides us with the best example of this. In areas like Bengal, geographical and climatic factors also influenced the character of the building work.

Regional architecture flourished in all the different sultanates of this period. Entire new cities came into being, such as Pandua in Bengal, Ahmadabad in Gujarat, Gulbarga and Bidar in the Bahmani kingdom and Mandu in Malwa. Some cities, formerly commissioned by the Tughluq rulers, were modernized and given a new face by the regional rulers. The Sharqi sultans of Jaunpur (1394–1479), for example, embellished the city originally founded by Firuz Shah Tughluq, and Ibrahim Sharqi celebrated the cultural glory of the city by building the great Atala Mosque in 1408.[62] Throughout this period the regional architecture is far more impressive than the imperial one at Delhi. Mosques, tombs and palaces were built, some of immense size and intricate designs. Many ruins of these areas testify to the vigour of the building traditions, of which we give two examples below.

From an architectural point of view, Bengal was a key region. With valuable lessons learnt over many millennia from floods, heavy rainfall, humidity and water-soaked ground, the ordinary Bengalis made good use of cheap, local materials such as thatch and bamboo. The curved roof of Bengali huts, for example, designed to meet the demands of climate, had become part of an indigenous convention; and the construction of numerous mosques under the various sultans from the thirteenth century onwards reflected this native tradition. Muslim architects also creatively used bricks and stones as building material.[63] The 1374 Adina Mosque of Pandua, the capital of the Ilyas Shahi rulers, is one of the largest mosques of the Indian subcontinent. Now roofless, widely shattered, in ruinous condition, and away from the tourist trail, it attracts few visitors; but its immense size, its numerous arches and originally over 360 domes, its *zenana* gallery and, above all, the grandeur of its setting, arguably make it the most iconic of Bengali buildings.[64]

Gujarat provides us with another example of interesting regional Islamic architecture. At first, with the advent of Muslim rule from 1296, the fine Hindu temple pillars were freely pillaged for the construction of mosques;[65] but this phase soon passed. The skilled craftsmen of the province began to adapt to new conventions by employing the pointed arch and leaving out figurative sculpture. The golden age of Islamic architecture began with the tolerant attitude of the Ahmed Shahi sultans from the early fifteenth century onwards. More than fifty mosques were commissioned by sultan Ahmed

Shah I (r. 1411–42) as part of his grand plan to build a new capital with a great gateway, Teen Darwaza (Plate 6.1) at the site which is now the great city of Ahmadabad.[66] A stupendous project of town planning, mosques and public buildings was put into motion and continued all through the sixteenth century. The Friday Mosque, the *Jami Masjid* (1424), lying at the centre of the city, has a vast courtyard covered by an arcaded cloister, the interior carried on 260 columns divided into fifteen bays, the rows of five domes in the roof and five *mihrabs*, the balconies and a variety of perforated screens, all in all creating a 'harmonious composition'.[67] Islamic architecture reached its apogee in the reign of sultan Mahmud I Beghara (r. 1459–1511);[68] and the glory of Gujarati Islamic buildings can be testified today by the province's surviving grand monuments with their plethora of exquisite *jalis*, lattice work, step-wells, etc. The Hindu and the Muslim worked together to achieve all this.

It would be in the later Mughal period that the regional architecture of the Deccan would truly flourish; but here we may take note of an impressive structure that testifies to the Persian and Central Asian tradition in the Bahmani kingdom. This was the great *madrasa* of Mahmud Gawan, built at Bidar in 1472.[69] Its massive court, arched portals and openings and one of its two cylindrical minarets rising in three stages with a domed summit at the top, all make it a typical Central Asian building. Countless generations of students learnt in its teaching rooms and libraries; and today the 'Mahmud Gawan Internet Café', just outside the entrance to the *madrasa*, provides internet access to young people eager for knowledge.[70]

The beginnings of the Mughal garden

Babur, the first Mughal emperor, pioneered the idea and execution of Timurid gardens in India. By Timurid gardens, we mean the gardens of Samarqand, left behind by Timur Leng and his successors, which had a great effect on Babur's aesthetic imagination. Not only the gardens of Samarqand but of all Central Asia at this time were copies of a Persian garden. There have been gardens in Persia from very ancient times, and a beautiful walled garden was for the Persians a *pairidaeza* (paradise). In its basic form, a Persian paradise garden was a rectangle divided into four quarters formed by the crossing of four water channels representing four rivers of life. This was called the *chahrbagh*.[71] Through the long history of Persia, we notice the modification or embellishment of the basic garden pattern and, with the coming of Islam, the octagon pattern with its symbolic squaring of the circle and its eight parts representing the eight divisions of the Quran brought greater geometrical symmetry into the designs.[72] Babur was most disappointed by both the lack of running waters in traditional Indian gardens or residences and poor irrigation skills of Indians generally.[73] His love of nature, technical understanding of sources and the channelling of water within the garden, and his botanical knowledge of flowers, fruits and shrubs, can be witnessed both in his own descriptions and some of the remains of his terraced garden palaces. Three particular gardens – the Bagh i-Wafa in Kabul, the Ram Bagh or Gul Afshan in Agra and the Lotus Garden in Dholpur – may be described as Babur's interpretation of a Persian paradise garden.[74] A recent historian of Babur has rightly said that Babur's conquest of India did not come to be expressed in religious monuments but 'pervasively as the imperialism of landscape architecture'.[75] Babur's legacy was continued by his great grandson, Jahangir, in the early seventeenth century.

The art of the book

An important strand of Islamic art from this period concerned the illustrated book and the art of the book. As the demand for Qurans soared among the Muslim populace, we see the emergence of a distinctive convention of Islamic book production in the sultanates from the late fourteenth century onwards. The copying of manuscripts of the Quran became a major business, and with that came the calligraphy of stylized scripts and headings illuminated by floral and vegetal patterns in a variety of colors.[76] The colourful decoration was meant to animate the script. The most outstanding example of such an illustrated book from this period is from the sultanate of Malwa. The *Nimatnama*, a cookery book from the decade 1495–1505, especially written for the delight of the eccentric Malwa sultan, Ghiyas ud-Din Khalji (r. 1469–1500), contains nearly fifty paintings showing preparation of recipes and including such images as those of the sultan, his servant, the landscape, the dresses, buildings, etc.[77] The paintings fuse the artistic traditions of Persia and pre-Islamic India. The artists belonged to the Jain school of painting in Gujarat whose artists at that time specialized in hybrid painting.[78]

Intermingling of literary styles

Arabic and New Persian (Farsi) were the two premier languages of Islam. Arabic was the *sine qua non* for those who wished to write anything on theology. But, as was pointed out in the last two chapters, the language of high culture in which the Muslim intelligentsia preferred to write was New Persian. The trend continued in this period, but this was not the age of the brilliant and original poetry of Masud Saad Salman or Amir Khusrau. The literary fashion in vogue was translations and compilations. Owing to the indigenization process within the wider society, indigenous Indian ideas and works were now receiving greater attention from Sufi writers and other Muslim intellectuals. Firuz Shah Tughluq, the sultan fascinated by the antiquity of Sanskritic knowledge, ordered an important Indian astronomical text to be translated into Persian. Conversely, his court astronomer, Mahendra Suri, introduced Central and West Asian astronomical ideas in his book *Yantraraja*.[79] King Zainul Abidin of Kashmir (r. 1420–70), well versed in Persian, Hindi and Tibetan, encouraged his scholars to translate into Persian the ancient Indian epic of *Mahabharata* and the history of Kashmir, *Rajatarangini*, by the famous Kashmiri author Kalhana.[80] The Delhi sultan, Sikandar Lodi, and his Prime Minister, Mian Bhua, assembled an array of specialists on various sciences; and out of their efforts came one of the well known medical texts of fifteenth-century India, the *Madan-i sifa–yi Sikandarsahi*.[81]

It was noted in Chapter 4 that Masud Saad Salman and Amir Khusrau had both introduced many Indian terms, literary forms, imageries and metaphors into their Persian works and thus initiated the process of indigenizing the Persian language in India.[82] Through the translation movement, Persian became increasingly Indianized, which caused a reaction among the guardians of the purist strand in Persian. One result of this was the development of Persian lexicography in India. While only four dictionaries were compiled in Iran between the tenth and the nineteenth centuries, sixty-six were compiled in India in our period alone, and largely produced before the Mughals.[83] These dictionaries were useful tools to study the language, and they set a standard of taste and refinement that the purists desired.

The trend towards writing literature in vernacular languages was also gathering pace in this period. Following the example of Amir Khusrau, Muslim writers were composing entire works in the vernacular of Hindavi rather than Persian. The particular medium of expression was Avadhi or eastern Hindavi, spoken further east from Delhi, in the Lucknow-Allahabad-Benares region. Here the Sufis developed a fine literary tradition by writing Avadhi romances in the style of the Persian *masnavi*s, emphasizing Islamic spiritual values through the mode of indigenous Hindu culture, religion and literature.[84] These romances, based on old Hindu or pre-Hindu folk tales, weave together Sufi mysticism with Yogic practices. They contain descriptions of the particular Indian locale and environment in which the stories were based and also numerous allusions to various aspects of Hindu culture: music, astrology, festivals, names of gods, stories from ancient epics like the *Mahabharata* and *Ramayana*. The popularity of Sufi romances among both Hindu and Muslim readers demonstrate the marked ability of their authors to empathize with their surrounding Hindu milieu.[85]

The blending of the Hindu Bhakti and Islamic Sufi values, so well brought out within the Avadhi Sufi romances, can also be noticed in other regional literary traditions. In Bengal, for example, the sultans were great patrons of those who composed the Vaishnava verses called *padavali*s and those who wrote plays and poems in Bengali. With the use of Persian and Arabic loan words, Hindu and Muslim writers created a new form of Bengali vernacular cosmopolitanism.[86] In the Bahmanid kingdom of the Deccan, the Sufi saint and scholar, Mohammed Gisudaraz (1321–1422), wrote works on mysticism in Dakhni Urdu (a mixed language containing elements from Hindavi, Arabic, Persian and some Marathi, Telugu and Kannada).[87] His contemporary, Fakr i-Din Nizami, produced a great secular poem about kingship and worldly issues, *Kadam rao padam rao*, in Dakhni Urdu heavily tinged with both Telugu and Sanskrit.[88] Muslim writers of the Gujarat Sultanate (1407–1573) wrote both in Gujarati and Hindavi/Urdu. Whether it was the Urdu poetry of a well known Gujarati Sufi, Sheikh Baha ud-Din Bajan (1388–1506)[89] or the village theatre tradition initiated by the Hindu playwright Asait Thakur,[90] we notice that both Hindus and Muslims were considered part of the wider Gujarati society.

Although there are few historical works from this period that can match those produced in the heyday of the Delhi Sultanate by historians like Barani, one splendid exception is the *Baburnama*, or *Tuzuk-i Baburi*, Babur's autobiography. He was endowed with a unique flair for writing, along with all the other mental and physical skills that he possessed. His account, first written in Chagatay Turkish, was translated into Persian at the court of his grandson, Akbar, in the late sixteenth century. It has also been rendered into other languages, including English. The nearly 700 pages of the autobiography contain evocative and poignant accounts of scenes and episodes in which Babur's friends, colleagues and warriors are intensely involved in all his activities. Many of the intimate passages in which Babur recalls incidents of his early life indicate his mastery of literature generally and literary lore in particular.[91] The autobiography also provides a great amount of information on the different lands with which he was familiar. In his Hindustan section, he sees himself as a warrior in the *ghazi* tradition.[92] While he stereotypically generalizes about its people (Excerpt 5.5), he is matter-of-fact in his observations on its climate, landscape, fauna and flora, etc.[93] We may well describe the *Baburnama* as a fine example of early Mughal literary taste.

Select excerpts

5.1 The sack of Delhi by Timur

Timur's psychopathic cruelty, his greed, his refusal to take personal responsibility for his actions and his schizophrenia are brought out in his own words in the following extract.

> All that day the sack was general. The following day, all passed in the same way, and the spoil was so great that each man secured

> from fifty to a hundred prisoners, men, women and children. The other booty was immense in rubies, diamonds, garnets, pearls . . . Gold and silver ornaments of the Hindu women were obtained in such quantities as to exceed all account . . . I ordered that all the artisans and clever mechanics, who were masters of their respective crafts, should be picked out from among the prisoners and set aside, and accordingly some thousands of craftsmen were selected to await my command . . . I had determined to build a masjid-i jami in Samarqand, the seat of my empire, which should be without a rival in any country; So I ordered that all builders and stone masons should be set apart for my own Special service . . . By the will of God, and by no wish or direction of mine, all the three cities of Delhi . . . had been plundered . . . It was ordained by God that the city should be ruined. He therefore inspired the infidel inhabitants with a spirit of resistance, so that they brought on themselves the fate which was inevitable. When my mind was no longer occupied with the destruction of the people of Delhi, I took a ride round the cities.
>
> (Elliot 1871/2004 reprint: 63–5)

5.2 *Iranian help to Humayun*

There is little doubt that without the help of Shah Tahmasp of Iran Humayun might have ultimately failed to recover his throne. However, a sentence in one of his letters to the Ottoman sultan may suggest greater private tension between him and the Shah. The authenticity of this letter in the Ottoman archives is somewhat disputed.

> If I were to describe the oppression and perfidy which that vile man (Tahmasp) has practiced against me, you would weep bitterly.
>
> (Humayun to Sultan Sulaiman II, October 13, 1548, in Islam 1982: 296)

5.3 *Two rival camps at the Bahmani court*

The rivalry between the native elite and the immigrant elite was real and bloody at the Bahmani court.

> The sultan Ahmad Bahmani II (1436–58) . . . ordered that, both at court and in cavalry formation, the Westerners (mostly Persian immigrants) should appear on the right side, while the Deccanis and the Ethiopians should be on the left. From that time to the

> present, an inveterate hostility has taken hold between the Deccanis and Westerners, and whenever they get the chance, the former have engaged in killing the latter.
>
> (From Muhammad Qasim Firishta, *Tarikh-i Firishta*, quoted in Eaton 2005: 59)

5.4 Africans in the Deccan

The presence of Africans in India is a subject of great curiosity among many historians. During the Middle Ages, most of them were Ethiopian slaves (Habshis) bought by Muslim merchants who sailed from the Deccani ports.

> Father Francisco Alvares (the first Portuguese Jesuit to reach the Ethiopian highlands in 1520) noted the enormous quantities of Indian silks and brocades consumed by the Ethiopian court, acquired both by gifting and by purchase. African demand for Indian textiles . . . appears to have been the principal engine behind Ethiopia's slave-extraction process . . . The Habshis were not intended to serve their masters as menial laborers but as elite military slaves.
>
> (Eaton 2005: 108–10)

5.5 Babur on Hindustan

Although Babur pays tribute to the Indians' capacity for producing prodigious wealth in their country he also betrays much ignorance of the true condition of India in the following excerpt.

> Hindustan is a country of few charms. Its people have no good looks; of social intercourse, paying and receiving visits there is none; of genius and capacity none; of manners none; in handicraft and work there is no form or symmetry, method or quality; there are no good horses, no good dogs, no grapes, musk-melons or first-rate fruits, no ice or cold water, no good bread or cooked food in the bazaar, no hot-baths, no colleges, no candles, torches or candlesticks.
>
> (Beveridge 1922: 518)

6

THE MUGHAL ASCENDANCY

Akbar and his successors (1556–1689)

Neither Babur nor Humayun (Chapter 5) ruled over North India long enough to influence the future shape of their fledgling kingdom. It was Humayun's son Akbar (Plate 4.2) who, during his long reign of forty-nine years (1556–1605), established institutions and created conditions that enabled his son Jahangir, grandson Shah Jahan and great grandson Aurangzeb to rule effectively over an empire of great majesty and awe. For 133 years after Akbar's accession, this empire remained a rock of stability for the vast majority of the peoples of South Asia, and Muslim power reached its apogee during those years. Aurangzeb reigned on until 1707, but after 1689 the empire lost its dynamism.

The four Mughals: a brief introduction

Akbar ascended the throne at fourteen and, for the first five years of his reign, was guided by two regents. From 1561, however, until his death in 1605, he, alone, controlled the destiny of the Mughal state. Throughout his reign, he continued to expand his territories with ambitious ruthlessness and occasional cruelty, reminiscent of both his Mongol and Turkish ancestors. He built up a strong and centralized state.[1] Wise enough to recognize the limits of brute power alone, he was prepared to use diplomacy and institutional structures to create a new type of imperial order. This gained him respect from his subjects. He also possessed much sensibility, curiosity and openness of outlook, qualities that he demonstrated in his handling of controversial issues of faith and belief.[2] To his court, he attracted an array of brilliant advisers, associates and protagonists of varied ethnic and faith backgrounds; while completely open to their counsel he remained his own man throughout his life.

Akbar's eldest son was Salim (Jahangir) whose Hindu mother, Jodha Bai, the daughter of the Rajput chief of Amber (Jaipur), had been his father's principal wife. A brief revolt by him before his father's death had not deprived him of Akbar's consent for his succession. However, in the Timurid tradition which all Mughals were proudly attached to, Jahangir could not by right expect to succeed to the throne. In fact, he faced opposition from his own eldest son, Khusrau, with the danger of a war of succession.[3] Khusrau's revolt collapsed but, as events were to turn out, the Mughal wars of succession were to become a marked feature on the onset of old age or the death of a number of succeeding Mughal emperors.

During his reign, Jahangir (1605–27) stayed close to Akbar's policies but displayed little original thinking about the art and practice of governance. Instead, there was substituted a pompous and elaborate court ritual which distanced him and his acolytes from his subjects. His most purposeful activity, that also happens to be his greatest legacy, was patronage of artistic creativity.[4] Power increasingly came to be exercised by his last wife Nur Jahan, a young and charismatic Persian widow. Ambitious, and skilful at exploiting political and economic resources, Nur Jahan created a close faction at the court, which included Jahangir, her father Itimad ud-Daulah, her brother Asaf Khan and Jahangir's third son, the ambitious and vigorous Prince Khurram whose wife was Mumtaz Mahal, the daughter of Asaf Khan.[5] This faction controlled the empire for over a decade.

Nur Jahan's guile and diplomatic skills failed her at the most decisive time when Jahangir's health worsened in the mid-1620s. In her desperate desire to retain influence at the court after his death she supported Khurram's younger brother who also happened to be married to her daughter from her first marriage. This was her undoing, since her brother Asaf Khan stayed loyal to Khurram. She had also underestimated the cruel ruthlessness with which Khurram, with the active support of Asaf Khan, plotted the death of his brother and other close relatives.[6] All the blood-letting happened between Jahangir's death in October 1627 and Khurram being proclaimed Emperor Shah Jahan in January 1628. Nur Jahan had no option but to retire with a generous allowance from Shah Jahan and income from her inheritance.

Khurram (Shah Jahan) reigned between 1628 and 1658, and was a more proactive monarch than his father. He fought many military campaigns and, despite some failures, increased Mughal territory. While less liberal in his religious attitudes than his grandfather, he emulated Akbar's institutional policies.[7] He also went far beyond Akbar and Jahangir in imposing a rigid Persian formalism at his court; and his dignified bearing, dress and demeanour epitomized the idea of a grand Mughal so fascinating to foreigners.[8] For us today, his greatest legacy is in the beauty and majesty of Indo-Muslim art, architecture and gardens, in pursuit of which he spent much time and expended a vast treasure. In the last nine months of his reign, a ferocious war of succession among his four sons consumed him in a most poignant tragedy.[9] Dara was his eldest son and a favourite for the succession; but in the Timurid tradition his other brothers could equally claim the throne, if successful. All four sons were men of prowess and ability; all wanted to succeed their father, and none wished to give way. In the end, the third son, Aurangzeb, emerged the ultimate victor. He made sure that, one way or the other, all his three brothers were put to death; and, without any compassion, held Shah Jahan himself under house arrest in Agra fort until the latter's death in 1666.

Aurangzeb's long reign of forty-nine years (1658–1707) can be divided into two periods. The first, included in this chapter, was the age of triumphalism that ended in 1689. Aurangzeb succeeded in completing the grand project of an imperial state that his great grandfather Akbar had espoused. By 1689, except for the very deepest south and Assam, most of the subcontinent and Afghanistan were in Mughal hands. Ironically, however, Aurangzeb undid Akbar's work during this period. He turned Sunni Islam into a dominant ideology, and his dogmas led him to enact discriminatory measures against Hindus, Sikhs and the Shia. While political ambitions and struggles over economic resources may explain some of the anti-Mughal resistance by diverse groups, the role of the religious animus that Aurangzeb engendered should not be underestimated.

The remaining eighteen years (1689–1707) proved to be a more difficult time for both Aurangzeb and the Mughal state. They therefore rightly belong to the next chapter which discusses the diminution of Mughal power in particular and Muslim power in general.

The Mughal imperial order

A pan-Indian empire

The Mughals built a multi-regional empire for a variety of reasons, of which the most important was economic power. A key index of this power was the securing of sufficient land revenue to sustain their entire political and military edifice.[10] The fertile lands of the Indo-Gangetic basin, from Punjab to Bengal, were therefore central to Mughal ambitions. Before Akbar's accession, these lands had gone through a tumultuous period of uncertain rule but, after what is known as the second Battle of Panipat (1556), the Mughal hold on the area became a reality.[11] By the time Akbar himself took full control in 1561, the great cities of Lahore, Delhi, Agra, Jaunpur and Benares were in Mughal hands; and from then on the four Mughal rulers of this period never lost their grip upon the Hindustan heartland. Akbar's defeat of the Malwa kingdom in 1574 and his annexation of Kashmir in 1586, followed by Jahangir's annexation of Kangra in 1620, completed the Mughal control of the north.

The Mughal ambitions, however, ranged in all directions, far beyond the north. From a strategic point of view Rajasthan was a great prize to capture, because it lay on the route to the Gujarat coast. Akbar also wanted the

Rajput monarchs to associate with his empire, in order that the Hindu and Muslim elites could work together.[12] He offered them some of the highest posts within the nobility in return for Mughal paramountcy. At first, only the ruler of Amber (Jaipur) accepted the bargain and, as noted earlier, offered his daughter in marriage to Akbar. Others held back, particularly after the ruler of Mewar, Rana Udai Singh, refused to give up his independence. It was then that Akbar resorted to force, destroyed the gigantic fortress of Chitor and gave orders to massacre its garrison and inhabitants (1568). When yet another fortress of Ranthambore fell, the Rajputs acquiesced by enlisting in Akbar's armies and accepting posts at the Mughal court. Rajasthan became a Mughal *subah*, with the Mughal resident based at Ajmer.[13] Mewar, nonetheless, continued to resist, until its final surrender to Jahangir in 1615.[14] The Rajputs generally stayed loyal to the Mughals until the 1670s when Aurangzeb started interfering in the affairs of the Marwar kingdom, a partner of Mewar, and upsetting their Hindu sensibilities.[15] The Rajput revolts of the late seventeenth century therefore dealt a blow to the multi-faith empire of Akbar's vision.

Gujarat and Bengal, renegade sultanates from the late fourteenth century (Chapter 5), were coveted by the Mughals for their plentiful land revenues and profits from external trade. Akbar incorporated Gujarat into the empire after two successive campaigns of the early 1570s.[16] Bengal presented greater difficulties, because the Afghan nobility there resented Mughal intrusions; and it was not until the 1580s that both Bengal and Orissa were taken. Jahangir followed up with further pacification in 1612, and Shah Jahan evicted the Portuguese from their fort at Hugli in 1631.[17] Firm Mughal control of Bengal was finally achieved in the 1660s under Aurangzeb.

It was in the Deccan that Mughal strength and patience were both tested to their utmost. The great plateau, with its land, trade, minerals and manpower, was essential to Mughal dreams of a pan-Indian empire. A major problem that faced them was that there were at least seven different centres of power to contend with: the sultanate of Khandesh; the five post-Bahmani sultanates of Ahmadnagar, Berar, Bidar, Bijapur and Golconda; and the Hindu Maratha family of Bhonsla in the western hills of Maharashtra. It took the Mughals over a century to control the six Muslim sultanates; the diplomatic ability and political shrewdness of Deccanis (like Malik Ambar, an Indian of Ethiopian origin and chief minister of Ahmadnagar), for long thwarted them.[18] Nevertheless, all six sultanates had been incorporated into the empire by 1689.[19] The Hindu Marathas proved a much more difficult enemy to deal with.

Language, faith and ethnicity shaped the Maratha identity.[20] From their hilly fastnesses in the Western Ghats, they had watched the steady erosion of Hindu authority over the Deccan from the thirteenth century onwards. Realizing the limits of their power, they devised a twofold strategy of sometimes serving the Muslim potentates and, at other times, attacking them

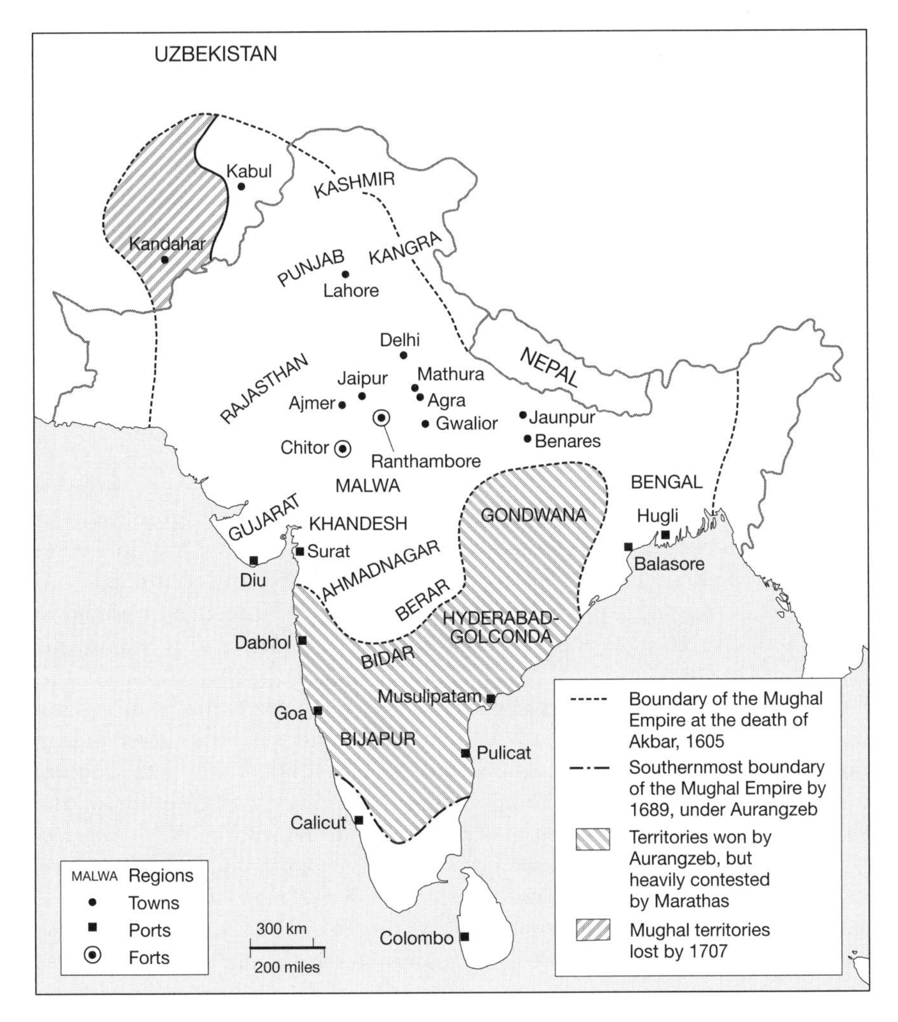

Map 6.1 The boundaries of the Mughal Empire: 1605–1707

separately. This strategy, perfected by a warrior chieftain, Shahji Bhonsla, during the 1620s and 1630s, secured the Marathas much autonomy in the lands straddling both Ahmednagar and Bijapur.[21] Shahji's son, Shivaji Bhonsla (1627–80), an able guerrilla fighter, perfected the strategy by constant attacks and quick withdrawals into the hills, thereby exhausting the Mughal armies. Aurangzeb, much agitated by these manoeuvres, had little success against him.[22] He did better after Shivaji's death in 1680, because the latter's son, Sambhaji (1657–89), lacking his father's military skills and political chicanery, could not keep his people united. Moving his entire court to the Deccan in 1682, Aurangzeb therefore decided to crush the Marathas once and for all. By 1689, Sambhaji had been defeated and executed,[23] but his death provoked his people into offering even greater resistance to Aurangzeb.

By 1689, the Mughal territory covered most of the subcontinent (Map 6.1). The territories in the far south kept their independence, although that did not unduly worry the Mughals. Assam, in the north-east, also remained outside Mughal control.[24] Much of Afghanistan, including Kabul, was retained; Kandahar, however, changed sides between the Mughals and the Iranians on numerous occasions.[25] The Iranians also denied the Mughals victory over their ancestral Timurid homeland of Uzbekistan; and the Afghan tribes were never entirely quiescent.

A stable empire

Notwithstanding periodic bouts of fighting and frequent agrarian revolts of various intensities, the legendary stability of the Mughal Empire between 1556 and 1689 was remembered by both the high and the low in Indian society for long afterwards. Even the East India Company continued with the fiction of Mughal supremacy almost until 1857.[26] Three centuries before 1857, Emperor Akbar's long reign was just beginning; and it was in his reign that the foundations of Mughal stability were established. Akbar and his advisers created an imperial system that was beyond the reach of any other competing authority; and what they left behind was little altered during the reigns of the next three monarchs and beyond. Three vital factors underpinned the stability of Akbar's system: the majesty of the emperor, the loyalty of the nobility, and bureaucratic institutionalization.

With his character and personality admired by so many, it was not long before Akbar came to be endowed with an aura and majesty that few dared resist or challenge. The legitimacy of his imperium, as articulated by his closest adviser, Abul Fazl, through his famous works, the *Akbarnama* and the *Ain i-Akbari*, was claimed from a long line of Timurid ancestors.[27] With Abul Fazl going as far as to state that Akbar's rule was divinely sanctioned,[28] it was inevitable that court painters portrayed him as a divinely illumined person.[29] His majesty came to be recognized increasingly by a series of elaborate court rituals, etiquette, nomenclatures and honorific symbols. The 'ideology of authority and legitimacy' was expressed through discussions at court, poetry readings, paintings and historical writings. It was not enough for Akbar to have his majesty and grandeur acknowledged at his court in one static place. He regularly changed his capitals, moving from Delhi to Agra, then to the newly built city of Fatehpur Sikri, then to Lahore, and so on.[30] This had the added advantage that no enemy could spring a surprise on him. He wanted all India to observe and wonder in awe at his grandeur and, for that reason, in his later years, his capital was often a mobile tented city anywhere in the empire.

Akbar's majesty would have been of little effect if he had failed to secure the loyalty of the nobility. He managed this by transforming the mindset of his courtiers: from a concept of loyalty based on family kinships, caste, region, etc., to that based on a pan-Indian imperial service.[31] A key element in this was the official policy of non-sectarianism, which greatly helped to secure the loyalty of Rajputs and many other non-Muslim and non-Indian groups. No person of ability from any ethnic group was barred from advancement by religious discrimination.[32] Second, under both Akbar and Jahangir, there developed an intimate ceremony of discipleship under which, regularly, a limited number of nobles, drawn from all communities, pledged a very special oath of allegiance and bonding to the emperor, in the way that a Sufi spiritual master and his disciples came together. Finally, the loyalty was cemented by a lavish system of gifts,[33] honours and assignments, based on meritocratic or functional ranking, within a centralized bureaucracy. Thus the idea of a *khanazad* was born: a nobleman ready to lay down his life for the emperor in the true spirit of a *ghazi*, a holy warrior.

Mughal political stability depended greatly upon a tight hierarchical bureaucracy. The Persian tradition that arrived with the Ghaznavids (Chapter 3) and survived for nearly five centuries was further fine-tuned and refined in the Mughal era. In his effort to control executive power Akbar sidelined the traditional office of the grand *wazir* and created four major central departments whose heads reported directly to him: finance; army and intelligence; justice and religious patronage, and the royal household.[34] Apart from this major innovation, most of the Persian machinery of rules and regulations, audits and controls, checks and balances, and rewards and punishments remained in place. There was, however, one particular feature of bureaucracy that may explain why the Mughal state may be considered as an early example of a proto-modern state. The quintessential quality of a modern state is that it is a rules-based state with strong, impersonal institutions run on a set of well formulated principles and criteria. We can see something of this in the Mughal *mansabdari* system.

The adjective *mansabdari* describes a system based on a *mansab*, which was a post, rank, office or status. A person who held the *mansab* was a *mansabdar*. The system involved a military-like ranking of all the great functionaries of the Mughals, whether military or civilian.[35] This ranking, split up into effectively thirty-three grades, depended upon the number of soldiers a functionary could provide for the emperor. It could be as few as ten soldiers or as many as 10,000. The number of soldiers provided determined the personal numerical rank, or the *zat*, of the Mughal *mansab*. This rank was further complemented by another rank, known as *suwar*, which depended upon how many heavy armed cavalrymen could be mobilized. The thirty-three grades of rank were divided into three classes. Any *mansabdar* who could raise both 5,000 infantry soldiers and 5,000 heavy cavalrymen with a requisite number of horses and all other military equipage was accorded a first-class ranking which, of course, entitled him to some of the most lavish privileges that the state could offer in terms of salary or lands or titles, etc.

We have to remember that the origins of the Mughal state lay in military feudalism, which meant that the aggregate royal power ultimately depended upon the number of soldiers and cavalry the monarch's feudal tributaries could supply. Different forms of military feudalism existed in the Middle Ages in different regions of Asia and Europe; but the Mughal model originated in the Islamic Middle East where military slaves or the *mamluks* had formed the armies of the varied sultanates. This was not therefore a system invented by Akbar; in India, even before the first Mughal had arrived, a version of it had been tried out by Ibrahim Lodi, the last of the Delhi sultans. Akbar, however, endowed the system with two striking features. One was that however high a rank that a *mansabdar* reached, he remained a servant of the imperial bureaucracy; his honour was personal to him, and his descendants could not claim any hereditary rights. Thus was reduced the scope for nepotism by family dynasties. Second, religion did not intrude into consideration for determining who became a *mansabdar*. Although the Muslims, owing to historical circumstances, occupied the highest ranks of the *mansabdars*, the number of Rajputs holding different grades within the system remained more or less constant under the great Mughals; and under Shah Jahan there were Marathas, too, in the higher ranks.[36] All nobles were *mansabdars*, but not all *mansabdars* were nobles. The lower class of *mansabdars*, along with an army of centrally employed clerks, accountants, auditors and other workers who kept the Mughal political and economic machinery in working order, were remunerated by salaries paid in cash that was widely available in a flourishing imperial coinage system. On the other hand, most of the great nobles, descending from the highest ranks of princes to Muslim *amirs* and Hindu *rajas*, were entitled to assignments of land, or *jagirs*, of varying sizes and values, with the right to collect their own taxes from the peasants. We explore this further in the next section.

The Mughal economy

The gigantic political and military machine of the Mughals could only be sustained by prodigious economic wealth. The four Mughal emperors had relatively little difficulty in amassing this. Throughout the sixteenth and seventeenth centuries, the Mughal Empire's wealth exceeded that of any other political entity, except perhaps China. By 1700, in fact, the empire's approximately 150 million people were producing the largest share, nearly a quarter, of the entire world's GDP.[37] Like in all pre-modern and pre-industrial empires, the wealth was mostly generated by peasants in the countryside, but trade, too, brought profits to individuals and the state.

Consumption mostly took place within urban centres, but further wealth was accrued there from the manufacture of crafts and sophisticated products.

Land revenues and rural hierarchy

The millions of peasants across the subcontinent were not merely an undifferentiated mass of exploited people. They enjoyed varied customary rights to own the land, rent it or even to sell it. However, since they were subject to payment of agricultural tax, they were obliged not to leave the land uncultivated.[38] The Mughal administration achieved the ideal of maximizing the state's take in taxation from land while still making it worthwhile for the peasant to bring more land under cultivation, thereby increasing productivity further. The tax was increasingly required to be paid in money rather than in kind, which had the desirable outcome of forcing the peasants into the market. In the tradition of Sher Shah Suri's work on the assessment of land revenues (Chapter 5), Akbar's financial guru, Raja Todar Mal, assiduously collected accurate data and standardized weights, measures and monetary values. All land was intricately surveyed, measured and categorized according to its fertility and cultivability; the varying yields in different types of lands were recorded over a long period of ten to fifteen years; the differences in the prices of different crops and products were also noted, again over many years. New revenue circles, incorporating *parganas* (created under Sher Shah), were consolidated, along with a revenue official known as *karori*; and the revenue assessment for each plot of land, each *pargana* and each revenue circle was worked out after much investigating, recording and averaging.[39] In the long run, this proved to be highly profitable for the Mughal treasury.

Within the agricultural pyramid there existed hierarchies of nobles and landlords. When Akbar ascended the throne in 1556, the power of the aristocratic families in most of his domains was overwhelming. Since 1351, when the authority of the Delhi Sultanate began to crumble, the *zamindars* had often arrogantly levied excessive taxes on their peasants, while avoiding as often as possible to make payments due to the Delhi authorities. Many of the *jagirs* or assignments of land granted during the heyday of the Sultanate had, contrary to customary understanding, become hereditary tenures. Now, Akbar and his successors refused to treat the *jagirs* as *iqtas* of the pre-Mughal period when the landlords held both political and administrative authority. The two functions were separated, with the result that Mughal *zamindar*s were left with certain autonomy, but which was not enough to challenge the central authority. Other imperial servants, working as collectors, military officers (*faujdars*), auditors, etc., were entrusted with the task of securing financial compliance from all sections of landed society. In this respect, the Mughals demonstrated a remarkable degree of modernity and efficiency that eludes many nations even today.

Trade: internal and external

Towns of varying sizes were the lynch-pin of market economy; and market centres ranged from the great urban emporia to small scale bazaars, fairs and isolated rural gatherings of peasants and middling traders engaged in bartering. By the standards of pre-industrial times, transport facilities in Mughal India were 'efficient and adequate';[40] on land, for example, different castes of transport carriers or *banjaras* conveyed goods in caravans of between 12,000 and 20,000 bullocks capable of carrying 1,600 to 2,700 tons of goods;[41] and on rivers and canals there were innumerable barges for the long distance transport of bulk goods. Responding to different needs of people, the markets were capable of performing different functions: as local markets, wholesale spot markets or wholesale forward markets.[42] Merchants with large and small capital, middling traders, tenant farmers and peasants, along with intermediary groups such as brokers (*dalals*), were all involved in the exchange of goods between urban areas and their rural hinterlands, between towns in different regions, and between ports and coastal hinterlands. Foreign observers were amazed to see market stalls bulging with foodstuffs and other goods even in the remotest areas of Mughal India.[43] The internal exchange economy generated profits for all who understood the rules and risks of trade; and the state also efficiently collected varieties of imposts as revenue.

The Mughals thought of themselves as essentially a land power, but they were not averse to opportunities to augment their wealth from sea trade. When the armies of Akbar and his successors reached Gujarat, Bengal and both the western and eastern coasts of the Deccan, sea ports like Cambay, Surat, Dabhol, Pulicat, Masulipatam, Balasore and Hugli came into Mughal hands. These ports were India's windows to the world.[44] The Indian Ocean trade was essentially a great Muslim enterprise.[45] Hadhramautis, Omanis, Persians, Indonesians, Malaysians, Swahilis – most of whom being Muslims – were joined in their ventures by Muslim mariners and traders of the coastal areas of India. The Indian Ocean trade was a peaceful activity, encouraging cooperation and partnership; and those who participated in it had few monopolistic or hegemonic ambitions to control it. A beggar-my-neighbour mercantilist mentality, a common feature of European trading policies during this period, was unnecessary when there was enough trade for everyone to enjoy a share of profits.[46] This must be considered as one of the great Islamic legacies to international economic history. On the Indian side, there were certainly many Hindu traders and financiers involved too, but without Muslim presence in a host of maritime and merchandising activities the Indian participation in the sea trade would not have been that significant. With their control of coasts and ports, the Mughals reaped a harvest of revenue from customs duties and the tolls on ports and ships.

Euro-centred histories tend to exaggerate the European role in the Indian Ocean trade from the time when Vasco da Gama arrived at Calicut in 1498. The sixteenth century is traditionally described as the era of Portuguese naval supremacy in Indian waters; and then, from the beginning of the seventeenth century onward, with the Portuguese elbowed out, the Dutch, British and French East India Companies are seen as the new naval super-powers of the Indian Ocean.[47] This is an over-simplification, because the pervasive Muslim presence in the Indian Ocean trade continued until the end of the seventeenth century.[48] The Portuguese, however, brought into the equation something new and highly consequential, which was the militarization of the Indian Ocean. With their capture of such places as Calicut, Goa and Diu in India, Ormuz in Iran, Colombo in Sri Lanka and Macau in China, the Portuguese felt emboldened enough to block off all access routes to and from the Indian Ocean against non-Portuguese vessels. They wished to have a complete monopoly of the spice trades, but they did not succeed. The Muslim traders sought out new land routes for access to European markets.

The Dutch and the English, while a little more subtle in their relationships with the Mughals, could on occasions be equally militaristic. Their main interest was Indian cotton cloth which sold profitably in Europe.[49] Since neither England nor Holland had valuable exports to offer for their imports, they were obliged to export gold and silver that they were plundering from the Spanish galleons and treasure ships plying between the New World and Spain. The Mughals appreciated the bullion windfall; they welcomed the European traders and permitted them to build the so-called 'factories', or goods and administrative depots, as long as they recognized Mughal sovereignty and paid handsomely in licence fees and other dues. As early as 1615, Sir Thomas Roe was appointed as the English ambassador at the court of Jahangir; other Europeans, too, sought favours from the Mughal court. None of the Mughal emperors, from Akbar to Aurangzeb, felt particularly intimidated by European warships in their waters. The situation changed only from the early eighteenth century onwards.

The essentials of a Mughal budget

Nearly nine-tenths of Mughal revenue was drawn from land tax, while the rest was raised from internal and external trade dues, gifts and other taxes. All revenue was received in cash form, either in silver rupees or in copper denominations known as *dam*s. Approximately 40 *dams* equalled a silver rupee. It has been estimated that between 1595 and 1709 the imperial revenues increased from 3,960 million *dams* (99 million rupees) to 13,340 million (334 million rupees).[50] Most of the increase can be accounted by enhanced land revenues and tributes from imperial expansion under Shah Jahan and Aurangzeb.

The revenues were showered on the Mughal military state and the maintenance of the dynasty itself. Thus, in 1595, out of a revenue of 3,960 million *dams*, 3,238 million were spent on the salaries and emoluments within the *mansabdari* system, 359 million spent on the so-called central military establishment, and 187 million on the imperial household.[51] This left a balance of 176 million *dams*, hoarded up in the treasury. The proportions would not have been much different in 1709. It is clear that, despite a measure of modernity that the Mughals showed in collecting revenue, there was nothing modern about a Mughal budget. The state showed no concern for our modern expenditure streams like health, education, social welfare, transport, etc. More or less everything went to the sustenance of the military. It is, therefore, no wonder that foreign observers have written about poor infrastructure and much wretchedness and poverty among the ordinary people of a nation that produced a quarter of the world's GDP in the seventeenth century.[52] On the other hand, the wealth of the nobles and *mansabdar*s generated demand for beautiful and elite products, which supported thousands of craftsmen and urban workers in jobs; and their creations attract even today the admiration of civilized and cultivated minds from all over the world. Perhaps we should not judge the Mughals too harshly, because the budgets of the British Raj were equally weighted towards the needs of the military and administration.[53] In fact, while the wealth of South Asia was much drained towards Britain during British rule, the Mughal wealth at least remained within the subcontinent.

Religion and state

Muslim rulers of India entertained few doubts about the legitimacy of their rule.[54] They thought of themselves as inheritors of the *ghazi* tradition of heroic Muslim warriors fighting for Islam. Mahmud of Ghazna was much extolled by court writers and poets as role model for a Muslim monarch. While Timur the Great had at first been reviled by contemporary Indo-Muslim intellectuals, with the passing of time he too had acquired the halo of a *ghazi*; and the Mughals drew inspiration from their Timurid background.[55] The question of legitimacy did not therefore trouble the Mughals or their predecessors. A more agonizing moral question for them was over their relationship with the majority non-Muslim subjects. As was pointed out in Chapters 4 and 5, the Delhi and provincial sultans had been highly realistic rulers, despite the pressure for anti-Hindu zealotry exerted upon them by the more orthodox and conservative *ulama*. They did not wish to alienate their non-Muslim subjects; they simply wished to manage their relationship. The policy was beginning to pay off, as we saw in Chapter 5, with Hindu and Muslim philosophers and intellectuals engaging in dialogue with each other and breaking at least some common ground through the *bhakti* movement and Sufism.

It was in Akbar's long reign that the blending process was taken a step further by his more liberal religious policy. At first, however, facing various challenges, he adopted the public face of a traditional Muslim ruler by following daily rituals, supporting many Islamic charities, funding the *hajj* pilgrims, and visiting the shrines of Sufi saints.[56] He encouraged many theological seminars at the court where he listened intently to various Muslim divines interpreting Islamic laws; in 1575 he commissioned an *Ibadatkhana* or the 'House of Worship'.[57] This marked the high-water point of his Islamic pieties. After becoming increasingly frustrated with the pedantic rhetoric of Muslim theologians he decided to turn the *Ibadatkhana* from an Islamic centre to a centre for inter-faith discourse. He invited the Hindu, Jain, Christian and Parsee-Zoroastrian religious leaders to various discussions, in order that not only he but the *ulama* too might learn about other faiths. He took great interest in Hindu beliefs and festivals and adopted some of the Brahmanic rituals at the court.[58] He greatly respected Jain views on non-violence, even to animals, and ordered special days when animal slaughter was banned.[59] He was drawn towards the ideas, symbols and miracles in Christianity in the way the Jesuits and the Portuguese Fathers explained to him. Although the orthodox Muslims were concerned that he might have become a Christian, they need not have worried because the Portuguese overplayed their hand and, despite all their efforts, failed to convert Akbar to Christianity. He was much too independent-minded for them.[60] He was also drawn towards Zoroastrianism after he had learnt from a Parsee priest about the symbolic importance of light from fire and sun representing all that was good.[61] He almost took to sun-worship (Excerpt 6.1).[62]

Akbar never abandoned Islam but, from the late 1570s onwards, he began to show a distinct coolness towards its more orthodox version. His disagreements with Islamic conservatives at the court ranged over a number of issues. While refusing to be indulgent in granting land rights to corrupt and fraudulent Muslim nobles, he granted lands to worthy non-Muslims and supported their places of worship.[63] He abolished the *jizya* tax in 1579 and thereby removed the discrimination felt by non-Muslims. In the same year, he proclaimed himself as the supreme arbiter in matters of religious dispute. Encouraged by his liberal supporters, such as Abul Fazl, the court chronicler and historian, he promulgated a set of religious formulations, known as *Din Ilahi*, the Religion of God or Divine Faith, which he got approved by a council of scholars, albeit reluctantly.[64] In theory, Akbar's ultimate goal was to establish a *sulh-i-kull* in India, the state of universal peace and toleration;[65] but in the *Din Ilahi* he also found a tool to forge a close and tight network of disciples and followers who could help to promote his political project of centralism.[66]

Akbar's attitudes did not go unchallenged. A number of distinguished *ulama* felt obliged either to speak out or attempt to subvert his plans. They paid a heavy price for their orthodoxy through loss of property, public humiliation, exile and even death.[67] The works of a historian, Badauni (1540–1615), and a Sufi religious scholar, Shaykh Ahmad Sirhindi (1564–1624), provide us with sufficient information about the anger and frustrations of the orthodox *ulama*.[68] From the hindsight of history, however, we can view Akbar's religious liberalism as a visionary project. By discouraging all forms of Islamic ascendancy at the centre of state life and by bringing the Hindu and the Muslim closer together Akbar attempted to create a unique pre-modern secular model.

Akbar's death in 1605 did not mean the end of the liberal regime, but a reaction was to be inevitable. Sirhindi's writings during Jahangir's reign sharpened the case for orthodoxy; despite experiencing a difficult personal relationship with Sirhindi, Jahangir was impressed by the latter's sincerity of beliefs.[69] He nevertheless continued with his father's policy of toleration. Shah Jahan was certainly more reactionary towards non-Muslims. He made apostasy a criminal matter, and ordered the destruction of Hindu temples in Benares and Orchha and a Christian church at Agra.[70] These actions did not, however, constitute a major deviation from the policies of Akbar. One of the reasons for his increasing restraint in his mature years was the benign influence of his eldest son, Dara Shukoh, a scholar-prince, an earnest student of Sanskrit literature, and opposed to ideas advocated by Sirhindi.[71] It was particularly unfortunate that Dara lost the succession struggle against his younger brother Aurangzeb, because religious liberalism was not high on the latter's agenda.

Aurangzeb was essentially an Islamic puritan. While he did not lack curiosity or interest in the scriptures of other faiths, the path of certainty for him lay in Sunni Islam. His sincerity, self-discipline and simple personal lifestyle were qualities admired by many of his contemporaries, but his Islamic partisanship made him a controversial monarch for not only his time but for posterity as well. He remains a hate-figure in today's Hindu world. The creation of an Islamic India was his goal; and for this he promulgated laws and ordinances that increasingly became unnerving for Hindus, other non-Muslims and for many Muslims too. His puritanical restrictions first affected those hallowed Hindu traditions at the court which his predecessors had retained as a way of demonstrating their goodwill towards Hindus. The *Jharokha-darshan*, an ancient ritual of the monarch appearing on a balcony in front of his people, was discontinued; and the use of Hindu astrologers in making almanacs and fixing auspicious dates for important events was discouraged. Other restrictions, enforced by *muhtasibs*, the officers of the 'morality police', affected both Hindus and Muslims. Thus, any outward signs of the joyousness and gaiety at various festivals, whether Hindu or Muslim, were disapproved; the great Shia festival of *Muharram* was stopped; lamps lit on the tombs of Sufi and other Islamic saints were banned. Even the average length of a beard was fixed at four fingers, and no more.[72] These petty and silly restrictions were born out of a particularly joyless understanding of Islam that Aurangzeb might have imbibed from his mentors. On the other hand, while there might have been a rational case for measures such as the prohibition of alcohol, narcotics, gambling and the restrictions on prostitutes, such measures bore little result in practice.[73]

Other measures were gravely resented. The re-imposition of the *jizya* in 1679, along with discriminatory customs duties between Muslims and non-Muslims, created two classes of citizenship based on religion. It may be said, in defence of Aurangzeb, that financial desperation caused by his incessant wars or the need to increase the incomes of Islamic and Sufi charities, rather than specifically anti-Hindu sentiment, might have driven him to levy discriminatory taxes.[74] For the Hindus, however, it was like returning to second-class citizenship in their own land (Excerpt 6.2). Again, the destruction of some of the most famous Hindu temples at such holy sites as Mathura and Benares and in the Rajput lands was a body blow to Hindu pride and respect. Quite likely, Aurangzeb destroyed the temples not out of anti-Hindu bigotry but out of a determination to punish defiant Hindu princes. An umbilical cord tied the royal Hindu states to their great shrines; the principal Hindu temple always symbolized the sovereignty of a Hindu state far more potently than any mosque did for a Muslim ruler. In this respect, Aurangzeb was perhaps no more culpable than most of the sultans before him; they desecrated the temples associated with Hindu power, not all temples. It is worth noting that, in contrast to the traditional claim of hundreds of Hindu temples having been destroyed by Aurangzeb, a recent study suggests a modest figure of just fifteen destructions.[75] This may be somewhat debatable, but we need to ensure that Aurangzeb is not unfairly demonized.[76] It is also a fact that he often gave grants for repairing old temples and even building some new ones.[77] Whatever the rights and wrongs, however, there is no doubt that the overall effect of Aurangzeb's actions on the Hindu psyche was highly negative.

Aurangzeb's theocratic mindset put him into direct confrontation with the small but increasingly dominant Punjabi community of the Sikhs. His grandfather, Jahangir, had had a difficult relationship with the fifth Sikh *guru*, Arjun, resulting in the latter's execution and martyrdom; but then the root of the problem had been political, not religious.[78] It was, however, Aurangzeb's religious policy that caused the friction between him and the ninth Sikh *guru*, Tegh Bahadur. In both Punjab and Kashmir the Sikh leader was roused to action by Aurangzeb's excessively zealous Islamic policies. Seized and taken to Delhi, he was called upon by Aurangzeb to embrace Islam and, on his refusal, was tortured for five days and then beheaded in November 1675.[79] Two of the ten Sikh *gurus* thus died as martyrs at the hands of the Mughals. The last *guru*, Govind Singh, who led the Sikhs for the next thirty-three years, began a course of open hostility against Mughal imperialism.

Courtly culture and imperial patronage

The court and the harem

The emperor was the centre of attention at the Mughal court, wherever it was located. An ensemble of musicians heralded his arrival with sounds of trumpets, drums and cymbals. It was an enormous privilege for courtiers and noblemen to be around the emperor, but strict rules of etiquette had to be observed. Each person knew his position and place, and was obliged to follow the court rituals.[80] A hierarchical positioning determined who could come closest to the emperor and who should confine himself to the outer perimeter. Bowing to the emperor and prostrating before him were compulsory for everyone, even for those who returned victorious from wars. One of the important ceremonies at the court was the giving and receiving of gifts.[81] Even the lowliest of *mansabdars* was expected to bring a gift, and foreign visitors were less than welcome when they turned up without expensive gifts. The court was also a place of entertainment; and music and dancing were an integral part of the courtly culture. A synthesis of Persian and indigenous Indian musical styles and instruments had been going on for nearly three centuries, but further refinements both in Hindustani classical music (*dhrupad*) and dance performances took place at the Mughal court.[82] Manuscript illustrations from this period provide rich evidence of this fusion. The most distinguished musician at Akbar's court was Tansen who originally worked in Gwalior, where earlier a brilliant musical tradition had been developed by the Hindu monarch Raja Man Singh of Tomar. Hundreds of male and female musicians and dancers relied for their livelihood on the patronage of emperors. Even the puritan Aurangzeb was not entirely immune from the enchantments of music.

The court was essentially male-dominated, but alongside it was the imperial harem for women – mothers, foster-mothers, stepmothers, wives, daughters, aunts, sisters, nieces, mistresses and concubines of the emperors, along with slave girls and eunuchs – who formed part of a huge and extended Mughal family.[83] The harem was not a place of licentious or orgiastic debauchery, as stereotypes would have us believe.[84] The women lived securely in comfort and style, befitting their seniority or ranks, within the seclusion of apartments and quarters of the grand palaces; but they were closely watched and guarded (Excerpt 6.3).[85] We do not have sufficient information about their state of mind engendered by life behind the *purdah*, nor are we certain about the freedom of movement they were permitted. Two particular facts should, however, be noted. First, since family is so much at the heart of Muslim social life, inter-gender bonds were strong.[86] Powerful emotional relationships tied the emperors to their mothers, wives, daughters and sisters.[87] Second, some of the women of the harem were formidable personalities who were prepared to break the bounds of conventionality.

For example, Jahangir's mother Maryam Zamani (Jodha Bai) and his wife Nur Jahan possessed sufficient economic knowledge to deal in international trade from behind the harem.[88] Other Mughal women, not necessarily connected with royalty, engaged themselves in education, charity work and aesthetic pursuits.[89] Many of them went out hunting, shooting and picnicking; and some even went to fight in wars. These examples of course only applied to Mughal women of higher rank; they were certainly ahead of their times. The experiences of ordinary women were very different.

Patronage of architecture

The world-renowned Mughal architectural tradition, that surpassed the legacy of all previous or contemporary Muslim kingdoms of the subcontinent, was due primarily to the munificent royal patronage offered to creative architects, builders and craftsmen by both the Mughal emperors and sub-imperial patrons in this period.

The Mughal achievement drew its inspiration from many sources.[90] The monuments of the Delhi Sultanate (Chapters 4 and 5), provided models for constructing courtyards, vaulted portals and domes, along with the decoration for calligraphy. The building materials and local architectural features of provincial monuments were also utilized by Mughal architects and builders. From the Iranian Timurid tradition came two central themes: first, the geometrical proportions for the construction of large central rooms surrounded by smaller chambers and interconnecting transversed arches, as found in the great *jami* mosques of Samarqand and Bukhara; and, second, the idea of a paradise-garden of the *chahrbagh* model (Chapter 5). There was also the influence of Hindu and Jain building traditions as found, for example, in the great fortress-palace of Gwalior built by Raja Man Singh Tomar *c.* 1500, with its flat-roofed rooms, the domed kiosks (known in Indo-Islamic architecture as *chattris*) and the gateways.

The grand forts built by Akbar and Shah Jahan were meant to project overwhelming Mughal majesty and power (Plate 5). Akbar's Agra fort, the most iconic stronghold built of red sandstone, had 22-metre high walls with a circumference of 2.5 kilometres; and, according to Abul Fazl's *A'in-i Akbari*, it contained 500 stone buildings.[91] Akbar also ordered the building of a new walled city without a fort at Fatehpur Sikri, near Agra. Now lifeless except for the tourists, Fatehpur Sikri offers us a glimpse into some of the classic contemporary state buildings and palaces that even now exude power and space fit for royalty (Plate 7).[92] Shah Jahan commissioned a new fort-city, Shahjahanabad (Plate 8), a fortified palace complex with 3-kilometre long red sandstone walls, with gardens and luxurious marble buildings, black and white with *pietra dura* inlays of precious stones (Plate 9.3), which were the envy of the world. Inside the fort lived 57,000 people working and producing everything from textiles to perfumes to serve the needs of the Mughal establishment.[93] Beyond the fort, the city extended with bazaars, pleasure gardens, canals and houses of the nobility.[94]

Conspicuous public dedication to Islamic faith was one of the motivations behind the Mughal rulers' patronage of grand mosques. Akbar's early religiosity and devotion to Sufism led him to plan a massive *khanqah* as part of the Fatehpur Sikri complex, adorned with an impressive gateway, the Buland Darwaza; inside the *khanqah* he built a great *jami* mosque and an elegant tomb to honour Salim Chishti, a local Sufi saint who had correctly prophesied the birth of Jahangir and his brothers.[95] The Moti Masjid at Agra fort is another of his exquisite buildings (Plate 10.2). Shah Jahan too was sufficiently imbued with Islamic fervour to build his monumental *jami* mosque in Shahjahanabad.[96] And South Asia's second largest mosque at present, the Badshahi in Lahore, was built by Aurangzeb who spared no expense over the most exquisite floral designs, arches and motifs outlined in white marble inlaid into the red sandstone surface of the entire building (Plate 10.1).[97]

The tombs were expressions of power, as can be seen in the majestic structures of Humayun's tomb at Delhi and Akbar's at Sikandra, Agra. They were also symbols of love for the departed. Neither time nor treasure was spared by Nur Jahan to build an exquisite tomb in Agra for her father Itimad ud-Daulah.[98] A small structure of pure white marble ornamented with an inlay of semi-precious stones, the tomb conveys 'the impression of a rich article of jewellery magnified into architecture'.[99] Nur Jahan also oversaw the building of her husband's tomb at Lahore (Plate 3.2). The immortal tomb of love, a crowning glory of Mughal architecture, is of course the Taj Mahal at Agra, built by Shah Jahan in memory of his wife Mumtaz Mahal. Much has been written about the Taj over the last 350 years; so anything here would be superfluous. All we can say is that the Taj is the final culmination of all the notable conceptions, works, designs and decorations that originated from different sources of Indo-Islamic architecture that were identified at the beginning of this section. A recent work by Ebba Koch, *The Complete Taj Mahal*, provides the most comprehensive survey of its construction.[100]

If power and religiosity were the unwritten themes behind the building visions of Akbar and Shah Jahan, it was the love of nature that guided Jahangir in his approach to architecture. Following in the footsteps of his great grandfather, Babur (Chapter 5), Jahangir found inspiration and comfort in gardens. It was in Kashmir, a land of valleys, clear streams and mountains, where he set up his summer residence and began his garden project. The central feature of a Kashmiri Mughal garden is a spring, of which the waters are collected in a canal that forms the main axis of the garden. The layout uses the undulating hillsides for terraces, pavilions, ponds and branch canals along the watercourse. The flowering plants, trees and shrubs, profusely cultivated in all their varieties, create a magical serenity of a Persian *bagh-i behesht*, the garden of paradise. Jahangir's own efforts were emulated by his courtiers and noble *mansabdars* who constructed such gardens at their riverside houses in Agra and other places.[101]

In this period of Mughal architectural excellence, it is worth bearing in mind that many provincial Indo-Muslim styles also flourished in different parts of the empire. Mughal and neo-Mughal influences made their imprint on traditions in Kashmir, Bengal and Gujarat. The post-Bahmani Deccani kingdoms were absorbed into the empire quite late, so we have an interesting mixture of Deccani provincialism and Mughal characteristics. Fine examples of the Nizam Shahi architecture of Ahmadnagar may be found in the Farah Bagh palace complex, the Hayat Behisht Bagh with its complex system of water flow management devised by Iranian experts, and the tomb of Malik Ambar, the great African soldier-governor of the province.[102] In the Adil Shahi kingdom of Bijapur, an exquisite mosque, the Ibrahim Rauza, was built during the memorable reign of its most illustrious monarch, Sultan Ibrahim II Adil Shah (r. 1580–1627),[103] while one of the most technically advanced domed structures may be seen in the grand mausoleum of Sultan Muhammad Adil Shah (r. 1627–56), the Gol Gumbad.[104] And in Hyderabad, founded in 1591, the Qutb Shahi ruler, Muhammad Quli (r. 1580–1611), planned two massive commercial thoroughfares whose crossing is marked by a monumental gateway of the Char Minar, a quartet of arched portals with four corner minarets (Plate 6.2).[105] In all these buildings, Mughal influences are to be noticed.

Royal paintings

Mughal paintings are prized possessions of well endowed museums and galleries or wealthy art collectors and connoisseurs of art. Their original inspiration came from the Persian miniature tradition. During his exile at the court of Shah Tahmasp in Iran (Chapter 5), Humayun had been much impressed by the artistic techniques and styles employed in the studios there. On regaining the Mughal throne, he invited Iranian artists who were disenchanted with the puritanical atmosphere in their native country.[106] His early death did not jeopardize the artists' position because his successor, Akbar, was equally impressed by their work. He provided patronage and funds for studios where teams of artists and skilled assistants were employed full time to produce works of art that he desired. Paper workers, sketch drawers of facial features, calligraphers, illuminators, gilders, illustrators, binders etc., all made their contributions in the traditional Indian way of working together. The final product was the effort of many, not just one brilliant artist.[107] There was no religious bar to the recruitment of Hindu artists and workers. Akbar looked for talent; ethnicity and religion did not matter to him. It was this mixed work force that created a magical intermingling of Persian and Hindu elements which are the hallmarks of the best of Mughal painting.

Manuscript illustration was the fashion of the day and, at first, manuscripts dealing with fables and legends were Akbar's favourites. A manuscript, known as *Tutinama*, contains short stories told each night by a parrot to his mistress in order to stop her taking a lover in the absence of her husband. These stories were complemented by illustrations.[108] Another manuscript, the *Hamzanama*, related the adventures of Hamza, an uncle of Prophet Muhammad, in fourteen volumes, with each volume containing a hundred illustrations, portraying characters like magicians, demons, dragons etc., and depicting love affairs, murders, kidnappings and battles.[109] When Akbar's interests moved to comparative religion, he commissioned translations of Hindu religious texts along with suitable illustrations. The traditional Hindu stories of the *Mahabharata*, for example, are illustrated in the Mughal style in a manuscript called the *Razmnama*.[110] European and Christian images, brought to India by travellers and Jesuit Fathers, also impressed Akbar greatly; and Mughal artists emulated Western use of perspective and modelling of three-dimensional figures.[111] The manuscripts that Akbar cherished most were those that incorporated suitable portraitures of scenes and heroic deeds of kings.[112] He himself liked to be portrayed in an energetic pose, either hunting or fighting.

The number of Mughal artists increased significantly under Jahangir (Excerpt 6.4). Most of them were in fact Hindus, a significant pointer to Mughal secularism. Jahangir generously commissioned individual pieces of work which were bound in albums, called *muraqqas*.[113] While Akbar was obsessed with themes of power and warfare, Jahangir loved nature and its creations in his paintings.[114] *The Turkey Cock*, by Mansur, is a fine example of this (Plate 9.1). Akbar had to show guile and courage to secure the throne, and his paintings therefore reflect much of what may be called activity; Jahangir came to the throne when the empire faced no enemies, and he is therefore portrayed in a secure setting.[115] In an iconic illustration, *Jahangir Preferring a Sufi Shaikh to Kings*, by Bichitr, now housed in the Freer gallery of Art, Smithsonian Institution, Washington DC, for example, a serene Jahangir is portrayed with a halo around his head, sitting on an hour-glass throne, and concentrating on a Sufi divine rather than the Turkish sultan or the English king.[116]

Shah Jahan, the great patron of architecture, loved all things of luxury and beauty, and especially diamonds, rubies and precious stones. He was also passionate about grand and formal ceremonies at his courts. We are able to get a glimpse of some of this from a manuscript, the *Padshahnama*, which chronicles most of his reign.[117] The ritual of court life is brilliantly evoked in the highly colourful illustrations accompanying the text. Shah

Jahan did not, however, have the same intensity of feelings towards the art of painting as his father and grandfather; and from his time onwards the quality of the classic and courtly Mughal painting declined substantially. Other traditions in the neo-Mughal style, however, continued to develop at regional Muslim courts of the Deccan.

It is in the Deccan that we must look for some excellent examples of art works that reflect both the Mughal and provincial Deccani influences. The Ahmadnagar School of Painting, for example, specialized in female portraiture, rare in that age.[118] Superseding the Ahmadnagar paintings are the treasures from Bijapur where the Adil Shahi dynasty held sway between 1490 and 1686. This kingdom was one of the very last that the Mughals conquered, and a truly independent artistic tradition existed there. Yet Mughal influences were never absent. The most illustrious ruler was Sultan Ibrahim II Adil Shah (r. 1580–1627) whose patronage of assorted arts was astounding.[119] A tolerant and eclectic person that he was, he was greatly fond of classical music and iconography; and he employed outstanding musicians, painters and calligraphers at his court. One of the evocative items in the Bijapur collection is a portrait showing him in a contemplative mood holding castanets and wearing beads of a mystic (Plate 11.1). Another, the highly sophisticated *A Prince Reading a Book*, has been described as 'among the most brilliant of all Deccani paintings' (Plate 11.2).[120] Finally, from the Deccan in the Mughal age, we also have some fine works from Golconda, such as the *Procession of Sultan Abdullah Qutb Shah Riding an Elephant* (*c.* 1650), now housed at the State Public Library, St Petersburg, Russia, and considered as one of the 'outstanding masterpieces of Indian art'.[121]

Literature and learning

One of the great delights in studying Mughal history is the availability of the extensive contemporary written material. Numerous European writers, for example, have left us useful and interesting accounts of their travels, their impressions of the subcontinent and their reception by all manner of people, from the emperors downward.[122] For our purpose here we shall concentrate on Islamic writings: and we need to start first with Persian works. Throughout this period, Iranian intellectuals and poets were seeking permanent settlement in India in order to write or express themselves freely. The Muslim rulers of India had for centuries welcomed the Iranians for their knowledge and expertise. In the sixteenth century, the best of them inevitably ended up at the Mughal court.[123] The Mughal rulers were welcoming because they much admired Persian culture and wished to emulate it. Akbar formally declared Persian to be the language of administration at all levels and ordered the introduction of Persian syllabi for different subjects in Islamic institutes of learning.[124] The Hindu middle class and the intelligentsia, with a long tradition of learning and 'panditry' behind them, also strongly encouraged their young people to learn Persian to secure lucrative posts under the Mughals.[125]

The elite Islamic literature, written mostly in Persian during this period, includes poetry, history and religious writing. The Persian poetry of India, which we have already referred to in Chapters 4 and 5, was composed in what is called *sabk-i Hindi* or the Indian style. This meant that within such Persian genres like the *masnavi* (narrative poem), *ghazal* (lyrical poem) or the *rubai* (the stanza of four lines) the Indian poetic sensibility, born out of the Indian natural environment or Indian imagination, was skilfully woven in. While this created controversy between the advocates of Persian poetry of Persia and the Indo-Persian poetry of India over the correct stylistic usage of Persian language, there is no doubt that the style developed in India reached a high level of achievement and became extremely popular.[126] One of the hallmarks of *sabk-i Hindi* was its ability to put in words the idea of accommodation and space for everyone, which Akbar and his successors encouraged. Among a number of Mughal poets, the two whose poetry best represented the Indo-Persian style were Faizi (1547–95),[127] the poet-laureate and brother of Abu'l Fazl, and Urfi Shirazi (1556–91).[128]

Islamic historical writing, developed from the earliest days of Islam, continued in Mughal times in the form of great narratives of the emperors. Babur had written a fine autobiography, the *Baburnama*, originally in Chagatay Turkish, which Akbar ordered to be translated into Persian (Chapter 5). Akbar asked his learned sister Gulbadan Begum (1523–1603) to write down from memory everything she knew about their father Humayun. Her work, *Humayunnama*, lost for a long time, has now been recovered, and is the sole surviving account by a Mughal royal female of the sixteenth century.[129] Akbar's own reign has been narrated in Abu'l Fazl's great volume, the *Akbarnama*, along with its appendage, the *Ain-i Akbari*.[130] Jahangir, like Babur, wrote his own autobiography, *Jahangirnama*, for the first seventeen years of his reign; his later years were covered by other court historians. One of the most reliable accounts of his reign can be found in the writings of Firishta (1560–1620), a renowned historian.[131] The early Western historiography of Islamic India relied much on Firishta. Finally, Shah Jahan's reign was recorded in the *Padshahnama* written by, among others, Abdul Hamid Lahori and his student Muhammad Waris.[132] The original copy of this beautifully illustrated manuscript is in the Royal Library at Windsor. Aurangzeb stopped the detailed annals after ten years; instead he commissioned one of the great Islamic legal texts of India, the *fatawa-yi Alamgiri*, a valuable work to understand the history of Mughal institutions;[133] his private letters are also a worthy source of historical knowledge.[134]

For the student of history, the most useful of all the works from this period is the *Akbarnama/Ain-i Akbari* of the historian and courtier, Abu'l Fazl. This work, which combines narrative history, personal opinions, minute and detailed accounts of various peoples and things, statistics concerning geography and economics, and much else, is an indispensable storehouse of Mughal knowledge.[135] One of its pervasive characteristics is the exaltation of Akbar in every situation. Abu'l Fazl was either unable or unwilling to evaluate Akbar with a critical eye; and that is why the *Akbarnama* cannot be considered an absolutely reliable primary historical source. Since Abu'l Fazl and his poet-laureate brother Faizi were strongly behind Akbar's liberal views on religion, both of them along with Akbar were heavily criticized by a conservative historian at the court, Abdul Qadr Badauni.[136] This serious Sanskrit scholar had translated the Hindu epics of the *Ramayana* and the *Mahabharata*: but he was no liberal in matters concerning Islam. He vehemently attacked Akbar, Abu'l Fazl and Faizi in his *muntahab al-tawarikh*, but did not publish the book until after Jahangir's reign had begun. It was then published uncensored.

The controversy about Akbar's liberalism and universalism in the historical works of Abu'l Fazl and Badauni had its echo in religious writings too. One of the most influential religious thinkers of the early seventeenth century was Shaykh Ahmad Sirhindi (1564–1624).[137] A follower of the Naqshbandi Sufi order, his aim was to curb what he saw as Sufism's excessive emphasis on universalism; he wished to integrate it within a Sunni framework.[138] His letters have had a great impact on many generations of Muslims in the subcontinent, and his ideas were particularly well received during the eighteenth and nineteenth centuries when Muslim power in the subcontinent was ebbing away. However, his championing of Islamic revivalism was bound to lead to poor community relations. The blossoming of such relations depended upon not treating the Hindu and the Muslim as two separate and mutually antagonistic categories. This is brought out in the mystical writings of Shah Jahan's eldest son, Dara Shukoh (1615–59), who was cruelly put to death after a sham trial decreed by his brother Aurangzeb. Dara was not the most earnest of princes, but he was an earnest Sufi.[139] He wrote a number of mystical works and letters which constantly emphasized the reconciliation between Muslims and Hindus; and the title of one of his books, *Majmua al-bahrain*, or the 'merging of two oceans', points to this ideal.

Muslim writings from this period reveal to us the high intellectual qualities of the respective authors. Their own education would have taken place in private schools, *madrasa*s and other higher seminaries of learning. This elite education was available to only a chosen few. Despite the fabulous wealth that the Mughals possessed, they never thought of using a portion of it on some form of universal primary education. The mass of the population had no opportunities for schooling; and whatever education some might have received was very basic and sectarian. Akbar was a great unifier of communities; if only he had thought of introducing a secular and practical education for all children between five and fourteen in his kingdom, the history of South Asia would have turned out very different indeed! It is quite likely that colonialism might have been avoided. This essentially was the tragedy of the second richest empire on earth in the sixteenth and seventeenth centuries.

Select excerpts

6.1 Two opposite views on Akbar's interest in other faiths

In the following extract Jahangir praises his father and points to a motivating force behind Akbar's interest in a multicultural approach to faith.

> My father always associated with the learned of every creed and religion: especially the Pandits and the learned of India . . . He associated with the good of every race and creed and persuasion, and was gracious to all in accordance with their condition and understanding.

Contradicting the above is the following opinion of Badauni, the historian, who blamed Akbar for uncritically accepting the ideas and opinions of non-Muslims and thereby harming the cause of Islam.

> crowds of learned men from all nations, and sages of various religions and sects came to the Court, and were honored with private conversations . . . everything that pleased him, he picked and chose from anyone except a Moslem . . .
>
> (Rogers & Beveridge 1909/1989 reprint, Vol. 1: 33 and 38; Lowe 1884/1976 reprint, Vol. 2: 263)

6.2 Aurangzeb's re-imposition of jizya

A very fine and comprehensive account of Aurangzeb's reign is contained in a historical work, *Muntakhabu-i Lubab*, by Muhammad Hashim, popularly known as Khafi Khan, who served Aurangzeb for a number of years. The following passage describes the re-imposition of *jizya* in a tone rather different to that adopted by the more partial medieval Muslim historians. The incident mentioned would have been one of numerous such events that ultimately destroyed Aurangzeb's reputation in the eyes of the majority Hindu population.

> Upon the publication of this order (the re-imposition of jizya) the Hindus all round Delhi assembled in vast numbers . . . on the river front of the palace, to represent their inability to pay, and to pray

for the recall of the edict. But the Emperor would not listen to their complaints. One day when he went to public prayer in the great mosque . . . a vast multitude of Hindus thronged the road . . . with the object of seeking relief. Notwithstanding orders were given to force a way through, it was impossible for the Emperor to reach the mosque . . . At length an order was given to bring out the elephants and direct them against the mob. Many fell trodden to death under the feet of the elephants and horses.

(Khafi Khan, *Muntakhabu-i Lubab*, in Elliot & Dowson 1867–77/2001 reprint, Vol. 7: 296)

6.3 The imperial harem

The following extract provides an insight into the strict procedures and protocols within the harem.

Though there are more than five thousand women, he (the Emperor) has given to each a separate apartment. He has also divided them into sections, and keeps them attentive to their duties . . . The salaries are sufficiently liberal . . . Attached to the private hall of the palace, is a clever and zealous writer, who superintends the expenditure of the Harem, and keeps an account of the cash and the stores. If a woman wants anything, within the limit of her salary, she applies to one of the cash-keepers of the seraglio. The cash-keeper then sends a memorandum to the writer, who checks it, when the General Treasurer makes the payment in cash, as for claims of this nature no cheques are given. Whenever Begums, or the wives of nobles . . . desire to be presented, they first notify their wish to the servants of the seraglio, and wait for a reply. From thence they send their request to the officers of the palace, after which those who are eligible are permitted to enter the Harem. Some women of rank obtain permission to remain there for a whole month.

(Blochmann 1873: 44–5)

6.4 Jahangir's expertise on paintings

Jahangir was a true connoisseur and patron of art and painting. His confidence and self-assuredness in the following extract stemmed from his deep knowledge and empathy with the art of the period.

As regards myself, my liking for painting and my practice in judging it have arrived at such a point that when any work is brought before me, either of deceased artists or of those of the present day, without the names being told me, I say on the spur of the moment that it is the work of such and such a man. And if there be a picture containing many portraits, and each face be the work of a different master, I can discover which face is the work of each of them. If any other person has put in the eye or an eyebrow of a face, I can perceive whose work has painted the eye and eyebrows.

(Rogers & Beveridge 1909/1989 reprint, Vol. 2: 20–1)

7

THE AGE OF MUGHAL DISINTEGRATION (1689–1765)

The year 1689 witnessed the high tide of Mughal fortunes. Emperor Aurangzeb reached the pinnacle of glory with his conquest of the Deccan (Chapter 6). No other Indian ruler, since the time of Emperor Ashoka, had ever ruled over a territory of the size of the Mughal Empire of the late seventeenth century. Yet, in just over three-quarters of a century, in 1765, the Emperor Shah Alam II lost all mastery over his domains when he ceded the authority of revenue collection in Bengal, one of the richest areas of his empire, to the British East India Company for its exclusive use, 'from generation to generation, for ever and ever'.[1] In return, he was to receive from the Company an annual pension of £260,000, a relatively trivial sum for a Mughal emperor, to maintain his household and support his imperial style. His long reign, which began in 1759, ended in 1806; and two further emperors were to continue the Mughal dynasty. But there was no substance left to imperial pretensions after 1765.

With imperial collapse being so swift and precipitate, the period 1689 to 1765 has traditionally been viewed as one of utter confusion, chaos and economic stagnation, a period which is also popularly viewed as the one during which the East India Company simply stepped in to fill the vacuum left by the dying and powerless empire.[2] There is much truth in this, but recent researches into eighteenth-century India also tell us that while the Mughal dynasty itself and the highly centralized empire of Akbar's creation were collapsing in this period, many of the Mughal traditions, institutions and cultural styles continued in other centres of power. These centres were controlled by new regional dynasties and elites, Hindu, Muslim or Sikh, which had once been part of the old Mughal *mansabdari* system.[3] Mughal influences pervaded the political economy and culture of the regions. New research also points to this period as not one characterized solely by economic stagnation.[4]

The disintegration of the central imperial structure meant that Muslim power in general was being eroded in this period. The beginnings of Islamic revivalism may be interpreted as a reaction by some of the orthodox to the perceived debilitation of the Mughal Empire. On the other hand, many regional Muslim rulers showed both flair and diligence in pursuit of statecraft. They created coherent state structures, gave support to Islamic institutions and also opened up spaces to many different communities to engage in trade and other activities. At the same time, Islamic cultural and literary activities continued to flourish both in the depleting Mughal heartland and the regions.

The rulers of the most important of the autonomous Muslim regional states – Bengal, Awadh, Hyderabad – were able to assert their authority by either drawing powers away from the Mughal centre or engaging with the rising power of the Marathas. For most of the first half of the eighteenth century, they were also able to maintain at least profitable, if not always equal, relationships with the respective East India companies of France and Britain, the two rising powers of Europe in the eighteenth century. It was in the mid-eighteenth century, however, that these autonomous rulers faced their moment of truth when the British began their active interference in the politics of Bengal. No Indian ruler, from then on, could be truly confident of maintaining his power base. The period 1689 to 1765 was therefore a curtain raiser to the British domination of India that was to follow.

Factors in imperial decline

The causes of Mughal decline from the late seventeenth century onwards have received much attention in historical literature.[5] Two of them were internal causes. First, Aurangzeb's Islamic orthodoxies and prejudices strained the loyalty of his majority non-Muslim subjects.[6] The Rajputs and the Marathas, once loyal supporters of the *mansabdari* system, turned against him.[7] Second, the vast expanse of the empire and its rich economy provided opportunities for those in power in the regions to assert their authority further. A third cause of decline was external, in the sense that the empire was unable to adapt to the fast changing world of European expansion in the eighteenth century. The Mughal rulers and their advisers could be seen as people out of touch with the forces of modernity that Europe was experimenting with from the mid-seventeenth century onwards. The same could also be said for other Islamic imperial systems and many Asiatic and African native governments of the time that faced threats from both European ideas and military might. In the specific Indian context, new commercial arrangements and alignments, for example, were beginning to be forged among social classes, often bypassing the imperial system.[8]

The immediate cause of Mughal decline can be traced to 1689 when, in the hour of his supreme triumph in the Deccan, Aurangzeb took the fateful decision to continue to stay there until he had crushed the last of his Maratha enemies. He thus condemned himself and his army to another eighteen years of arduous warfare that sapped the strength of the empire.[9] His wiser predecessor, Akbar, in similar circumstances, would have paused, taken stock of the situation, attempted to conciliate the battered Marathas and thereby consolidate his conquests. Aurangzeb, instead, went on to execute Sambhaji, the captive Maratha leader (Chapter 6). The Marathas, safely ensconced within the formidable hills of Mahrashtra, refused to be intimidated (Excerpt 7.1); their lightly armed but swift moving bands of soldiers harried and harassed the worn-out Mughal armies across the Deccan. In the years 1702–4 they attacked Hyderabad, creating a massive trade disruption and famine throughout the surrounding countryside. The situation remained both perilous and stalemated for the Mughals until Aurangzeb died, a disappointed man, in 1707.

The war in the Deccan proved to be extremely costly. The Marathas extorted from the peasants the taxes that were meant to be collected by the Mughal-appointed *zamindar*s and revenue officers. Aurangzeb attempted to resolve the crisis by declaring large tracts of land as crown lands (*khalisa*) in order that taxes could be directly collected by crown officers.[10] This, however, resulted in a shortage of lands that were needed as *jagirs* for *mansabdars*. The expansion of infantry forces had required a corresponding increase in the number of military *mansabdar*s; but the shortage of *jagirs* de-incentivized them. The system depended upon a balanced equation between the loyalty of nobles to the emperor and his generosity in assigning them *jagirs*. The steady decrease in the number of *jagirs* meant serious financial hardship for the nobles. Most of them could not maintain the level of troops and cavalrymen as would have been agreed and, with little monitoring of military deficiencies, the entire Mughal military machine increasingly became ineffective.[11]

The Deccan campaign also created another crisis of confusion for the nobles. The *mansabdars* based in the north were immune to southern problems, but felt cut off from the court; the southern or the Deccani nobles were near to the ageing emperor but were all too aware of his impotence in the face of the Marathas.[12] There was a large contingent of Maratha *mansabdars*, but their credentials and motives were increasingly suspect. They remained a 'large and unsettling'[13] force among the nobility. Thus, a combination of factors – a lack of a decisive victory over the Marathas, the shortage of funds and *jagirs*, the differences in strategies, etc. – led to the breaking up of the corporative spirit of the *mansabdari* system. With bribery, corruption and disloyalty spreading through both the ranks and the officer class, each of the Mughal nobles was cast adrift to look after himself.

The crisis worsened after Aurangzeb's death. As revenues decreased, the Mughal officers increasingly resorted to what is called revenue-farming; that is, they awarded contracts to individual merchants, bankers and large landlords to collect tax from the peasants and remit an agreed sum to the treasury in return for a commission and local political privileges and powers.[14] Unwittingly, therefore, the Mughals were creating a new class of rural gentry whose power was to grow. Each *mansabdar*, too, attempted to increase his income by means other than those traditionally agreed upon.[15] The concept of *khanazad* loyalty to the emperor thus began to dissolve.[16] In that atmosphere, the nobility at the very centre of power became fragmented into self-interested disparate groups; and it was only a matter of time before the greatest of the Muslim nobles, originally the most loyal of the *khanazads*, would consolidate their own power bases away from the centre. We see this taking place in the 1720s when three Mughal governors of Bengal, Hyderabad and Awadh established their own autonomous states, while yet preserving the fiction of Mughal sovereignty. The rulers and courts of these principalities held attention at a time when imperial power was decaying.

The tendencies towards localism, factionalism and secessions could have been checked if the centre had held its nerve under some able emperors. This was the missing factor in this period. All seventeen Mughal rulers between 1526 and 1857 craved for power, but only the first six knew what to do with it. Those six left an enduring legacy, although Aurangzeb's Deccan wars after 1689 proved disastrous. But with all Aurangzeb's mistakes, the empire had survived nonetheless. Of the eleven emperors who followed Aurangzeb, nine ascended the throne within the short space of fifty-two years, during this period. Compare that to the first six emperors who ruled between them for 181 years. All the nine emperors, except perhaps Bahadur Shah I, lacked character and vision. They provided no leadership and had no cohesive programme of renewal or rejuvenation to offer to their subjects. They were either overawed by powerful courtiers or were themselves engaged in treacherous conspiracies against the courtiers. This can be demonstrated by studying how their reigns began and ended.

Aurangzeb's eldest son, Bahadur Shah I, could take the throne only after he had defeated and killed two of his brothers in a brutal war of succession.[17] His successor, Jahandar Shah, became emperor after three of his brothers were first disposed of by his *wazir*, Zulfikar Khan.[18] Jahandar's nephew, Farrukhsiyar, incensed by his father's death, led a revolt with the assistance of a family of two brothers, Sayyid Husain and Sayyid Abdullah, better known as the Sayyid brothers. This revolt succeeded and, within a year, both Jahandar Shah and Zulfikar Khan were killed.[19] After taking the throne, Farrukhsiyar turned against his protectors, the Sayyid brothers, who then joined up with the Marathas, the enemies of the Mughals, and got him blinded, strangled and executed.[20] After promoting two young princes for the crown, each of whom died of illness after a few days' reign, the Sayyid brothers finally settled on one of the royal princes, Muhammad Shah. He, however, disliked their interference and, conspiring with another courtier, Nizam ul-Mulk (later of Hyderabad), brought about their demise within a year.[21] Muhammad Shah ruled for three decades (1719–48) and died peacefully in his bed, but his two successors, Ahmed Shah and Alamgir II, came to the throne more or less as puppets of Nizam ul-Mulk's son and grandson, Ghazi ud-Din Khan I and II. The fate of both emperors was sealed once they were of little use to their patrons: Ahmed Shah and his mother were blinded, and Alamgir II was assassinated.[22] The last of the nine emperors of our period, Shah Alam II, secured the crown only on the recommendation of an Afghan invader, Shah Abdali.[23] During his long reign between 1759 and 1806, he suffered many humiliations: the granting away of the Bengal *diwani* to the British in 1765, his own blinding by an Afghan chieftain in 1788, and the British occupation of Delhi in 1803. Two more emperors would follow him, but they were mere pensioners of the British.

The later Mughals lived at a time of great changes and uncertainties. Their empire, if it was to survive, needed them as effective custodians of power. Personalities do matter in history. John Stuart Mill observed that 'the initiation of all wise or noble things, comes and must come from individuals; generally at first from some one individual'.[24] He also correctly said that 'the worth of a state, in the long run, is the worth of the individuals composing it'.[25] The institutions of a state can be effective only as long as there are capable individuals to sustain them. The greatest of institutions and structures can fall apart without the support of worthy individuals. The fall of the Mughal Empire was in part a function of the ineffectiveness of the later Mughals.

Imperial meltdown

Successive internal and external waves of aggression sapped the Mughal strength.[26] The first crisis arose in the north among the Jat agriculturists dissatisfied with punitive imperial taxation.[27] Originally a pastoral people in the lower Indus basin, the Jats had been migrating northwards, centuries previously, towards Punjab; and from there they had spread along a long corridor of land up to the middle Gangetic basin.[28] The cities of Delhi and Agra were situated within their home lands. A homogeneous community of farmers, peasants and some pastoralists, the Jats were, however, divided by faith: Muslim Jats in West Punjab, the Sikhs in East Punjab and Hindu Jats further to the east. The first to rebel against the Mughals were the Hindu Jats who carved out a small statelet, Bharatpur, for themselves. Among the Sikhs of East Punjab too a spirit of rebellion was ever present, owing to a heady brew of agricultural grievances, pride in a strong religious identity and the memory of the martyrdom of some of their leaders at the hands of the Mughals.[29] After 1708, when their tenth *guru*, Gobind Singh, was assassinated, they were led by a millenarian figure, known as Banda Bahadur, who raised large peasant armies not only to defy the Mughals but also to carve out a Sikh principality in north-west India.[30] This was a precursor to the rivalry between the Sikhs and the Muslims in Punjab generally. Banda's rebellion was eventually suppressed, but at great cost to the Mughals.

One of the serious aspects of the Jat-Sikh rebellions was the growing power of local *zamindars* refusing to follow imperial dictats.[31] They sought increased autonomy. While religious grievances certainly played a part in arousing the Sikhs, it is important to note that the primary factor connecting the Jat and Sikh rebellions was the rural political economy in which *zamindars* and peasants both had a stake. Although the Mughals suppressed a number of smaller northern rebellions, it was more difficult for them to halt the march to autonomy of the great Rajput rulers and *mansabdars*.

The continuing Maratha aggression throughout this period further drained the Mughals. Rapid movements and lightning raids by their young soldiers, along with the policy of cutting off Mughal supply lines, were the classic Maratha military tactics.[32] After 1713, their overall policy was increasingly conducted from the office of a chief minister, the *peshwa*. The first *peshwa*, Balaji Vishwanath, led a march on Delhi to force the Mughals to cede taxation rights in the Deccan. His dynamic son, Baji Rao I, who became the *peshwa* in 1720, coordinated moves by heads of different Maratha families to secure sections of the Mughal provinces and set up their own civil and fiscal administrations.[33] Having arrived at the gates of Delhi by 1737, he forced the Mughals to surrender not only Gujarat and Malwa but practically all lands between the rivers Narmada and Chambal.

No sooner had they ceded a large swathe of their territory to the Marathas, the Mughals were attacked in 1739 by the Persian ruler, Nadir Shah.[34] Within the course of a single day 30,000 citizens of Delhi were killed; and the massacre was stopped only after Emperor Muhammad Shah personally begged for Nadir Shah's mercy. By the time the Persian invader returned home, in May 1739, with all the loot, including the priceless Peacock Throne, all carried by 1,000 elephants, 7,000 horses and 10,000 camels, Delhi and Shahjahanabad were left in a ruinous state.[35] While in 1737, in their hour of victory, the Hindu Marathas under Baji Rao I had felt a lingering sense of honour and decency to constrain themselves from entering the gates of Delhi, the Muslim Iranians under Nadir Shah felt no such qualms.

One can hardly talk about the Mughal Empire after 1740 (Map 7.1). The emperor remained on the throne, propped up by one faction of the nobility or the other. A period of civil war ensued between the factions; and the autonomous rulers of the principalities of Awadh and Hyderabad had their own court parties at war with each other. The Marathas continued with their plunder and pillage and, in the 1750s, their territorial sway reached up to Punjab in the north-west. Parallel to their expansion was the aggression unleashed by an Afghan leader, Shah Abdali, who attacked the north-west and the north on ten different occasions between 1748 and 1763.[36] On his penultimate attack in 1761, he inflicted a crushing defeat upon the Marathas

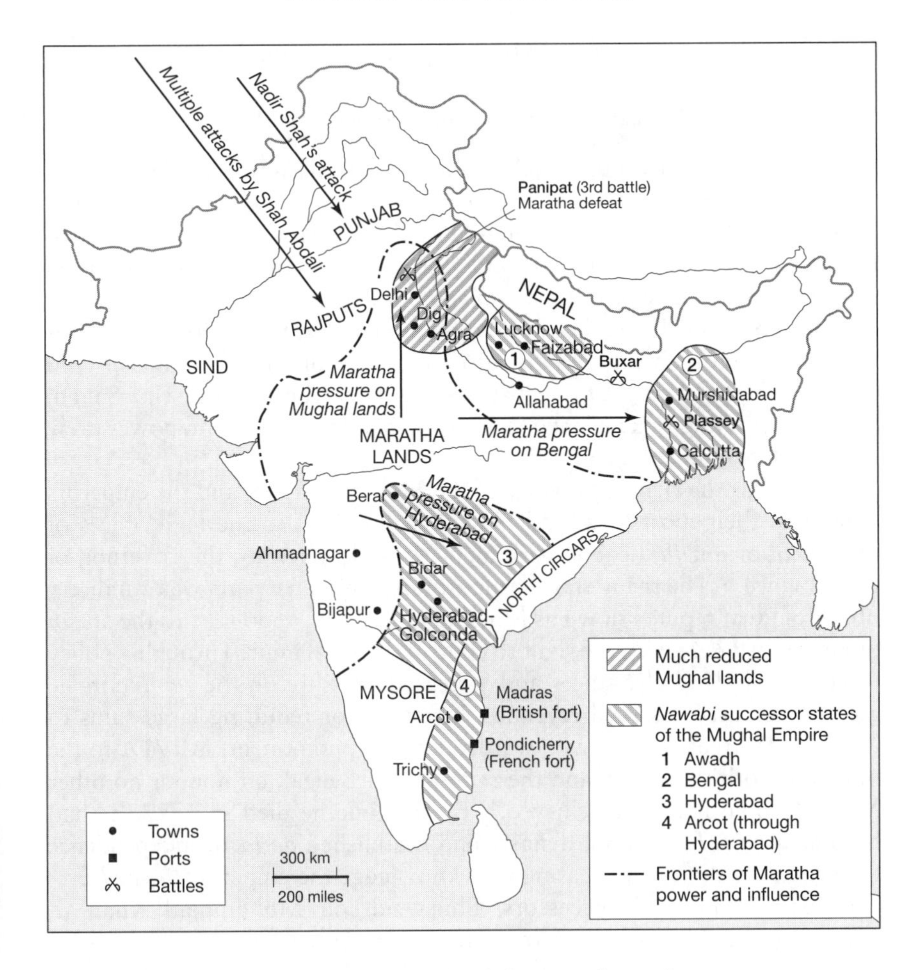

Map 7.1 Diminution of Mughal power and the rise of *nawabi* successor states: mid-eighteenth century

at what is known as the third Battle of Panipat (the first and the second battles are referred to in Chapters 5 and 6).[37] The Maratha defeat made no appreciable impact upon the fortunes of the Mughals. The Emperor, Shah Alam II, confined himself to the Red Fort of Delhi, while occasionally Mughal forces attacked the local recalcitrant Jats. The Sikhs became a cohesive force in the Punjab and the Rajput princes went their own way. Although most sources of finance had dried up, the imperial treasury still received a substantial financial tribute from the autonomous principality of Bengal, a region rich in resources and trade. The family name Mughal and its Timurid ancestry continued to retain a magical hold on the imagination of all in this period. Even the Marathas paid ritual respects to the emperor.

The post-Mughal political scene

Nawabi *rule in Muslim successor states*

In some key provinces of the Mughal Empire – Bengal, Awadh and Hyderabad/ Carnatic – a form of neo-Mughal rule continued in the wake of imperial disintegration. This rule was exercised by the *nawabs*, or princes, who had once been loyal to the emperor and who had worked as his governors in the provinces.[38] The first great *nawab*s and their immediate successors enjoyed a brief spell of real power during a very short period from the 1720s to 1760s, before being drawn into the circle of British influence and domination after 1765. Although at least one later Muslim ruler, Tipu Sultan, would offer a mighty challenge to the British, effective Muslim power drew to an end around the 1760s.

Bengal was the richest province of the Mughal Empire, and the emperors appointed their most loyal and trusted *mansabdars* to the high offices of the *diwan* or *nazim*, dealing with finances, and *subahdar*, the governor, of the province.[39] The rising star during Aurangzeb's later years was a military officer of great repute known as Murshid Quli Khan, appointed as the *diwan* of Bengal in 1701, with the right to control the local mint. Through a policy of large scale land clearances and increased pressure on the *zamindars* he greatly increased his land revenues. In return for remitting large sums to the imperial treasury he was rewarded by his appointment, in 1717, to the offices of both the *diwan* and the *subahdar* of Bengal, an honour no other Mughal noble had ever achieved.[40] By the time he died in 1727, he had become a *nawab*, set up a dynasty and established *de facto* independence for Bengal, without ever ceasing to acknowledge the emperor's formal pre-eminence.[41] His two successors, Shuja ud-Din Muhammad Khan (r. 1727–39) and Alivardi Khan (r. 1740–56), essentially continued with his policies. All three *nawabs* also established a complex and profitable relationship in overseas textile trade with Hindu bankers and British and Dutch traders.[42] This relationship unfortunately ended when the fourth *nawab*, the impetuous Siraj ud-Daulah, confronted the British from 1756 onwards.

While Bengal was a long way from Agra and Delhi, the Mughal province of Awadh sat on the very doorstep of imperial heartland. If any part of India could be considered as the Mughal monarchs' very own, it was Awadh. Its nobles and *mansabdars*, drawn from varied Muslim backgrounds, considered themselves as the elite of the Mughal court. When mutual rivalries and intrigues of various factions destabilized the court, the Mughal emperors often appointed the most difficult of nobles to high posts elsewhere. When a nobleman of Iranian origin, Saadat Khan, fell into disgrace at the court, the Emperor Muhammad Shah appointed him as the *subahdar* of Awadh.[43] Like Murshid Quli Khan in Bengal, Saadat Khan increased his revenues by confronting the mighty *jagir* holders, including the powerful Muslim lobby of *madad-i maash*, the holders of benefits and assignments from the revenues of land granted to them as religious leaders or scholars.[44] Proximity to the centre and their great wealth enabled him and his successors, Safdar Jang and Shuja ud-Daulah, to meddle at the imperial court with the aim of securing the position of *wazir* of the empire.[45] The rulers of Awadh indeed carried the title of *nawab-wazirs*. The Awadhi armies also cooperated with, contested or checkmated the Marathas and the Afghans. Until the early 1760s, Awadh's military position remained robust, but her defeat by the East India Company at the Battle of Buxar in 1764 presaged an uncertain future (Chapter 8).[46]

Hyderabad was the third great *nawabi* state. Aurangzeb's victories in the Deccan included the provinces of Khandesh, Berar, Aurangabad-Ahmednagar, Bidar, Bijapur and Golconda-Hyderabad. It was imperative for the Mughals to resist the Marathas who had the land revenues of these provinces in their sights; but in the absence of any possibility of delivering a knock-out blow the only alternative was to contain them by both low intensity warfare and statecraft. This policy came to be perfected by an able Mughal *subahdar* of the Deccan, Qamar ud-Din Khan, popularly known as Nizam ul-Mulk of Hyderabad (1671–1748).[47] Having served Aurangzeb on the battlefield from the age of seventeen onwards, the Nizam had risen high at the Mughal court; disenchantment with royal politics, however, forced him to withdraw to Hyderabad in 1724. For the next twenty-four years, he remained a major force in both Deccani and Mughal politics. He dealt with Maratha aggression sometimes by ceding them the right to collect taxes and sometimes by challenging them on the battlefield.[48] In time he lost many of the provinces of the Deccan and also the coastal area of the Carnatic, where a sub-*nawabi* came into existence under two separate dynastic lineages; but he consolidated his power base over Hyderabad-Golconda sufficiently strongly to set up, with the blessings of Emperor Muhammad Shah in Delhi, his own Asaf Jahi dynasty that carried on ruling Hyderabad state until 1948 and whose rulers called themselves *Nizams* rather than *Nawabs*. Nizam ul-Mulk never renounced Mughal sovereignty and, although he did not send a single rupee to the imperial treasury, he attempted to help the emperor on various occasions, such as when Nadir Shah invaded Delhi. While firmly opposed to any French or British territorial ambitions, he saw advantages in permitting their trading posts on his coastline. He, however, failed to foresee that the Anglo-French rivalry would first begin in his lands and that Hyderabad would become just a pawn in the game played by other powers after him.

Certain features were common to all the *nawabi* states.[49] The *nawab*s secured their autonomy when they were in reality the Mughal governors of their provinces during the early eighteenth century. As governors, they first won the right to appoint their own administrative officials without reference to the centre. From there, they proceeded to appoint their successors, thus establishing their own dynasties. They found ways of diverting revenues due to the imperial treasury for use in their own regions, and conducted diplomatic and military activities. They invested considerably in their regional capitals, such as Murshidabad in Bengal and Faizabad in Awadh, where their private courts resided. These were substantial steps towards autonomy, but they desisted from forcing mosque authorities to declare their independence at the Friday *khutbah* prayers. Ultimate Mughal sovereignty therefore remained unchallenged, at least in name. Also, Mughal courtly culture with its elaborate manners and etiquette continued under the influence of nobles, scholars, artists and poets who left Delhi and Agra for greater security in the provinces. Considering themselves as ideal Muslim rulers, the *nawabs* offered patronage to iconic Muslim institutions like the mosque, *madrasa* and the Sufi *khanqah*. Mughal-style religious liberalism, in the best tradition of Akbar's reign, was also demonstrated by employing Hindus to highest offices in their kingdoms. Hindu generals led the *nawabi* forces, and Hindu bankers oiled their finances.[50] Hindu shrines and temples were offered support.

European interventions

Away from the courts of the Mughal emperor and the *nawabs* a crucial development on the south-east coast of India was to foreshadow the rise of a new political power in the subcontinent. Until the mid-eighteenth century, the British and French trading companies, operating from their trading stations licensed by Mughal governors, had engaged in competitive but relatively peaceful trade along the coastline. The Coromandel coast was part of the Carnatic region of the autonomous Hyderabad kingdom. In the same way as Hyderabad under Nizam ul-Mulk had broken from imperial control, the Carnatic gained its autonomy from Hyderabad under a ruler known as the Nawab of Arcot.[51] The British trading post of Madras and the French one at Pondicherry were thus, technically speaking, part of Nawab of Arcot's possessions. When the European wars of the mid-eighteenth century – the War of the Austrian Succession (1740–8) and the Seven Years' War (1756–63) – transformed the Anglo-French commercial rivalry in India into one for political and military superiority, the two European antagonists proactively intervened in the affairs of both the Carnatic and Hyderabad by supporting rival claimants to their thrones and provoking conflicts.[52] The British and the French demonstrated their ability to make and unmake local Indian *nawabs* by proxy wars. Ultimately the French master strategist Dupleix was trumped by the British, and the French were eliminated from the Carnatic; that aside, the weakness of Indian armies became glaringly evident when, for example, the French stormed Madras in 1746 and the British captured Arcot in 1751.[53]

What had been tested out in the Carnatic was replicated in Bengal, except that it was mostly the British who did the running. The young *nawab*, Siraj ud-Daulah, who came to the throne in 1756, found it difficult to reconcile himself to the British presence in Calcutta and other trading posts in Bengal. A series of mishaps and imbroglios, including the infamous, and still a much argued about, incident of the Black Hole of Calcutta where many British people were suffocated, inevitably led to warfare (Excerpt 7.2). The British supremo, Robert Clive, used his guile to hatch a plot against Siraj, in conspiracy with his many enemies, including members of his own family and the Hindu bankers whose friendship Siraj had not valued enough, as in the days of his predecessors, Murshid Quli Khan and Alivardi Khan.[54] Once Clive had won the ensuing Battle of Plassey (1757), although more by default and treachery than by military brilliance, he became the main power-broker of Bengal.[55] Siraj's throne was given to his uncle, Mir Jaffar, in return for the latter agreeing to pay a massive indemnity to the East India Company and granting Clive personally an enormously valuable *jagir* which was to greatly enhance his wealth. When Mir Jaffar could not keep up with the payments the British replaced him with his son-in-law, Mir Qasim. He, too, fell out with the British; but he also plotted with the Nawab of Awadh, Shuja ud-Daulah, and Emperor Shah Alam II in order to fight the British.[56] In the Battle of Buxar (1764), the British forces, under Major Hector Munro, won a resounding victory, although both the Company's army and that of the two *nawabs* and the emperor were evenly matched. Mir Qasim's defeat made the hapless Mughal emperor change sides, seeking protection of the British, which was exactly what they wanted. Clive clarified the confusing political scene, in August 1765, with two bold moves: first, by recognizing the importance of Awadh as a buffer state and restoring to its *nawab* most of his lands, except Allahabad district, on payment of a large sum of money; second, by ceding the Allahabad district to the emperor, along with a pension, in return for Shah Alam II agreeing to grant the entire *diwani* of Bengal, Bihar and Orissa, to the East India Company. No greater humiliation for the Mughal Empire could be imagined than the decision of Shah Alam II to stay on in Allahabad rather than return to his court in Delhi.

Some trends in Indian Islam

The progress of Islam in the subcontinent was to be affected by three interesting, and somewhat contradictory, developments in this period of imperial disintegration and the rise of the successor states. The first of these was the increasing influence of the Shia among the newly emerging Muslim ruling class. The second was an attempt by some Muslim rulers to encourage syncretism between the Hindu and Muslim faiths, in the tradition of Akbar. The third was the articulate re-assertion of Sunni orthodoxy as a reaction to the first two. Out of these three developments would emerge some of the religious and political issues confronting South Asian Muslims in the nineteenth and twentieth centuries.

Shia political and cultural influence

Shia communities of Ismailis, Khojas, Bohras and Ithna Asharis had been settled for many centuries in Gujarat, Sind and Punjab. Until the beginning of the eighteenth century, with the exception of early Ismaili control in Sind and two sixteenth-century Deccan dynasties in Golconda and Bijapur, all Muslim monarchs of India were Sunni. The majority of the *ulama* and the Sufis, who gained patronage from these monarchs in return for bestowing on them an Islamic legitimacy for their rule, were also Sunni. Yet, as we have noticed in the previous four chapters, Muslim rulers felt themselves to be part of the Iranian cultural milieu. They looked up to Iran and Iranian/Persian civilization as sources of high culture of Islam. For centuries, therefore, Iranian writers, poets, skilled craftsmen, painters, astronomers, mathematicians and theologians had been welcome at the courts of the sultans of Delhi and Mughal and Deccani rulers.[57] Many of them were eclectic Sufis but adjusted themselves to the Sunni orientation of Muslim India. However, with the rise of the Safavid dynasty in Iran in the early sixteenth century, Shiism was established as the state religion of Iran; and increasingly the Iranian migrants during the sixteenth and seventeenth centuries were Shia in outlook. A Shia Iranian court party always vied for favours at the Mughal and Deccani courts.

The Mughal governors who carved out autonomy for themselves in the three provinces of the empire during the early eighteenth century – Murshid Quli Khan in Bengal, Saadat Khan in Awadh and Nizam ul-Mulk in Hyderabad – were all Shia noblemen of Iranian origin. Shia influence and self-confidence in India grew in, and out of, their provinces.[58] Flourishing trade relations between these provinces and the Iranian world yielded great profits for the treasuries of the *nawabs*. The collapse of the Safavid dynasty in 1722 did not mean a cessation of contacts; many Iranians went to Iraq, and contacts between India and Baghdad were established.[59] Besides the movement of trade and money, there was also a fertile exchange of ideas. For example, the theological arguments among the Shia of Iran and Iraq were also mirrored within the Shia community in Awadh.[60] With Shia princes holding power in the provinces, it was inevitable that the Shia religious establishment desired to gain ascendancy by putting in place Shia-oriented judicial and educational institutions. The Friday congregational prayers also became an important tool of Shia dominance and assertion.[61] The effusive public celebration of the month of *Muharram* became institutionalized. While the Sunni sultans and emperors, except Aurangzeb, had never placed restrictions on the *Muharram* processions, they had obviously disliked any overt anti-Sunni messages being propagated. This now became more common in the Shia dominated provinces, leading to tension between the followers of the two branches of Islam.

Syncretist practices

Since the mid-sixteenth century, Hindu participation in the affairs of the Muslim state had become more pronounced. Hindu administrators, accountants and bureaucrats were part of the state machinery. The post-Mughal princes relied heavily on Hindu bankers and financiers to help them with their wars or civic projects.[62] They could not afford to alienate the Hindus. What we therefore witness in the eighteenth century is that, in both Hindu and Muslim successor states of the Mughal Empire, the rulers were encouraging forms of reconciliation and syncretism rather than religious division. Two examples, one from the north and the other from the south, will demonstrate the particular ways in which Muslim rulers of this age adapted to new currents in society.

The composite culture of Awadh flourished greatly under the third *nawab* of Awadh, Shuja ud-Daulah (r. 1754–75). In his time Faizabad was the capital, while Lucknow was the commercial centre. Large numbers of people of different classes and faiths were attracted to both cities, and Shuja ud-Daulah's policy was to facilitate as much cosmopolitan mixing as possible.[63] This was particularly noticeable during the month of *Muharram* when many Hindus participated with much feeling and emotion in the mourning rituals. Sufi *pir*s and Hindu syncretists served as 'cultural mediators' and transmitters of symbols from one group to another, thereby helping their followers to learn to empathize with each other.[64] Their work was akin to a form of inter-cultural education strategy. Shuja ud-Daulah and his two predecessors had also taken steps to protect the ancient Hindu *devi* temples in and around Lucknow. These temples, devoted to the worship of *devi*, the consort of Lord Shiva, were centres of pilgrimage. The idols, broken by the earlier iconoclastic marauders, were re-consecrated alongside the new ones.[65] This symbolism of reconciliation was to play a major part in the flowering of Awadhi culture in the early nineteenth century.

A good example of syncretism practised by Muslim rulers in the south of India is provided by the actions of the *nawab*s of Arcot who had broken away from the suzerainty of Hyderabad. These *nawab*s, from the two dynastic lines of the Navaiyats and the Walahjahs, first equipped themselves with strong armies and created an effective state revenue apparatus.[66] They next proceeded to fulfil a responsibility which, in the eyes of their mostly Hindu subjects, was considered an essential attribute of kingship in South India. The royal benevolence towards the great Hindu temples and shrines was considered to be the hallmark of a good ruler. Among the many holy southern cities the sacred site of Trichy, with its massive rock-fortress and the temple of Srirangam, was held in great affection by the Hindu masses. Trichy was also the city where there were many *dargahs* of Sufi *pirs* and saints who had undertaken their arduous mission into the heart of the Hindu south.[67] The most well known of these *dargahs* was that of a *pir* known as Nathar Wali. By the early eighteenth century, Trichy was therefore a centre of both Hindu and Muslim pilgrims who, while worshipping either Shiva or Allah, felt a common spiritual bond with each other. It was for this reason that the most famous Nawab of Arcot, Muhammad Ali Walahjah (r. 1749–95), directed his most lavish acts of royal piety towards the shrines of the Srirangam temple and the Nathar Wali *dargah*.[68]

It is easy to exaggerate the importance of syncretism in the context of Indo-Muslim history. The two examples cited above and many others did not truly lead to any lasting reconciliation between the Hindu and Muslim in the subcontinent. They were therefore worthy but somewhat tokenistic measures. Vested interests and historical memories on both sides were never going to make the task of the syncretists any easier. Hindu–Muslim riots were already beginning in this period.[69]

Orthodox yearnings of Shah Wali Allah

By the mid-eighteenth century the glory of Delhi was no more, but there was more at stake than the ruination of the capital city. The rapid disintegration of the Mughal Empire struck at the very foundations of Muslim power in India. Muslim authority was now fragmented and dispersed in certain regions of the subcontinent, but in the absence of strong metropolitan suzerainty life itself seemed unsafe and insecure.[70] It was in this atmosphere of fear and moral panic that the eighteenth century's most influential Muslim religious scholar articulated, clearly and concisely, the case for a return to the fundamental principles of Islam for the salvation of his people.[71] His ideas inspired many thinking Muslims in the nineteenth and early twentieth centuries; and even today his arguments resonate in many Muslims' hearts and minds (Excerpt 7.3). His name was Shah Wali Allah.

Born in 1703, Shah Wali Allah was from an early age inculcated into Sufi doctrines. During his higher education in Islamic Studies in the Hijaz (modern Saudi Arabia), he came to be inspired by the works of a medieval Syrian thinker and theologian, Ibn Taymiyyah, who believed without any qualification in the supremacy of the Quran and the *Sunnah* of Prophet Muhammad and the early Muslim community. On his return to India, he began a career of teaching and writing. Although he showed a degree of understanding towards other faiths, his primary concern was the welfare of the Muslim community of India.[72] He strongly felt that the Muslims needed help and succour at a time when they faced existential threats from non-Muslims such as the Marathas, Jats and the Sikhs.[73] He therefore appealed to the Emperor Muhammad Shah and other Muslim grandees like

Najib ud-Daulah, a chieftain of the Rohilla people, Nizam ul-Mulk of Hyderabad and Shah Abdali of Afghanistan to build up a united Muslim front against non-Muslim armies; he was to be disappointed by both their response and capacity. The Mughal state went on disintegrating.

In his intellectual works, Shah Wali Allah located the roots of Muslim disarray and confusion of this period in the moral degeneration of Indian Islam. Although it may appear arcane to our modern minds, one of his first concerns was the highly philosophical question of the nature of God and how ordinary Muslims perceived God.[74] In this, he was guided by the writings of Shaykh Ahmad Sirhindi (Chapter 6) who, although himself a Sufi, had severely attacked a very common Sufi idea of what is sometimes referred to as 'ontological monism', popularized by a medieval Spanish Sufi, Ibn al-Arabi, in his doctrine of *wahdat al-wujud*, or the Unity of Being, meaning that God and His creation were all one. Strongly suspecting this belief to be very close to the Hindu concept of pantheistic monism, Sirhindi promoted the idea of the Unity of Appearance or *wahdat al-shuhud*, implying that while God existed and was unique, no created object could be part of him. Sirhindi wanted the masses to proclaim 'All is from God', not 'All is God'. The Muslim creed, he affirmed, was 'There is no God, but God', not 'There exists nothing but God'.[75] Although Wali Allah attempted a compromise between the two ideological positions, this important philosophical distinction was central in his view. He appealed to his readers to go back to the basics of Islam: to proclaim the uniqueness of God and to unswervingly follow the Quran. Extraneous ideas from Hindu or other faiths that had percolated into liberal Sufi thinking had to be expunged, and Sufi disciplines needed to be rationalized and integrated with the fundamental tenets of Islam.[76] For him, too close an association with non-Islamic ideas was a corrupting influence on Muslims. This insular form of fundamentalism was one of the unfortunate legacies of Shah Wali Allah's writings. Yet, in other ways, he was liberal and inclusive in his thinking. While the Quran and the *Hadith* were sacrosanct, the formulations of religious law as interpreted by the *ulama* and jurists of the four Islamic schools of law were in his view relative, and he reaffirmed the important Islamic legal precept of *ijtihad* or legal reasoning.[77] He was also prepared to go against the grain of tradition by translating the Quran into Persian, since he strongly believed that the Word of God had to be understood by people in their own tongues. Later on, his sons translated the Quran into Urdu.

In his works and sermons, Shah Wali Allah greatly emphasized the moral duties and obligations of Muslims in both their private and public lives. He appreciated the usefulness of the communitarian aspects of Muslim life, such as the Friday prayers, the celebration of festivals, paying homage to the holy shrines of Sufi *pirs* (as long as there was no veneration of tombs), the *Hajj* pilgrimage, etc.; but he also believed that the ideal Islamic state required a universal Islamic *khilafat*, presided over by a just and all-powerful

Islamic monarch, in the tradition of the first rightly guided caliphs of Islam who succeeded Prophet Muhammad.[78] Today, nothing is more irksome to a modern Western person than the mention of a *khilafat* in some of the literature propagated by anti-Western Muslim fundamentalist organizations. This is understandable because the West has undergone many painful revolutions in order to achieve democracy, rule by the people through their representatives. The idea of a *khilafat* exercised by a strong and supposedly just ruler is, for Western thinkers, more akin to enlightened despotism that prevailed in Europe in the late eighteenth century. The West has come a long way since then and considers the *khilafat* as an outdated method of government. On the other hand, since Shah Wali Allah had no knowledge or understanding of what was happening in Europe in his own time, let alone what the West was to achieve in the arts of government during the two succeeding centuries, it is not surprising that the idea of a *khilafat* seemed to him to be a model form of government in pre-modern India paralysed in his time by Mughal breakdown.

Cultural developments of the later Mughal period

Muslim culture did not crumble in the age of imperial disintegration. We notice both continuities and newer trends within this period of turmoil; and to understand the context of these we choose three particular areas of cultural development: literature, education and architecture.

Literature: a gentle transition from Persian to Urdu

By the end of the seventeenth century, the importance of Persian in Indian life was an established fact. Ordinary people continued to use their regional vernaculars for daily conversation, but Persian was considered as the principal language of politics, commerce and culture. Over many centuries the language had been cultivated and nurtured by an array of Muslim writers, poets and philosophers (Chapters 3 to 6); but when Emperor Akbar insisted that children should learn the Persian alphabet and basic vocabulary he was, in a sense, pioneering a form of standardized literary education, at a time when there was no provision for universal education.[79] With his exhortation, Akbar began the trend of 'Persian for many' rather than 'Persian for a select few'. Anyone who wished to enter the large Mughal bureaucracy as an accountant or a scribe had to be well qualified in Persian, since all papers and imperial orders (*firmans*) were written in that language. The elders of the Hindu castes such as Kayasths and Khatris, who were professional scribes, encouraged their children to learn Persian; and Hindu writers in Persian increased greatly in numbers through the eighteenth century.[80] Since many aspects of the Mughal bureaucracy were adopted by

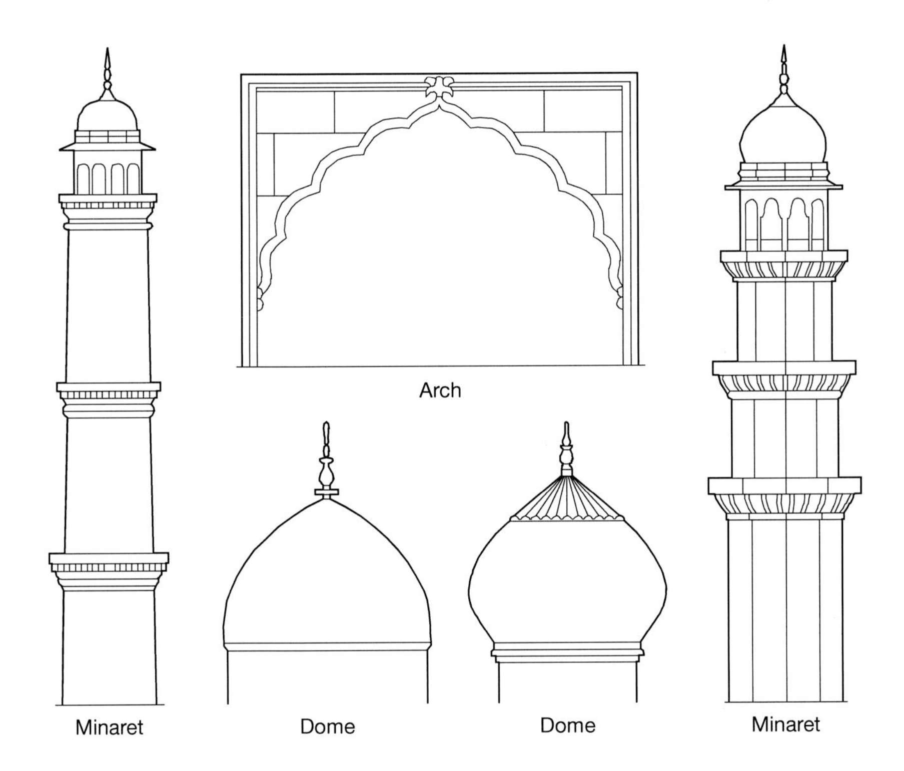

Plate 1 Islamic architectural styles

Source: Mr Ismail Lorgat, Blackburn, UK, private album

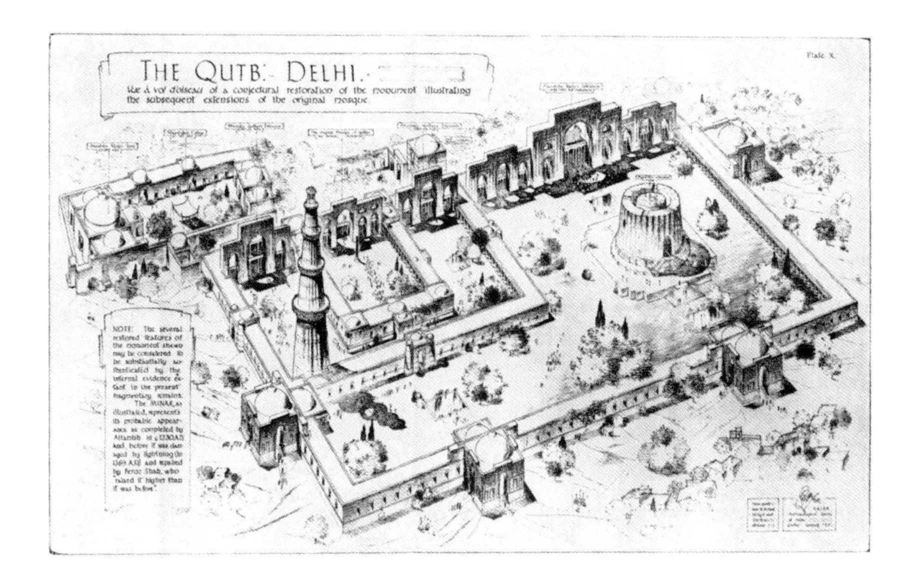

Plate 2 An artist's model of the Qutb complex in Delhi

Source: Sir Wolsey Haig, editor: *The Cambridge History of India*, Vol. 3, Cambridge University Press, 1928

Plate 3 Two styles of royal tombs: Sher Shah and Jahangir

3.1 Source: Sir Richard Burn, editor: *The Cambridge History of India*, Vol. 4, Cambridge University Press, 1937

3.2 Source: Mr Akhtar Hussain, Manchester, UK, private album

Plate 4
Two Mughal emperors:
Babur and Akbar

4.1 Source: Dr Peggy Woodford Aylen, *Rise of the Raj*, Midas Press, 1978

4.2 Source: Dr Peggy Woodford Aylen, *Rise of the Raj*, Midas Press, 1978

Plate 5
Two Mughal forts: Lahore and Agra

5.1 Source: Mr Akhtar Hussain, Manchester, UK, private album
5.2 Source: Mrs Maharukh Desai, Goa, India, private album

Plate 6
Two gateways: Teen Darwaza (Ahmadabad) and Char Minar (Hyderabad)

6.1 Source: Miss Azzmin Mehta, Ahmadabad, India

6.2 Source: Dr George Michell, co-author: *Architecture and Art of the Deccan Sultanates*, Cambridge University Press, 1990

Plate 7
. . . 'space fit for royalty' – a Mughal palace and a garden

7.1 Source: Mr Imtiaz Patel, Blackburn, UK, private album

7.2 Source: Mrs Maharukh Desai, Goa, India, private album

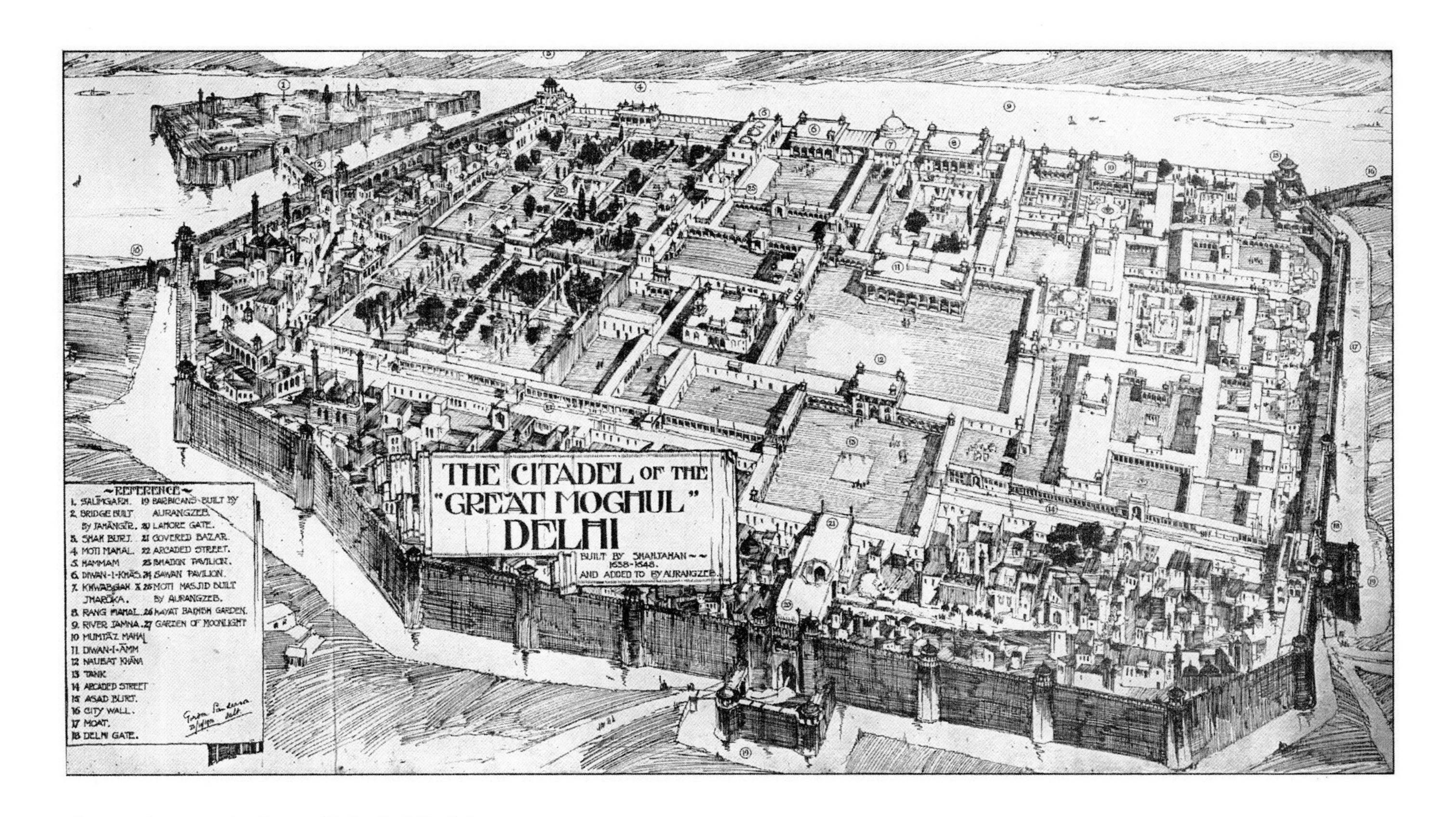

Plate 8 An artist's view of Mughal Delhi

Source: Archaeological Service of India

Plate 9 Examples of Mughal art and aesthetics

9.1 Source: Victoria and Albert Museum, London, UK

9.2 Source: Mrs Maharukh Desai, Goa, India, private album

9.3 Source: Mrs Maharukh Desai, Goa, India, private album

Plate 10 Two Mughal mosques: the Badshahi (Lahore) and Moti Masjid (Agra)

10.1 Source: Mr Akhtar Hussain, Manchester, UK, private album

10.2 Source: Mrs Maharukh Desai, Goa, India, private album

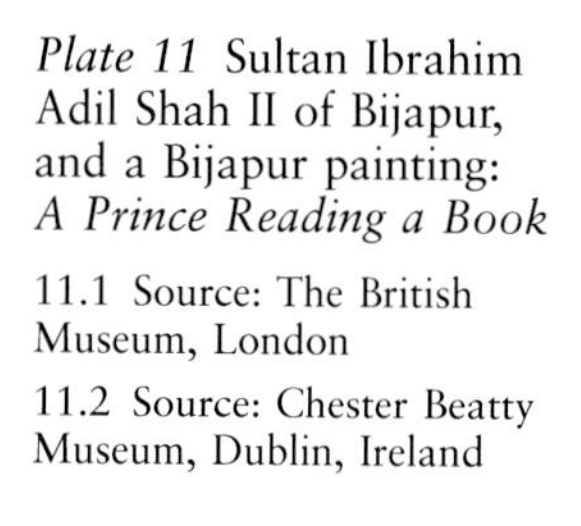

Plate 11 Sultan Ibrahim Adil Shah II of Bijapur, and a Bijapur painting: *A Prince Reading a Book*

11.1 Source: The British Museum, London

11.2 Source: Chester Beatty Museum, Dublin, Ireland

Plate 12
Three nineteenth-century Muslim educators: Sir Sayyid Ahmad Khan and his colleagues, and a twentieth century feminist: Jahanara Shah Nawaz Begum with her husband at the London Round Table Conference, 1932

12.1 Source: Professor Nizami, Aligarh Muslim University, India, *The History of the Aligarh Muslim University*, Vol. 1, Delhi, 1995

12.2 Source: Rozina Visram, *Women in India and Pakistan: The Struggle for Independence from British Rule*, Cambridge University Press, 1992

Hindu, Muslim and Sikh successor states and their rulers, there was ample opportunity for those fluent in Persian to find posts in regional civil services.

Few historians or poets writing in Persian were officially celebrated in this period, partly because neither Emperor Aurangzeb nor his successors gave much encouragement for the two arts to flourish at the court. The popular anthologies and biographical handbooks, or *tazkirahs*, however, provide us with copious information about those who were writing books, diaries, memoirs, letters, poems and essays in Persian of varied quality.[81] Two of them deserve a special mention. The first was the poet Mirza Bedil (1644–1721), whose mature works fall in this period. Living in a cosmopolitan world of Hindus and Muslims, Bedil was a syncretist at heart. He was also against the injustices of the feudal system, the hypocrisy of the clergy and the sycophancy of fellow poets.[82] Another was Siraj ud-Din Ali Khan Arzu (d. 1756) who wrote in both Persian and Urdu. Affected by the general collapse and confusion in Delhi at this time, he championed the Persian language as a symbol of a larger trans-local and trans-regional pan-literary identity; and through his dictionaries and grammars he encouraged a standardized Persian literary style which set norms for the learned in Persian.[83]

Alongside Persian we notice the steady growth in the use of the Urdu language, particularly in poetry. Two particular forms of the language had been evolving since the thirteenth century: first, the *hindavi* Urdu based on a mixture of many northern dialects, whose simplified form was called *khari boli* or standard language, along with Persian, Arabic and Turkish etc. and, second, the *dakhni* Urdu consisting of a mix of *hindavi*, some southern dialects, Persian, Arabic and Turkish.[84] The principal foreign ingredient in both forms was Persian. The man who fused the two forms together was Shams ud-Din Wali (1668–1744) of Aurangabad who, after his two visits to Delhi in the early eighteenth century, adopted the northern *hindavi* in his poems and lyrics, and thus accelerated the development of modern Urdu.[85] Wali adopted a style that had earlier been pioneered by Amir Khusrau (see Chapter 4), known as *rekhta* ('mixed' or 'broken'). It meant that the poet wrote alternating Persian and *hindawi* lines in his verses.[86] This form of poetry was much patronized at the court of the *nawabs* of Awadh.

The skill of the poets was generally tested at gatherings known as *mushairas* which were hosted by eminent scholars, poets or nobles.[87] Often the invitation would include a line of verse, and those who participated would be expected to recite a poem incorporating the line in rhythmic harmony. Commendations were praised by participants with fulsome acclamations, while criticisms were made gently and sensitively. It could be at the *mushairas* of this period that five noteworthy Urdu poets came to the attention of the literati. Their verses set the standards for the literary golden age that was to follow at the court of Lucknow from the late eighteenth century onwards (see Chapter 8). The first was Shah Hatim (1699–1791) who used 'easy, elegant and fluent' Persian and Arabic words to produce a mix of Urdu poetry.[88] He wrote movingly about Delhi (Excerpt 7.4). Another was Mirza Sauda (1713–81), a poet of exceptional brilliance, who wrote satirical verses about his contemporaries.[89] The third poet, Khwaja Mir Dard (1719–85), a mystic and a Sufi, wrote about philosophical themes and struggles of daily life.[90] The fourth one was Mir Taqi Mir (1722–1810) who wrote some of the finest *ghazals* in Urdu on a variety of themes: distress in war, the melancholia of death or the beauty of a rose. Emotionally attached to Delhi, he too, like Mirza Sauda, left the city for greater safety in Lucknow.[91] He compiled a magnificent *tazkirah*, *Nikat al-shuara* (The Points about the Poets) in which he evaluated the works of many poets writing in Urdu and Persian.[92] Lastly, there was Mir Hasan (*c.* 1727/8–86), who wrote romantic *masnavis* of which the most famous is the *Sihr ul Bayan*, the Enchanting Story of the adventures of Benazir, the prince. The verses of this *masnavi* accurately observe and describe in detail such things as gardens, dresses, etiquettes, daily conversation and many other elements that can be imagined in an idealized setting of the more peaceful parts of Muslim India.[93]

Education: two curricula of Islamic learning

Islam exhorts Muslims to revere learning and seek knowledge. The pre-modern Muslim educational curriculum consisted of two main strands of knowledge: *manqulat* or transmitted sciences and *maqulat* or rational sciences.[94] The six branches of transmitted sciences were Grammar and Syntax (*sarf wa nahw*), Rhetoric (*balaghat*), Jurisprudence (*fiqh*), Principles of Jurisprudence (*usul al-fiqh*), Traditions (*hadiths*) and Quranic Exegesis (*tafsir*). The rational sciences were those of Philosophy (*hikmat*), Logic (*mantiq*), Theology (*kalam*), Mathematics and Astronomy (*riyaziyyat*), and Mysticism (*tasawwuf*). From the varieties of knowledge imparted from within both strands we can deduce that the Muslim educational curriculum was in no way inferior to the curriculum that was in use in Europe until the eighteenth century. Just like in pre-modern Europe, however, there was little notion of democratic education. Only the nobility and the professional classes had both the motivation and the means to get their children educated. Repressive sexism prevented most girls from getting any formal education whatsoever.

During the early centuries of Islam, there was a balance maintained between the *manqulat* and *maqulat*. This was a time when there was great curiosity about the sciences, geography and the wonders of nature.[95] From about the thirteenth century, however, the spark of curiosity that marked the earlier period seemed to wane; and the ecclesiastical voice of the *ulama* became a dominant force in the Islamic world. The only people who were able to challenge the orthodoxies of the *ulama* were some exceptionally dissident Sufis. This meant that the *manqulat* tradition came to dominate the educational curriculum. Since Islamic rulers needed an army of legal professionals trained in the Quranic studies and the principles of jurisprudence in order to maintain their authority over the populace, the *ulama* enjoyed a free hand in the educational process. This was the general position in Muslim India before Akbar's time. In Iran, however, the educators had maintained a balance between the two strands of the curriculum; and with the arrival of many more Iranians into India in the sixteenth century, the Muslim curriculum was to become more balanced.[96] Through the influence of a major Iranian scholar at Akbar's court, Fadl Allah Shirazi, greater interest began to be taken in the subjects that were part of the *maqulat* tradition.[97] Akbar himself was highly curious about the rational sciences; but an orthodox reaction that set in after him revived the *ulama*-dominated *manqulat* curriculum. The best known *madrasa* of the eighteenth century, for example, founded in Delhi by Shah Abd al-Rahim, the father of Shah Wali Allah, gave its students a thorough training in all the subjects of the *manqulat* curriculum. A number of Muslim revivalists of the later period came out of this institution, known as the Madrasa-i Rahimiyya.

New ideas from Iran, Central Asia and, to a lesser extent, the Ottoman Empire kept up the interest of the educated classes in the rational sciences of the *maqulat* throughout the seventeenth century; and, in this connection, we need to refer to another institution in Lucknow, not Delhi. Around Lucknow and throughout the Mughal province of Awadh, there were innumerable *madrasas*, mosques and *khanqahs*, in charge of scholars, Sufis and noblemen holding *madad-i maash* grants of land from imperial authorities.[98] Quite often they faced opposition from local *zamindars* who resisted imperial control. A teacher, named Mulla Qutb al-Din Sihalwi, a follower of Fadl Allah Shirazi, was brutally murdered in 1692 by some *zamindars* at his *madrasa* in Sihali; and his library of 900 books was burnt.[99] Moved by this tragedy, Emperor Aurangzeb assigned to his family a European merchant's house in Lucknow, in which the family started residing around 1695. In this house, that came to be known as Farangi Mahal (the foreigner's house), the third son of Mulla Qutb al-Din, Mulla Nizam al-Din, started teaching the various rational sciences of the *maqulat* curriculum. Drawing on the contents of textbooks written in the Timurid tradition by such Samarqand authors as Sayyid Sharif al-Jurjani and Saad al-Din Taftazani, he devised a new syllabus, Dars-i Nizamiyya, with strong emphasis on logic and science.[100] A large number of educated, articulate and politically active Muslims of the nineteenth and twentieth centuries were trained in the Dars-i Nizamiyya syllabus. They were an important section of the

Muslim elite in British India. Many foreign students also studied at Farangi Mahal.[101]

Architecture: new patrons

The pinnacle of Mughal architecture was reached under Emperor Shah Jahan; after him there could only be anticlimax. Aurangzeb did not have his father's vision or enthusiasm in this area, although some notable religious buildings were constructed during his reign, such as the monumental Badshahi Mosque in Lahore (Chapter 6). His long and costly involvement in the Deccan campaign resulted in the shortage of funds for architectural patronage. His immediate successors, embroiled in their own succession struggles and wars with the Marathas and Jats, also had little time to devote to architecture. The main casualty of this lack of concern was the grand capital of Shahjahanabad in Delhi. Yet Delhi somehow continued to maintain its pre-eminence throughout the first half of the eighteenth century; and its glory and reputation allured both men and women of means to sponsor their own private initiatives.[102] A powerful nobleman at Emperor Muhammad Shah's court, Raushan al-Daula Zafar Khan, spent his huge ill-gotten fortune upon constructing many buildings, of which the most distinguished is the attractive Sunahri or Golden Mosque in Chandni Chowk, built in 1721–2.[103] Another nobleman's widow, by the name of Kaniz-i Fatima, also entitled Fakhr-i Jahan, commissioned in 1728–9 the building of the Fakhr al-Masjid, a beautiful red sandstone mosque faced with white marble. She thereby continued the older tradition of Mughal court ladies providing patronage for religious buildings.[104] And Emperor Muhammad Shah himself, a pleasure-seeking monarch with a penchant for astronomy, was able to persuade Sawai Jai Singh Kachhwaha (1699–1743), the Rajput ruler of Jaipur who was busy transforming his city, to invest in a unique observatory known as the Jantar Mantar.[105]

Delhi suffered terribly from Nadir Shah's invasion of 1739, and was to continue to suffer from the depredations of intruders;[106] but the recovery was relatively quick. New patrons from among the rich maintained their confidence in the city. In the short reign of Ahmad Shah the Queen Mother Qudsiya Begum, a woman of means and influence, commissioned a beautiful palace and garden complex, called the Qudsiya Bagh.[107] It was also in Delhi, and not in Faizabad, that the *nawab-wazir* of Awadh, Shuja ud-Daulah, decided to bury his father Safdar Jang (d. 1754) and to build a grand tomb. This tomb, set within a walled *chahrbagh* complex, is the finest example of late Mughal tomb architecture with some specifically novel eighteenth-century characteristics.[108]

Architectural syncretism in this period can be demonstrated by two particular structures. The one with a dominant Hindu trait is the palace complex of the rulers of Bharatpur, Badan Singh (r. 1722–56) and Suraj Mal (r. 1756–63), at their capital in Dig. The beauty of the palace pavilions, recalling Hindu temples, is enhanced by their grand setting in a Mughal *chahrbagh* garden.[109] The traditional Islamic building of the Katra Mosque at Murshidabad in Bengal reveals another form of syncretism. Nawab Murshid Quli Khan built this enormous mosque to make a statement of his autonomy from the emperor, but the building contains a mix of classic Mughal styles and pre-Mughal indigenous Bengali features.[110]

Select excerpts

7.1 The terrain of Maharashtra

In a letter to Aurangzeb's 'officers and counsellors' Shivaji affirms his faith in the hardy terrain of the Maratha hills to protect his kingdom, and compares it favourably to the sites of his enemies' armies.

> My home, unlike the (Mughal) forts of Kaliani and Bidar, is not situated on a spacious plain, which may enable trenches to be run (against the walls) or assault to be made. It has lofty hill-ranges . . . everywhere there are nalas hard to cross; sixty forts of extreme strength have been built, and some (of them are) on the sea coast.
>
> (Gordon 1993: 71)

7.2 Controversy over the Black Hole of Calcutta

The incident of the Black Hole of Calcutta, in which 146 British inhabitants were confined on the night of 20 June 1756, and from which only twenty-three came out alive the next day, was graphically described and narrated by John Zephaniah Holwell, the governor of the city's Old Fort William. The incident remained transfixed into British imagination for a very long time; and as late as 1901 Lord Curzon, the Viceroy of India, restored the much dilapidated Holwell Monument in Calcutta as a 'sacred trust to be guarded'. Historians today are less sure of the veracity of the conventional British version of the event, in the light of what a distinguished Indian historian, Sir Jadunath Sarkar, wrote in the mid-twentieth century:

> the number of victims afterwards given out and accepted in Europe (namely 123 dead out of 146 confined) is manifestly an exaggeration . . . a floor area of 267 square feet cannot contain 146 European adults. This point was established by Bholanath Chunder, who

> fenced round an area 18 feet by 15, with bamboo stakes and counted the number of his Bengali tenants, who could be crammed into it; the number was found to be much less than 146, and a Bengali villager's body occupies much less space than a British gentleman's . . .The true number was considerably less, probably only sixty. It is a very reasonable supposition that all the former British residents of Calcutta whose manner of death could not be clearly ascertained . . . were afterwards set down as 'perished in the Black Hole' and their names were blazoned on Holwell's monument. Holwell and some other leading servants of the Company were . . . were set free, and joined the English fleet at Falta.
>
> (Sarkar 1973: 476–7; Busteed 1908/1972 reprint: 30–56)

7.3 Thoughts of Shah Wali Allah

Shah Wali Allah's writings are contained in some twenty-five different volumes, very few of which have been translated into English. The following selection hints at his powers to articulate on varied issues:

On the meaning behind any pilgrimage:

> All people have a place of pilgrimage; it may be a sanctuary or a river like the Ganges where the Hindus go on a pilgrimage, or it may be a tree, a semi-desert plain, a tomb or a porch upon which wonderful signs appeared. Crowds gather there in order to be filled with the beneficent virtue of the venerated object. This is not so much a matter of custom or habitual practice; it is rather a universal proneness to single out an object of worship.

On the harm done by gambling and the charging of interest:

> Gambling and interest (charged) are unlawful gain, and are not in keeping with civic spirit and mutual aid; for, as a rule, the borrowers are people fallen into a severe state of indigence, mostly not able to pay their debt in time . . . When this way of earning money takes root, it leads to the abandonment of agrarian trades and skilled crafts which are fundamental means of earning a living.

An example of his Islamocentrism:

> If the (Muslims) forsook jihad and co-operated with non-Muslims they would suffer severe humiliations and people of other religions would subdue them.
>
> (Baljon 1986: 184, 186, 192)

7.4 Shah Hatim on Delhi

Delhi has been the subject of praise and lament by many Urdu poets. Two years before Nadir Shah's brutal attack in 1739, the poet Shah Hatim wrote about the luxury and culture of the city in the following verses:

> I had a sudden impulse to say something in praise of Delhi,
> Delhi is not a city but a rose Garden
> Even its desert is more pleasing than an orchard,
> Beautiful women are the bloom of its streets and markets.
> Shy coquettish beauties are found at every step
> While every place is adorned with greenery and lovely cypress trees.
> And one living in Hindustan in this age is the Shahjahan of his times.

Some years later, after the city had suffered repeated loot and pillage by Afghan, Maratha and Jat invaders, Hatim lamented thus the turn of fortunes in so short a period:

> How can one be happy even for a moment in this age
> That is devoid of all that brings happiness:
> The cup bearer, festivity and peace of mind
> The constant flow of tears from my eyes
> In fact comes from my heart
> A cupful of sorrows brimming over.

(Haque 1992: 21–2)

8

MUSLIMS UNDER THE EAST INDIA COMPANY (1765–1858)

The English East India Company was formed by a group of London merchants who received a royal charter to trade in eastern waters from Queen Elizabeth I on 1 December 1600.[1] The Company ceased to function when the charter was revoked by the British parliament in 1858. Over this long period, the Company's role in the subcontinent changed in three successive phases. During the seventeenth century, it acted as a great corporate trader, in competition with other European East India companies and the native Indian traders, under the protection of the Mughal Empire. It gained certain commercial and judicial privileges from the Mughals, but essentially its role was confined to that of trade.[2] During the first half of the eighteenth century, when the empire was disintegrating, the Company played both a political and military role in addition to trade. This was the second phase, when it dealt with the Mughal successor states and defeated the French rivals (Chapter 7). After the successes of the battles at Plassey (1757) and Buxar (1764) it went on to conquer over half the subcontinent, exercise great power over its peoples and to inaugurate Westernization touching many aspects of Indian life. This third phase, which is the subject of this chapter, coincided with overwhelming British military might born out of the technological successes of the Industrial Revolution.

At first, during this period, most Indians considered the Company as just another player in the politics of the subcontinent and were happy to work with it. Their resentment increased when the Company began to impose itself upon Indian society and mould it. The changes wrought by economic and social measures initiated by the Company's representatives along with the domineering and imperious attitudes of some of the governors-general disturbed a considerable number of people, from all classes of society, whose interests were adversely affected; and the great Indian rebellion of 1857 (the Indian Mutiny) saw many Hindus and Muslims come together in a remarkable act of unity to offer a sustained and bloody challenge to the Company within the narrow belt of the Gangetic basin. Their attempt failed, but the Company's nemesis was its own dissolution by the British government.

The supremacy of the Company

By the Treaty of Buxar in 1765 (Chapter 7) the East India Company strengthened its hold on the *nawabi* of Bengal. The office of the *nawab* continued to exist, but after 1772 it was merely a showcase behind which all true power was exercised by the Company's governor-general. The Bengali capital was transferred from Murshidabad to Calcutta; and, in 1798, the Marquess of Wellesley commissioned a palatial residence fit for the company's premier governor-general in India.[3] The military success at Buxar also enabled the Company to wrest much of Bihar from Shuja ud-Daulah, the *nawab* of Awadh.[4] At the beginning of our period, therefore, the Company possessed Bengal, Bihar and, from the earlier campaigns, Madras and some forts on the Coromandel coast. By 1857 the Company's direct control extended over 62 per cent of the area and three-quarters of the population of the subcontinent.[5] The Company exercised a so-called 'Indirect Rule' over what later came to be called the Princely States, those that occupied the other 38 per cent of the area of the subcontinent.

The enormous territorial gains of the Company were not achieved as a result of a deeply thought-out planned strategy in its London headquarters; rather they came about through bold initiatives, resourcefulness and some good luck on the part of the Company's men on the ground in India. In any disagreement between London and Calcutta over further British expansion, it was the latter that seemed to have the upper hand. Indian princes of every faith bent to the will of the Company throughout this period. Before we examine the chronology of acquiescence by Muslim states, it is important to ask three general questions: What were the motives behind the Company's desire for more territories? What means did it employ to further its ambitions? Why was it so successful?

Until a more rigorous analysis of the Company motives began to be explored by historians after the independence of India and Pakistan, the British public had felt comfortable with the notion that India had been conquered by eccentric and absent-minded imperialists, a sentiment somewhat mistakenly associated with a memorable sentence written in 1883 by a late Victorian historian, Sir John Seeley, who said: 'We seem, as it were, to have conquered and peopled half the world in a fit of absence of mind'.[6] There were, however, certain specific reasons which impelled the Company to carry on expanding. First, there was the economic gain to be made out of India by a monopolistic stranglehold on its resources and trade through territorial control, generating undreamt of profits for the metropolitan economy.[7] Second, there was the need to eliminate headstrong Indian rulers who had the potential to challenge the Company by building alliances with dangerous external enemies, like France for example[8], and later Russia. Third, it was thought necessary to expand in order to enhance liberal values. A large number of annexations were defended on the grounds that the Company rather than the native rulers had the best interests of their people at heart.[9] Fourth, the complex problems arising out of the Company's push towards the frontiers of native states could, it was concluded, be resolved only by further expansion. Stability for the Company was always just 'one annexation away'.[10]

The force of arms was not the principal method employed by the Company in its desire for greater territorial and political control. While that remained the ultimate weapon, the preferred method was diplomatic enticement, known as Subsidiary Alliance.[11] The success of this method was due to the mistrust among the Indian successor rulers of the Mughal Empire and their fears about each other's intentions. Through a Subsidiary Alliance, the Company offered protection to each ruler who promised to pay for the cost of maintaining its troops in his territory. This process was overseen by a high representative of the Company, known as an Agent or a Resident, posted at the ruler's court. This person inevitably became the Trojan Horse for the Company. Through him, gradually, greater limitations were placed on a ruler's own independent army; communications in and out of the state were controlled; and the ruler's freedom of action was circumscribed.[12] The success of this arrangement, however, depended upon the ruler making large regular payments to the Company.[13] Since the classic Mughal system of raising land revenues had long been superseded by unstable forms of revenue farming, it was extremely difficult for Indian rulers to maintain unbroken payments; and that weakness reduced the value of the Subsidiary Alliance for the Company. In that situation, annexation became the Company's chosen tool of control. Another circumstance that was used as an excuse for annexation was when an Indian ruler died without a male heir. This was done by a device known as the Doctrine of Lapse, used most frequently by Lord Dalhousie, the governor-general between 1848 and 1856.[14] If annexation was not agreed willingly by the ruler, then the Company threatened him with the very army that he had been paying for. The Company thus made fools of Indian rulers in this period.

One reason for the Company's success lay in its ultimate military might. With the exception of the armies of Mysore and the Sikh kingdom in Punjab, most of the Indian armies of this period were not modernized enough to challenge the Company's regularly paid, well drilled and well disciplined native infantry and cavalry, with access to the finest musket technology then available.[15] However, a more crucial reason for the Company's success lay in Indian cooperation. There were Indians who were prepared to serve the Company for personal and family interests. Indian traders, bankers and financiers, used to wheeling and dealing, worked happily with British traders, agents and administrators and had little to gain by any form of patriotic opposition.[16] This was made easier by the fact that, during the earlier decades of its power, the Company's representatives had genuinely wished to learn about India in addition to ruling it. They showed enough humility for many Indians to warm to them and ultimately help them in their goal. This was a collaboration born out of respect for the particular sort of Indophile Briton in authority in the late eighteenth century.[17] The relationship cooled at the turn of the century, however.

Termination of Muslim political authority

The Company's motivation, methods and the reasons for its success can be studied case by case in relation to each of the Indian states, but, for our purpose here, we shall concentrate on six important Muslim authorities: the Mughal emperor, and the rulers of Hyderabad, Arcot, Mysore, Awadh and Sind (Map 8.1).

For the Company, the first Muslim person of any consequence was still the Mughal emperor because, in the minds of millions, he embodied in his person the sovereignty of all India. This period witnessed the reigns of the last three emperors: Shah Alam II, Akbar II and Bahadur Shah II. Each of them was highly cultured, and possessed fine literary talents: but these qualities were of little use in the political challenge they faced from the British. The Company took control of Delhi in 1803 after ejecting the Marathas; and, from then on, each emperor was reduced to worrying about the size of his pension, the continuity of his legal suzerainty vis-à-vis the Company's government and the question of succession to the throne. Their correspondence with the Company's governors-general consists mostly of these three topics; the last two monarchs even sent emissaries to London, including the Hindu intellectual, Raja Ram Mohan Roy, to plead for their case not just with the Company directors but with King William IV and Queen Victoria.[18] Their efforts failed conspicuously. Little by little, the emperors were reduced to helplessness. For a short while in 1857, when the great revolt was spreading through Delhi in his dynasty's name, Bahadur Shah II might have felt a flicker of optimism: but that was just a chimera. His incarceration in Rangoon after the revolt was the final end of the Timurid dynasty.

Three Muslim states in peninsular India – Hyderabad, Arcot and Mysore – had close encounters with the Company, but with different outcomes for each of them. For many decades, Hyderabad's affections veered between the French and the British; but in the end the *nizams*, terrified by both the Hindu Marathas and two headstrong Muslim rulers of Mysore, threw in their lot with the British.[19] The Company, realizing the limits of its power, behaved with restraint and desisted from making unwelcome encroachments on Hyderabad's sprawling territory. An amicable coalition of interests therefore kept the *nizams* and the British in a friendly embrace until 1947.

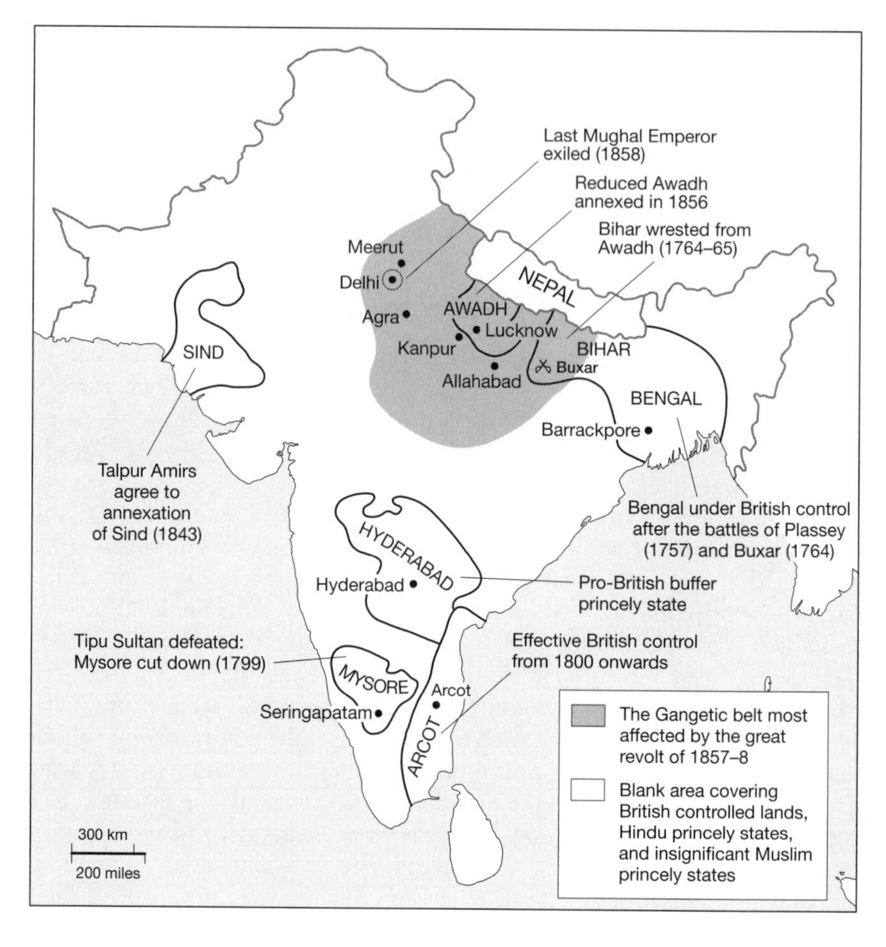

Map 8.1 The end of Muslim power in South Asia: 1765–1858

The *nawabs* of Arcot in the Carnatic were also extremely responsive to the Company's needs and demands, but they received a different treatment. The Company's long held control over Madras (Chapter 7) and the sea lanes around the coast emboldened it to demand greater tributes for the supposed benefits of the Subsidiary Alliance that had been agreed in 1765. Between 1765 and 1801, despite various protestations from the *nawabs*, the Company continued interfering without exercising direct control.[20] Finally, the hard-line governor-general, the Marquis of Wellesley, insisted in 1801 on the *nawab* ceding direct control to the Company, in return for a Company pension and the retention of royal title. In 1855, however, when a *nawab* died without a male heir, Lord Dalhousie applied the Doctrine of Lapse and proclaimed the end of the Walajah dynasty. An appeal to Queen Victoria made no difference.

Unlike the rulers of Hyderabad and Arcot, the father and son rulers of Mysore were clearly anti-British. Originally a Hindu kingdom, left over as a remnant of the medieval Vijaynagar Empire, Mysore had never been part of the Mughal Empire. A Muslim soldier, Haidar Ali, seized power from a Hindu king in 1761 and started the modernization of his state with the help of the French.[21] The upgrading of the state's military capability was seen as a great threat by the Company which saw independent Mysore as an impediment to its effective control of the peninsula. Between 1767 and 1784, two Anglo-Mysore wars were fought. During the second war, Haidar's son, Tipu, crushed the Company forces at the famous Battle of Pollilur (1780). He became the sultan of Mysore after Haidar's death in 1782; and his drive for modernization with French help was even more intense than that of his father. Although for seventeen years he thwarted the Company's plans for expansion in the south, he failed to secure a strong anti-British alliance with local Indian kings. The Company's diplomacy proved to be more adroit; and Tipu's own end came in 1799 when the Company forces finally stormed his capital, Seringapatam.[22] Mysore was then cut down to size, and a Hindu dynasty was restored on its much diminished throne (Excerpt 8.1).

Unlike Haidar and Tipu, the rulers of Awadh had shown good friendship towards the Company. A Subsidiary Alliance was in place since 1765, and the *nawb-wazirs* continued to finance the Company lavishly for their supposed protection by its army. Awadh's revenues became a cash cow for the Company; yet the Company perceived the state as a weakness on the frontier and a block to its own further expansion. By its coercive pressures, the Company forced the *nawab-wazir*, Saadat Ali Khan, to cede half of his kingdom to it in 1801.[23] Nearly fifty years passed by after this first annexation, during which time the *nawab-wazirs* took care not to alienate the Company. Interpreting this as effete weakness, the Company continued to make impossible demands from them. The relationship deteriorated markedly after Lord Dalhousie arrived as governor-general in 1848. Determined to annex Awadh one way or the other, he plotted for eight years with his two henchmen Residents at Lucknow, William Sleeman and James Outram, to demoralize and finish off the *nawabi* once and for all (Excerpt 8.2). The *nawab*, Wajid Ali Shah, was personally targeted and branded unfit as a ruler, although, despite his artistic idiosyncrasies and personal eccentricities, he had not been an oppressive ruler. His refusal to sign away his kingdom did not stop the annexation of 1856.[24] The distasteful way he and his family were treated by the Company reveals to us some of the ugly nature of Victorian imperialism.[25]

The last important Muslim authority that the Company disposed of was that of the Talpur Amirs of Sind. They were *de jure* under the sovereignty of Afghanistan, but enjoyed considerable autonomy. The Company, during its mainly trading era, had never shown any great interest in the province.

Its policy changed during the 1820s and 1830s owing to concerns about Russia's intentions in Afghanistan. The strategic and commercial importance of the River Indus was then exaggerated, and the Company was determined to become more proactive. By a treaty signed in 1834, the Amirs were forced to accept a British Resident, whose sole task was to keep an eye on them and to keep them as divided as possible.[26] At this time, the affairs in Afghanistan were coming to a head, and the first Anglo-Afghan War was fought between 1838 and 1842. Its results were inconclusive for the Company, but Sind became a victim of British paranoia. The year 1842 saw the career advancement of two Company men: Lord Ellenborough becoming the governor-general, and Sir Charles Napier assuming the command of British armies stationed in Sind under the 1834 treaty. The two destroyed the Amirat of Sind by fair or foul means; and the annexation of the province took place in 1843.[27]

The six Muslim political authorities extinguished by the Company joined the numerous Hindu rulers, the most important of whom were the Marathas who were finally crushed in 1818, and the Sikh kingdom of Punjab, defeated twice in 1846 and 1849. By 1857, the Company had had 100 years of supreme glory in India. It is time, therefore, to examine how this affected Muslim consciousness in general.

Muslim resentment and resistance

Millions of people, whether they liked it or not, became subject-citizens of a military dominion exercised by the East India Company across the length and breadth of the subcontinent. They belonged to varied classes and faiths, and they all had to learn to adjust to the realities of being ruled by an alien authority whose prime interests lay in promoting the power and welfare of Great Britain, not India. Life was good for those who worked with and for the Company. They lived mostly in cities and appreciated the peace and order established by the Company, which was basically to protect government, military and business infrastructures. The Hindu merchant and literary classes, particularly in Bengal, were the Company's greatest admirers; and so were the Parsees of Bombay. Most Muslims, however, remained at best circumspect. The extinction of all their own rulers' political authority left the Muslim body politic reeling with shock and bewilderment. A huge class of people that had enjoyed unparalleled power and privilege over many centuries was now left with a feeling of hopelessness and helplessness. Muslim embitterment simmered over many decades; and in 1857 Muslim fighters along with others played a leading role in the great revolt against the Company. The causes of the revolt were numerous, and some of them were identified by the nineteenth century's most distinguished Indian Muslim statesman, Sir Sayyid Ahmad Khan.

The causes of resentment and Sir Sayyid Ahmad Khan's analysis

Issues concerning land were the first major source of grievance. In its determination to collect as much land revenue as possible to pay for its army and administration, the Company encouraged massive commercialization of Indian agriculture and established new rules on ownership of property.[28] Rural India was profoundly transformed by deforestation, logging of timber and the planting of cash crops which fetched high prices on world markets.[29] In addition to long term ecological damage, millions of people like the hunter-gatherers, nomads, tribesmen and landless labouring peasants were uprooted from their environments.[30] Newly stratified peasant communities, much influenced by money lending and trading groups, were encouraged to grow cash crops like cotton, tea, tobacco, opium, coffee, etc. To a certain extent, no doubt, the peasantry prospered, but its fortunes were tied up with the uncertain price cycles of commercial crops, which in turn led to the peasant's marked dependence on the banker and the money lender.[31] The huge Muslim peasantry in Bengal suffered much as a result of this. Also, with insufficient attention paid to food crop production and the increase in the number of famines, price inflation adversely affected the urban people, many of whom were Muslims.

The Company also established a new system of property rights in land by regulations concerning the ownership of property, obligation to pay the land revenue, and the sale of land on the open market in the event of non-payment of land revenue.[32] Under the so-called Permanent Settlement of Bengal of 1793, bankers and money lenders could obtain decrees from the civil courts and have the bailiff auction lands to satisfy the decrees. To Sir Sayyid Ahmad, this British system of revenue settlements was particularly burdensome and unjust for the Muslim landed gentry.[33] They had been used to the system of revenue collection based upon the differential values of varied qualities of land, as first laid down in Emperor Akbar's reforms (Chapter 6). The cash value of the revenue, then, was determined by reference to current levels of the market price of produce. The Company assessments, on the contrary, were permanently fixed without any concern for contingencies. Some of the great magnates of Awadh, who became active in the revolt of 1857, had become victims of this arrangement. Other members of the Muslim gentry who had, in previous times, secured rent-free lands from the sultans and the Mughals, for services to the state or Islam, faced confiscations under the Company rules. Muslim religious and educational institutions suffered great hardships as a result.

Another set of grievances concerned the employment opportunities for Muslims. A traditional pathway for many young Muslims – military service – was much restricted, since there were no longer any independent Muslim principalities with their own armies. The advent of the British-imposed peace in the countryside had severed the links between rural power and military service. A loss of status in their localities stoked the resentment of many Muslim aristocrats against the British.[34] The Muslim rulers of India had created employment for different groups and classes of people, at the court, in the army and in towns. A person endowed with skills could rise to the very top; and even many foreigners from Iran and Central Asia had greatly prospered under a non-discriminatory system. The East India Company, however, had institutionalized discrimination almost from the beginning. Under the regulations of Governor-General Lord Cornwallis, no Indian could be employed in the service of the Company to any post which paid a salary of more than £500 a year.[35] The adversities faced by members of the Muslim higher nobility badly affected in turn the fortunes of the Muslim artisan class in urban areas. The production of luxury articles and goods, formerly in great demand, now slumped; and unemployment increased. The large scale import of British manufactured goods was to ruin Indian artisans and craftsmen, including many Muslims.[36]

From their own cultural point of view, Muslim upper and middle classes especially regretted the end of the old Mughal order. The Mughal rulers had been great patrons of art and culture. Sir Sayyid Ahmad noted how the rulers had expended funds in rewarding 'faithful servants, victorious generals, scholars, holy men, poets, recluse mystics, and indigent people'.[37] There was a time in Bengal, during the governorship of Warren Hastings (1773–84), when the Company had shown much multicultural liberality, and Hastings himself had collected around him a coterie of British scholars who might be called truly Indophile. They attracted some of the most intelligent Hindus and Muslims to work for and with them. Many such Britons continued with their Indian interests in later decades, but some of the newer groups of officers and civil servants who arrived in India in the early nineteenth century were imbued with ideas of utilitarianism and showed disdain for Indian culture.[38]

The Company's social and educational policies were resented on specific grounds. Historians have dwelt at length on such policies as the abolition of the custom of *sati* and the fight against thugs in the countryside. These were positive policies; but their role should not be exaggerated. It is worth remembering that the Company was run on fiscally conservative principles of minimal government and parsimonious social budgets.[39] There was never enough money set aside for a truly dynamic social policy. Nevertheless, even within this limited area of activity, misunderstandings were bound to arise. The vast majority of Muslims found it inexplicable that English should replace Persian, the language of Muslim governments for six centuries. They also resented anti-Islamic rhetoric of Christian evangelists; and Sir Sayyid Ahmad himself noted that the Indians were alarmed to witness, during the 1837 famine, Hindu and Muslim orphans being raised in the Christian faith by the missionaries.[40]

Sir Sayyid Ahmad provides many other examples of British neglect of Indian feelings, but perhaps the most telling reason he lists for Muslim discontent was the lack of any meaningful participatory system of government under the Company. He argued that the rebellion of 1857 was spawned by the subject culture superimposed on Indian society by the British. He argued for a transformation of this into a 'mixed subject-participant political culture', without necessarily implying a full-blown democratic system, which he was shrewd enough to appreciate could only lead towards complete independence. In his memorable words, Muslim discontent arose because the Company had 'kept itself as isolated from the people of India as if it was the fire, and they the dry grass'.[41]

Fighting for the crescent

Long before the great conflagration of 1857, the unpopularity of the Company had led to numerous small-scale riots, revolts and rebellions by different groups of people – tribes, nomads, peasants, *zamindars*, *taluqdars*, urban artisans – in different parts of the subcontinent.[42] Most of the uprisings had local economic origins; and, like other economically dissatisfied groups, some Muslims also took part in them. The duration of these uprisings was very short, and they were normally suppressed quickly and forcefully. An early movement of resistance among the poorest of Bengali Muslim peasants, against the Hindu *zamindars*, was led by Mir Nathar Ali, popularly known as Titu Mir.[43] Another so-called *Faraizi* movement in Bengal, started by Haji Shariat Ullah, was also an anti-*zamindar* movement, but developed into a more violent campaign under his son, Dadu Miyan.[44] Islamic fervour influenced both movements; and the *Faraizis* particularly adhered to a strict set of Sunni beliefs. With the Company so well entrenched in power in Bengal, and supported by the rich native landed and commercial interests, these rebellious movements stood little chance of success.

A small circle of Muslims, led by the Sunni *ulama*, were, however, determined on a more ambitious plan of resistance. They appreciated the limits of their physical power as far as military resistance was concerned, but they were convinced that they could compensate for that weakness by awakening and revitalizing the Muslim masses with the traditional message of Islam based on Sunni principles and instilling in them the keenness for a *jihad*. A modern Western tendency is to equate the term with an unreasoning, frenzied and mad form of Islamist extremism.[45] This is understandable in the context of suicide bombing carried out by Islamist extremists in the name of *jihad*. Many Muslims, however, consider *jihad* as the sixth pillar of Islam and for them it is either a struggle against the evil temptations that face them or a struggle for their moral elevation. Warlike *jihad* is thought of as a weapon of last resort against what Muslims might perceive to be unjust subjugation. The idea of a *jihad* against the Company first arose in the writings of a scholar and preacher, Shah Abd al-Aziz (1746–1824), the son of Shah Wali Allah (Chapter 7). Although he had generally excellent relations with the British, he nevertheless proclaimed that under the Company the subcontinent was no longer *dar-ul Islam*, a land where Islam enjoyed sovereign authority and political power, but *dar-ul harb*, a land where Islam was not free.[46] He was not strictly correct, because the Company did not persecute Islam in any way; but his stricture may have helped to legitimize some form of armed resistance against it.

Ironically, it was not the Company that was the first target of the *jihad* movement; it was, in fact, the Sikh kingdom of Punjab.[47] This kingdom, which had expanded and re-organized itself under a brilliant leader, Rana Ranjit Singh, and which was allied to the Company, put pressure on the Pathan tribes who inhabited the extreme north-western areas of the subcontinent. It was in order to help them that a dedicated disciple of Shah Abd al-Aziz, Sayyid Ahmad Shahid (1786–1831), recruited a band of *mujahids* or fighters for Islam.[48] His campaign failed completely and he died fighting the Sikhs in vain; but the *mujahid* idea lived on. Time and again, in the 1830s and 1840s, the *mujahid* bands of fighters, full of high spirits and strong rhetoric but lacking all other resources, threw themselves against British cannons and were slaughtered.

The revolt of 1857

The major landmark in the futile story of armed resistance against the Company was the great Indian uprising or rebellion of 1857. The British called it the *Sepoy* Mutiny, owing to the fact that it was the native troops (the *sepoys*) of the Company's Bengal Army who were the instigators and who offered the greatest resistance. These troops nursed many grievances.[49] One was the enactment of the General Service Enlistment Act of 1856 which required them to serve abroad. Another that angered the Muslim military class in Awadh particularly was the Company's annexation of the state in 1856. Issues around recruitment and pay added to the grievances. The most immediate cause of the mutiny was the issue of the new Lee Enfield rifles with cartridges greased with animal fat, which rightly or wrongly inflamed the religious passions of some of the soldiers, both Hindu and Muslim. It was they who started the mutiny, first in Barrackpore and then in Meerut.[50]

The Muslim role in this dramatic event cannot be ignored. The revolt was at its most intense in the traditional heartlands of Muslim elites: Delhi, Agra and Lucknow. Bahadur Shah II, although completely powerless, was still nominally on the throne in Delhi, and the name of his family invoked memories and emotions among Muslims. In the immediate aftermath of the revolt, many British commentators and historians attributed the origins of the revolt to a Muslim conspiracy.[51] That was certainly an exaggeration, but the British had been clearly unnerved by Bahadur Shah's support to the rebels. The conspiracy theory was strongly denied by Sir Sayyid Ahmad Khan as part of his attempt to mollify the victorious British and to lessen their hatred towards his people;[52] but in his attempt to bring about a *rapprochement* between the British rulers and Muslim subjects he might have helped to gloss over the extent of Muslim participation in the revolt.

Different groups of Muslims were in the vanguard of revolt. First, there were the soldiers (*sepoys*) of the Bengal Army who possessed guns and heavy artillery which they had acquired by breaking into gunpowder arsenals, called the *topkhanas*. They, along with their peers from other faiths, caused the greatest losses to the British. The second group was made up of the influential *ulama* and the Sufis of the Chishti and Naqshbandiya orders, some of whom being linked to the followers of Sayyid Ahmad Shahid and Shah Abd al-Aziz. One such person was Maulvi Ahmad Allah of Faizabad, a Sufi preacher, who went around the countryside rallying the people to the cause of resistance; later, he displayed greatest courage and common sense on the battlefield, and only met his death through the treachery of a local *raja* who had invited him to his palace.[53] The preachings of the otherwise peaceful Sufis fired up thousands of *mujahids* and *jihadis* who fought hard and died for their cause. Another group consisted of disaffected Muslim administrators, landed gentry, members of the old nobility and royalty, etc., who did not take part in actual fighting but ran the civilian machinery in some areas just for a short period wherever the British were ousted.[54] Behind these groups were Muslim civilian mobs, the more naïve of them being aroused by millenarian *fakirs* and soothsayers who preached that the Company rule was destined to end in 1857, the hundredth anniversary of the Battle of Plassey in 1757.[55]

As a complete episode, the revolt lasted for two years, from May 1857 until April 1859; in some centres it ended earlier than in others. The most critical period for Delhi was between 13 May 1857, when Bahadur Shah II was proclaimed the new Mughal emperor, and 21 September 1857, when he was captured by the British at Emperor Humayun's tomb. He could not control the events that unfolded, as he was essentially a weak monarch.[56] In Lucknow, Hazrat Begum, the Queen Mother of Awadh, the wife of the deposed king Wajid Ali Shah, did her best to avoid sectarian divisions and also to build a strategic alliance with the Hindu ruler of Kanpur, Nana Sahib, and his adviser, Azim Ullah Khan.[57] The Lucknow revolt began with the siege of the British Residency on 30 June 1857, which lasted until 21 March 1858. In Kanpur, the other great centre of Hindu–Muslim revolt, the rebels' siege of what was known as Wheeler's Entrenchment, a defensive British fortification, began on 5 June 1857; there many gruesome events followed, which shocked Victorian sensibility, and it was only in December 1858 that Kanpur was retaken and Nana Sahib fled to Nepal on 4 January 1859.

The murder of so many innocent British and Europeans in various places was both monstrous and senseless;[58] but it is important to remember that thousands and thousands of Muslims (along with Hindus) also died between 1857 and 1859 for what they believed to be a righteous cause; and most of them have disappeared from historical records. The avenging British forces spent months scouring the Gangetic countryside; the rebels were rooted out, and hanged or placed at the mouths of cannons to be blown apart.[59] Both Hindu and Muslim rebels went to their deaths without any fear.

The revolt failed for a number of well known reasons. Punjab and Bengal, the two key areas for the British, remained safe in their hands. The trouble in the south was minimal; and in Gujarat and Central India the ordinary people had been reconciled to the Company's rule after new revenue assessments had been carried out. The British controlled the Gangetic transport system; they had the railways and telegraph system; and, above all, they had the resources of their rich mother country behind them. They suffered grievously, but they won in the end. The Muslims could have shown an even stronger defiance than they did, but for some critical shortcomings: the sectarian suspicions among themselves, the less than total support given by many of the Muslim gentry who might have hoped for a quiet life under the British, the unusually docile attitude of the Pathans and, above all, the lack of a continuous supply of heavy armaments.

Queen Victoria's proclamation of 1 November 1858, ending the East India Company's rule, symbolically closed a terrible chapter of Indian history. To the great credit of Lord Canning, the governor-general, the British retribution against Muslims was measured. Certainly, all those who had carried out or incited the murder of Europeans during the revolt were put to death. Others participating in the revolt were either transported or given long jail sentences; and rebellious members of the Muslim nobility lost many of their *jagirs* and pensions. The Muslims in Delhi suffered grievously. While all this happened in the immediate aftermath, there was no sustained vendetta against Muslims in the long run.[60] Most of the Muslim *zamindars* continued to retain ownership of their lands and properties; and a circular from Canning prohibited any discrimination against Muslims in civil and judicial appointments. Not for nothing was the term 'Clemency Canning' coined, perhaps as a form of mild mockery, by those imperialists bent on vengeance at all costs.[61]

Continuity of Muslim social order

Away from the political conflicts and socio-economic turbulence engendered by the power of the East India Company, the more traditional aspects of Muslim life carried on unperturbed in this period. Muslim society in general remained static, with little social mobility, in a modern sense, among its groups and classes. Increased social mobility could only be achieved through modern mass education, but this only became possible in the second half of the nineteenth century. Certain individuals displayed aspects of modernity in private and public life, but overwhelmingly there was a continuity of older social categories with their traditional concerns which had prevailed through many centuries of Mughal and sultanate rule in the subcontinent.

Very few Muslims at this time could be described as belonging to the middle classes; the primary social division lay between those who worked with their hands and those who formed the elite. The former consisted of peasants and manual workers. Muslim peasantry was concentrated in Bengal, Punjab and Sind, but less so in the imperial heartlands, the Gangetic Doab and the south. While all peasants had to bear the brunt of paying land taxes to avaricious *zamindars* and revenue-farmers, both Hindu and Muslim, the Bengali peasants were particularly crushed by the East India Company's land policies that seemed to enrich the commercial classes at their expense.[62] Muslim manual workers included skilled and unskilled craftsmen, artisans, tailors, bricklayers, masons, barbers, washermen and washerwomen, domestic servants, etc. In Delhi, the term generally applied to such workers was *karkhanadar*, someone who worked in the numerous workshops inside and outside the royal fort, serving the needs of the Mughal royal family.[63] Such *karkhanadar*s were found in all towns and villages where demand for their work and skills existed. Both the peasants and workers, with their large families, lived a life of hard work and prayers five times a day. Their entertainments were few, and their care-free days were limited to festive occasions. Their children, with few opportunities for education, followed them in their fields, works and trades, in the mould of the Hindu caste system. There was, however, less of the seclusion of the *purdah* among proletarian Muslim women.

The East India Company might exercise its power over many aspects of Indian life, but for the vast majority of Muslims at this time the traditional elite groups that mattered most were the Muslim monarchy, nobility, professional people and the religious establishment. At the apex of the elite classes, or *sharifs*, was the Mughal emperor. The elaborate formalities and codes of etiquette which governed his relationship with individuals and groups had been in place for nearly three centuries; even in this period of decline, the Mughal court fascinated all people, native and foreign.[64] The presentation of *nazrs*, or gifts from an inferior to a superior, was a ceremony of great significance for the emperor. The Company followed the convention until the late 1840s. The Mughal monarch was meant to be a patron of cultural projects and religious benefactions; but none of the three emperors of this period possessed enough wealth to emulate the style of their predecessors. The monarchs of the Mughal successor states of Awadh, Hyderabad and Arcot were, however, animated and rich enough to promote and patronize architectural, artistic or literary projects along with religious endowments with a certain degree of flair and style. The Awadh capital of Lucknow, for example, was adorned with splendid buildings and monuments.[65] Unfortunately, surrounding any Indian royal family, principally in Delhi, but also in other states, there was the constant presence of innumerable relations and dependants who were proud enough to boast of their royal connections but were too proud to put in an honest day's work. Over 2,000 of such parasites hung on in Delhi's Red Fort, expecting to subsist from a share of the emperor's pension.[66]

The second category of the elite classes consisted of the nobility. The *mansabdari* system (Chapter 6) had by now long gone; but from the late seventeenth century onwards many a noblemen at the court, along with various *zamindars*, *taluqdars* and assorted revenue-farmers, had managed to acquire hereditary land rights from the Mughal emperors or the successor state kings in return for military and administrative services. This was the origin of a new form of nobility or service gentry.[67] Their members held vast lands and palatial buildings both in towns and in the countryside. A corporate unity held this service gentry together by such devices as tight cross-cousin marriage alliances and strong attachments to specific places and settlements, often connected to the shrines of Sufi saints. Their influence at royal courts was substantial, but their standing within the small Islamic townships, popularly known as the *qasbah*s in North India, was even more significant. Their power was drawn from their patronage and support of Islamic institutions like the mosque, *madrasa*, *dargah* and *khanqah*. Their education, deportment, local knowledge and shrewd instincts at manipulating power also made them valued servants of the East India Company. This was particularly so during the late eighteenth century, when the Company displayed much multicultural sensibility, and when both Hindus and Muslims belonging to service gentry were able to secure privileged positions in its service. Among Muslims, the Shia intellectuals of Bengal and Awadh were particularly prized for their contributions to the Company's administration.[68] While Hindus were securing more financial posts with the Company, most of the judicial appointments were made from Muslim ranks. One of the most distinguished Muslim advisers of the Company, Muhammad Reza Khan, who served both Clive and Warren Hastings, might have been 'perhaps the last Indian to hold high office in British India until the early twentieth century'.[69] There were others, too.

A section of the service gentry consisted of those who may be described as practical intellectuals or indigenous knowledge specialists. Three professional groups, whose work was not limited by any sectarian boundaries, enjoyed enormous respect and influence within the wider community. Those who could wield their pen with skill were known as *munshis*, who served as teachers, clerks, letter-writers or linguists. They were also much sought after by those Britons or Europeans who wished to learn Persian, Urdu or

Hindustani in order to expand their intellectual horizons or to secure promotion and career enhancement in the service of the East India Company.[70] A second service group included Muslim physicians, known as *hakims*, who had specialized in the Greco-Arabic medical system, known as *Unani Tibb*, but who were also able to combine that with the Hindu *Ayurveda* tradition of healing. Every social group valued the work of *hakims*, some of whom could command high remuneration from people of wealth. There were medical colleges, called *tibbkhanas*, in cities like Hyderabad, Delhi and Lucknow, where training was given to those who wished to become barber-surgeons, bone setters or cataract removers, among other skills. Women were taught to deal with childbirth and termination of pregnancies.[71] A third professional service group consisted of astronomers and astrologers, called *jyotishs*.[72] The science of astronomy has a long ancestry in India, but Muslim civilization has also held this science in high regard. A very practical way the astronomers displayed their knowledge was by determining the time of the new moon for the month of Ramadan. Their astral knowledge was also valued by people high and low who commonly believed in the power of astrology. The Hindu masses are well known for their faith in astrology, but the subcontinent's Muslims have also been accultured in this tradition; and both Hindu and Muslim *jyotishs* wielded much influence in society.

In their search for religious inspiration and guidance, the Muslim masses looked up to two important sources of authority: the *ulama* and the Sufi masters. Under them came the more junior ranks of cantors, memorizers of the Quran and mystics in various states of seclusion from worldly affairs.[73] A major controversy in theological circles revolved around the question of what authentic Islam meant. For those among the Sunni *ulama*, who uncritically followed the teachings of Shah Wali Allah and his son Shah Abd al-Aziz, the purity of Islam lay exclusively in the *Sunnah* and the *Shariah*. The *Sunnah* encompassed established customs, precedents, codes of conduct and cumulative traditions, typically based on Prophet Muhammad's example, and the divine revelations of the Quran. The *Shariah* was the ideal Islamic law derived essentially from the *Sunnah*. Everything else was considered as *bidah* or harmful innovations. Hindu customs were thus to be condemned wholesale, along with some of the Shia rites of passage and festivals. The most ardent followers of Shah Abd al-Aziz, such as the group around Sayyid Ahmad Shahid who fought the Sikhs, undertook a massive propaganda campaign in the name of the reform of Islam.[74] From hindsight, these so-called reformers or revivalists might be considered responsible for sowing the seeds of mistrust between Hindus and Muslims generally and between Sunni and Shia particularly. Their effort was never going to produce the desired outcome, for the simple reason that the subcontinent is a land of immense cultural diversity. Muslims had, for centuries, imbibed the immemorial traditions and customs of the Hindus; and it was just not possible to persuade the Sunni masses to simply follow normative Islam. The presence of the Shia was pervasive; and in the areas dominated by them there was little opportunity to re-educate them out of their cherished beliefs.

Part of the reason why this particular campaign of revivalism could also not truly succeed was the influence of the Sufis. The Sufis could be ardent champions of the *ghazis* and could adopt a war-like tone; but all their teachings were essentially based on the ideas of peace, contemplation and reconciliation. The shrines of their *pirs* were sometimes supported by Muslim monarchs, but generally the Sufis were keen to retain their own autonomy. They understood that if shrines were dependent on landholdings rather than donations of their followers, they were exposed to the state. In the extreme south, however, where Hindu culture was extensively prevalent, both Hindu and Muslim monarchs remained eager patrons of the shrines whose political significance they took care not to underestimate.[75] The shrines were iconic symbols of piety and reverence for both Muslims and Hindus, rich and poor; but especially for the poor and the desperate, they remained the last refuge of hope and salvation. The Sunni reformers called it veneration and a form of *bidah*; but, short of using violence, they could not stop humble Muslims from visiting them. Their task was to be more difficult when even the East India Company extended its support to the maintenance of shrines. But perhaps the most important reason for the shrine tradition to continue was that the Sunni activists themselves were not as puritanical as the Wahabis of Arabia. They were the followers of such people as Shah Wali Allah, Shah Abd al-Aziz and Sayyid Ahmad Shahid, all of whom had been influenced by the teachings of Sufism. It is, however, fair to say that some of the theological quarrels started by them in this period have continued to haunt and trouble the Muslim population of the subcontinent ever since.

Cultural and intellectual ferment

There was no mass literacy in India during this period, but many people were conscious of the importance of literacy. Discourses and debates were not confined to the intelligentsia alone.[76] Issues concerning politics, literature and morals were among the popular subjects discussed, while intellectuals marshalled learned arguments in the discussion of logic, philosophy and mathematics. Muslims and non-Muslims, whether they be office-holders, jurists, Sufis, Brahmans, elders in various states of alertness, musicians, courtesans, all engaged in dialogues with light-hearted vivacity and quick wit. Discourses among Muslims took place in mosques, ordinary homes, the all-night *mushaira*s or the splendid settings of *darbar*s. The Muslim intellectual world was not dormant, and the indigenous wit and wisdom were well and alive.

The flowering of Urdu poetry

After nearly two or three centuries of maturation, Urdu came of age during this period. Essentially it was the age of Urdu poetry; Urdu prose was a late developer. The main form of Urdu poetry was the *ghazal*, a brief lyric poem, generally made up of two-line verses, both romantic and mystical in tone, and essentially concerned with the passions of lovers. What further animated the *ghazal* at this time was the development of innovative word-play.[77] In this, the essence of imaginary, abstract or elusive themes was captured by the use of a clever word. This was known as *khiyal bandi.*

The *mushaira* increasingly became the main arena of competition as well as a technical workshop. It was also a place of gossip and social rivalry. It was not restricted to Muslim poets, since there were many Hindu poets who excelled in this art. Poets of humble background were also not excluded, and neither were women.[78] While a *mushaira* was generally called by a patron of some means, the gathering itself was controlled by the poets themselves. A noticeable feature at the gathering was the presence of many students or acolytes of established poets. They were known as *shagirds*, and they had close relationships with the master poets, who were called the *ustads*. A *shagird* may recite a verse or two, while his *ustad* approved his composition after changing one or two words. *Ustad-shagird* circles traced their own lineages through deceased *ustads*, thus providing a sense of historical continuity to the community of poets. Each poet claimed to have derived his skill from a past *ustad.*[79]

Many Urdu poets considered Delhi as their natural home, because they could secure imperial patronage. Unfortunately, with Mughal fortunes at their nadir, there was little opportunity for royal support in that city. There were, however, other Muslim courts of Mughal successor states, in places like Murshidabad, Lucknow and Hyderabad, where they were welcome. The *nawab-wazirs* of Awadh were exceptionally generous patrons of all artistic endeavours; and, with greater stability and safety in their state, they were able to attract a large number of Delhi poets to Lucknow.[80] Poets like Mirza Sauda and Mir Taqi Mir (Chapter 7) had both left Delhi for Lucknow. Mir was feted as the doyen of Lucknow poets, and he was the first Urdu poet whose work was typeset and printed. Since the publication of his collected works, *Koolliyati Meer Tyquee*, in 1811,[81] his reputation in Urdu literary circles has remained high.

The royal court of Lucknow was a place of much gaiety and sometimes childish frivolity. The Lucknow School of Poetry included many poets of distinction whose works reflected the relatively permissive and pleasure-seeking atmosphere that the *nawab-wazir*s encouraged in the state.[82] The poet Saadat Yar Khan Rangin (1756–1834), for example, wrote poems describing his amours with courtesans and dancing girls.[83] The Lucknow rulers, being Shia, encouraged Shiite rituals and ceremonies. The commemoration of the martyrdom of Imam Husain at Karbala in Iraq on the tenth day of *Muharram* was the occasion of solemn mourning and the recitation of poems of struggle, resistance and lamentation, called the *marsiya*.[84]

The most distinguished poet of this period did not live in Lucknow, but in Delhi. He was Ghalib (1797–1869), a poet of genius whose *ghazals* and mystical compositions are still cherished by people of literary taste in the subcontinent. Modern in his thinking, he had a high regard for the British, in whose praise he wrote a number of *qasidahs* or eulogies.[85] His immense knowledge, poetic skills and liberality of outlook were noted by them, and he was granted both honours and a pension (Excerpt 8.3). Such was his reputation that his pupils included princes and *nawabs*, high ranking officers of the East India Company and Hindu nobles. Although he was based at the Mughal court he remained strictly neutral during the great uprising of 1857, which helped him later to restore his relationship with the British. He was, however, profoundly upset by their indiscriminate revenge against Delhi in the immediate aftermath of the rebellion, and wrote some of the most moving lines on the wretchedness of the great city.[86] He died a broken man, broken in heart and spirit. His poetry lives on. Another broken heart was that of the last Mughal emperor himself, Bahadur Shah II, who was a poet of some renown, and whose poems reflect the frustrations and rage he felt by his impotence and weakness in the face of British might.[87]

Learning to understand the West

Just as some Europeans and Britons of this period showed empathy with Indian or Indo-Muslim civilization, Indian intellectuals in turn showed a keen curiosity about the West. While an overwhelming majority of Muslims veered between apathy and antagonism in their feelings towards Western culture, a minority of their intelligentsia both appreciated and admired it. As a first step they learnt English. Then, seeking deeper understanding of the progress of the West, they undertook long sea journeys to Britain. Mir Muhammad Hussain, for example, visited Britain in 1776 in order to study Western advances in astronomy and anatomy. He came to the conclusion that an outlook based on scientific methods and approaches was the key to Western success.[88] Mirza Abu Talib, who worked for the East India Company in Awadh, stayed in Britain from 1799 to 1804 and wrote a vivid travelogue (Excerpt 8.4).[89] Another Muslim intellectual, Lutfullah (1802–54), visited Britain in 1844, as part of a delegation; contrasting the behaviour of those at the top in the subcontinent, he remarked that 'in England you will find those who are highest in rank are the politest in society'.[90] Lutfullah was perhaps the first Muslim from the subcontinent to write his autobiography in English.[91]

A number of progressive Muslims prepared the way for future modernization of their community by their support of the Western educational curriculum, particularly the inclusion of scientific education. Traditional Islamic centres of learning, like the Madrasa-i Rahimiyya in Delhi or the Farangi Mahal in Lucknow, while providing sound Islamic education in some of the humanities, mathematics and astronomy, did not consider the academic study of such sciences as chemistry, physics or engineering as appropriate. This meant that there was little opportunity for young minds to investigate or experiment. There was enthusiasm, therefore, when the East India Company proposed that the Delhi-based Madrasa-i Ghazi ud-Din be modernized and re-named.[92] Thus was born the Delhi College in 1825. Science education was to be the first priority of this new college; in 1830 a patron from Awadh donated a large sum for use 'solely on the instruction of the sciences in Arabic and Persian'.[93] Science education then became very popular among students who often took their experiments home to complete. A spirit of free inquiry by students was encouraged at the college by its joint British/European and Indian staff. The more conservative Muslims found the idea of free inquiry deeply unsettling; and the college came to be seen by them as a means of securing a government job rather than as a place of true learning which, for them, had to be Islamic in nature. Opposing this trend were some very earnest young Muslim academics, such as Zaka Allah (1832–1910), Muhammad Husain Azad (1833–1910) and Nazir Ahmad (1836–1912), all of whom being involved with the college.[94] They were joined by even those who had not attended the college, such as Altaf Husain Hali (1837–1914) and Sayyid Ahmad Khan (1817–98). All these men, who led Muslim opinion in the second half of the nineteenth century, advocated Western-style modern education.

An important question that confronted the academics of Delhi College was over the medium of instruction. There was a genuine feeling among many educationists and college staff that all subjects should be taught in the vernacular, preferably Urdu. This could not be acted upon in practice because, following the famous Minutes of 1835 written by Lord Macaulay, the Company discouraged the vernacular medium in all government-sponsored institutions. English, not Urdu or Sanskrit, was to be the paramount medium of instruction. In response, some of the staff members founded a body called the Delhi College Vernacular Translation Society for the purpose of translating all useful works from English into Urdu.[95] The experiment did not last long because, first, the college was engulfed in the cataclysmic events of 1857 when it was closed and, second, by the time it was merged with the Government College in Lahore in 1877, the rising Indian middle-class intelligentsia had started clamouring for their children to be taught in English. People like Sayyid Ahmad Khan who, when young, were ardent supporters of the Delhi College outlook, became the new champions of English education. Much of the mental and emotional distance separating today's academics from the mass of the people in the subcontinent is a result of this incessant desire to learn English. On the other hand, it may be argued that the subcontinent is better placed to withstand competitive pressures in today's world dominated by the English language.

Challenging the missionaries

Until about the mid-eighteenth century, the Christian presence in India was confined to places like Kerala, the home of the Syrian Orthodox St Thomas Christians, or Goa and parts of the west coast of India, where Catholic influence prevailed. With the establishment of British rule in Bengal from the 1760s onwards, Protestant missionaries increased in numbers. Although keen to convert the Indians to Christianity, they remained focused on educational work.[96] After having learnt Indian languages, they went on to translate the Bible, with the aim of propagating Protestant Christianity. The East India Company, mindful of the religious sensibilities of both Hindus and Muslims, displayed an ambivalent attitude towards Christian missions; and, while not hindering the work of existing missions, forbade until 1813 further new missionary ventures within its own territories in India.[97] The rising tide of Christian evangelicalism in Britain in the early nineteenth century, however, forced it to relent after that date; and this led to intense and fervent missionary activities in India. Vernacular Bibles began to be distributed in their thousands; educational and medical work spread outwards from Bengal; and conversions were attempted on a larger scale. We noted earlier that Sayyid Ahmad Khan mentioned the conversion of Indian children in famine-ravaged areas in the 1830s as one of the popular grievances against the East India Company.

Convinced of their own superiority in matters of faith, some of the more zealous Protestant missionaries were keen to engage with the religious leaders of Islam, Hinduism and Zoroastrianism, with the purpose of refuting and disproving the theological validity of non-Christian faiths. Some Hindus in Bengal and the south, along with a few Parsees in Bombay, converted to Christianity; but in both communities there was much resistance too.[98] The missionaries faced greater challenges with the Muslim community. Although Islam and Christianity both belong to the Abrahamic religious family, serious differences have existed between the two faiths over doctrine and social matters. The believers of both faiths have long memories of their acrimonious relationships since, first, the Moorish conquest of Spain and, later, the Crusades. With their dominance over a great part of India for nearly six centuries, the Muslim elite had never felt any sense of disadvantage or inferiority vis-à-vis the Christians. The one moment when they were seriously worried was when Emperor Akbar invited the Jesuits to his court and began showing interest in their arguments (Chapter 6). But that was to pass. By the late eighteenth and early nineteenth centuries, however, the situation had changed, because the British were in a position of dominance. It could be said that the balance of psychological advantage lay with

Christian missionaries in an India controlled by the British in this period. The most active among the Protestant missionaries was a German, the Reverend Carl Gottlieb Pfander, a scholar and linguist.[99] He had a good command of Urdu; and he wrote many Urdu works to spread the message of the Gospel. Another churchman, Thomas Valpy French, a scholar of Islam who knew Arabic well, was to become Pfander's intellectual partner.[100] It was Pfander's works that motivated Muslim intellectuals to mount a challenge against his ideas.

An Islamic public debate using questions and answers, and held before an authority, was known as a *munazara*. Originally theological in nature, such a debate in later times was also concerned with issues of law, science and philosophy. Some of the earliest *munazara*s were conducted in Baghdad in Abbasid times;[101] and in the sixteenth century Emperor Akbar's *Ibadatkhana* (Chapter 6) became the arena of a great *munazara*. A *munazara* could only be truly stimulating when the two sides are thoroughly matched in intellectual rigour and are prepared to challenge each other with robust arguments. While missionaries like Pfander were certainly knowledgeable about both Christianity and Islam they underestimated the intellectual strengths of their Muslim challengers. We have already noted that some Muslims were proficient in English long before Macaulay's educational changes. A number of them were also familiar with the religious and secular literature of Britain and Europe, which was increasingly to be found in the libraries of institutions like the Delhi College or the Hindu College in Calcutta. The Muslim counter-attack came with the publication of tracts and pamphlets written by their scholars; and in 1854 a great *munazara* was held in Agra, at which Pfander and French were publicly challenged by two Muslim luminaries – Rahmat Allah Kairanawi and Dr Muhammad Wazir Khan – who were perhaps the most knowledgeable Indian Muslims of their day in both Islam and Christianity.[102] All matters to do with doctrine, such as the question of the Trinity, the corruption of texts or the prophethood of Muhammad, were argued over at great length, in front of vast crowds of people and the high officers of the East India Company. The Muslim contestants gave a good fight to the two missionaries who failed to secure a knock-out blow to Islamic arguments.[103]

Select excerpts

8.1 Tipu Sultan and the Hindus

Tipu Sultan was a true nationalist who understood the dangers that the British posed to all Indian sovereigns. It is not surprising that he was much demonized by the British. More controversial was his relationship with the

Hindu majority in his kingdom. They, too, nurtured great resentments against him over his land policies and occasional Islamic zeal. There was, however, another side to him, as can be seen from the following quotations. He was a great organizer and innovator; and in his Revenue Regulations we witness his meticulous application to details. Regarding the valuables in Hindu temples, he gave the following order to his inspectors:

> You are to examine the jewels, clothes, copper and brass utensils etc, which belong to all the Hindu pagodas throughout your district, and have an account taken of them by the Serishtadars and Shamboges, with the description and weight of each article; and you are to deliver them over to the charge of the Shamboges, with directions to allow the use of them at times when they are wanted in celebrating worship, and afterwards to put them away with care.

Tipu also had great respect for the Hindu monastery at Sringeri and its Swami. In 1791 he wrote to the Swami:

> You are a holy personage and ascetic . . . We request you to pray to God along with other Brahmans of the matha (monastery), so that all the enemies may suffer defeat and take to flight and all the people of our country live happily, and to send us your blessings.

When the Swami requested funds from Tipu to re-consecrate the goddess after the Marathas had vandalized Sringeri in 1792, Tipu replied thus:

> People who have sinned against such a holy place are sure to suffer the consequences of their misdeeds at no distant date in the Kali Age . . . (After) having consecrated the goddess and fed the Brahmans, please pray for the increase of our prosperity and the destruction of our enemies.
>
> (Brittlebank 1997: 120, 128–30)

Tipu's reference to Kali and his seeking of blessings from a Hindu Swami suggest that, far from being a tyrannical Muslim ruler, he was behaving in the traditional manner of a South Indian Hindu king.

8.2 *The annexation of Awadh*

The person who did most to bring about Lord Dalhousie's dream of annexing Awadh for the East India Company was his appointee, Colonel

Sleeman, the British Resident at Lucknow. Sleeman, who had a great reputation as the destroyer of the countryside dacoits, wrote a completely biased report on conditions in Awadh, which became the tool that Dalhousie needed to find a justifiable excuse for annexation. Two years after Sleeman's resignation in 1854, the annexation became a reality through the joint efforts of Dalhousie and the successor agent, James Outram. The official explanation for the annexation was that the Company needed to liberate the people of Awadh from the incapable and effete monarch, Wajid Ali Shah. This version was first repudiated in 1857 by Samuel Lucas in his book, *Dacoitee in Excelsis or the Spoliation of Oude by the East India Company*, in which he argued that Dalhousie had made up his mind before he appointed Sleeman as the Resident and that Sleeman was pressured to write a damning report on Awadh and her king. With biting sarcasm Lucas wrote thus:

> It might be said of Colonel Sleeman, in his most composed moments, that he was ever on the look-out to capture a thief . . . Set a thief to catch a thief, is an old maxim which by no means applied to so excellent a person as the Colonel. But it was an ingenious thought of Lord Dalhousie, that the converse might hold true, and that a gentleman who had caught dacoits in unexampled numbers was the instrument to assist him in an act of dacoitee unrivalled. Colonel Sleeman was, as it were, made to Lord Dalhousie's hand; but before he went, that there might be no mistake, and no option of impartial scrutiny, he bound him to effect his object by a very definite commission . . . (whose purpose) was the inculpation of Oude Government, thus prejudiced and precondemned in the mind of Lord Dalhousie.

Samuel Lucas's accusation of Dalhousie's chicanery is to a certain extent borne out by the discovery of a letter that Dalhousie wrote to a friend on 18 September 1848, just two days after he had appointed Sleeman as the Resident.

> Meanwhile I have got two other (in addition to Punjab) kingdoms on hand to dispose of – Oude and Hyderabad. Both are on the high road to be taken under our management – not into our possession; and before two years are over I have no doubt they will be managed by us. You will laugh doubtless . . . to think of (me) sitting here and bowling about kings and kingdoms as if they were curling stones! But although one does laugh, it is anxious work, I can tell you.
>
> (Fisher 1996: 261, 264–5)

8.3 Ghalib seeking appointment at Delhi College

Ghalib's ego was legendary. His encounter below with the Principal of Delhi College is an example of his feudal craving for recognition even amidst ordinary circumstances of life. It is also a sign of his self-assuredness and courage to speak up for himself, a rare ability denied to most Indians in colonial times.

> In 1842, when Delhi College was reorganized on new principles, Mr. Thomason, Secretary of the Government of India, who later became Lieutenant-Governor of the North Western Provinces, came to Delhi to interview the candidates. A teacher of Arabic had already been appointed at a salary of Rs.100, and Mr. Thomason wished to make parallel appointment for Persian. The names of Ghalib, Momin Khan and Maulvi Imam Baksh had been suggested to him, and Ghalib was the first to be called for interview. When he arrived in his palanquin . . . Mr. Thomason was informed, and at once sent for him. But Ghalib got out of the palanquin and stood there waiting for the Secretary to come out and extend him the customary welcome. When some considerable time had passed and Mr. Thomason found out why Ghalib did not appear, he came out personally and explained that a formal welcome was appropriate when he attended the Governor's durbar; but not in the present case, when he came as a candidate for employment. Ghalib replied: 'I contemplated taking a government appointment in the expectation that this would bring me greater honors than I now receive, not a reduction in those already accorded me'. The Secretary replied: 'I am bound by regulations'. 'Then I hope that you will excuse me', Ghalib said, and came away.
>
> (Russell & Islam 1969b: 62–3)

8.4 Mirza Abu Talib's travelogue

The following very accurate description of a great British institution is an example of Mirza Abu Talib's powers of acute observation. The whole of his book is irresistible reading.

> (To the art of printing) the English are indebted for the humble but useful publication of newspapers, without which life would be irksome for them. These are read by all ranks of people, from the prince to the beggar. They are printed daily, and sent every morning to the houses of the rich; but those who cannot afford to subscribe for one, go to read them at the coffee-rooms or public houses. These papers give an account of everything that is transacting, either at

> home or abroad; they contain a minute description of all the battles that are fought . . . the debates in the Houses of parliament . . . the price of grain . . . the birth or death of any great personage; and even give information that, on such a night, such a play will be performed, or such an actor will make an appearance.
>
> (Taleb 1814: 294–5)

9

STIRRINGS OF A MUSLIM MODERNITY UNDER THE RAJ (1858–1924)

A new phase in India's history opened after the 1857 rebellion and the consequent dissolution of the East India Company. The era of the colonial Raj began with Queen Victoria's proclamation of 1 November 1858. This benign document set a new tone of authority and conciliation.[1] The queen promised, first, no further annexations of the remaining native states. This came too late for many *nawab*s who had by now been dethroned. The Mughal emperor had been exiled to Rangoon, Burma, and his successors had been executed. Although nearly 500 princely states, big and small, survived, relatively few large ones were under Muslim rulers. After 1858, the vast majority of Indian Muslims lived within the directly controlled territories of British India or within the Hindu princely states, the most prominent of which, Kashmir, was where a Hindu prince ruled over a Muslim majority (Map 9.1). The queen's proclamation also promised that, with the exception of those landlords who had taken an active part in the rebellion, all the *zamindars* and subordinate categories of landlords could retain their lands and possessions. This meant a further lease of life for Muslim landed gentry, particularly in Punjab, Sind and the United Provinces (incorporating the pre-1856 *nawabi* of Awadh).[2]

The period of the colonial Raj was only a short phase in India's history, barely lasting nine decades. It was a time of enormous political, social, economic and cultural change which awakened a spirit of nationalism in the peoples of India. From a Muslim perspective, it is convenient to divide this period into two parts. This chapter covers the years between 1858 and the mid-1920s, when a modernized Muslim intellectual and political leadership came to define and articulate essential Muslim civic and political requirements under the Raj, at the same time as an incipient nationalist movement was being forged by a wider middle-class elite challenging British absolutism. The tension between Hindus and Muslims began to emerge in this period. When it became greatly aggravated after the mid-1920s, an almost irretrievable gap opened up between what we might call Indian

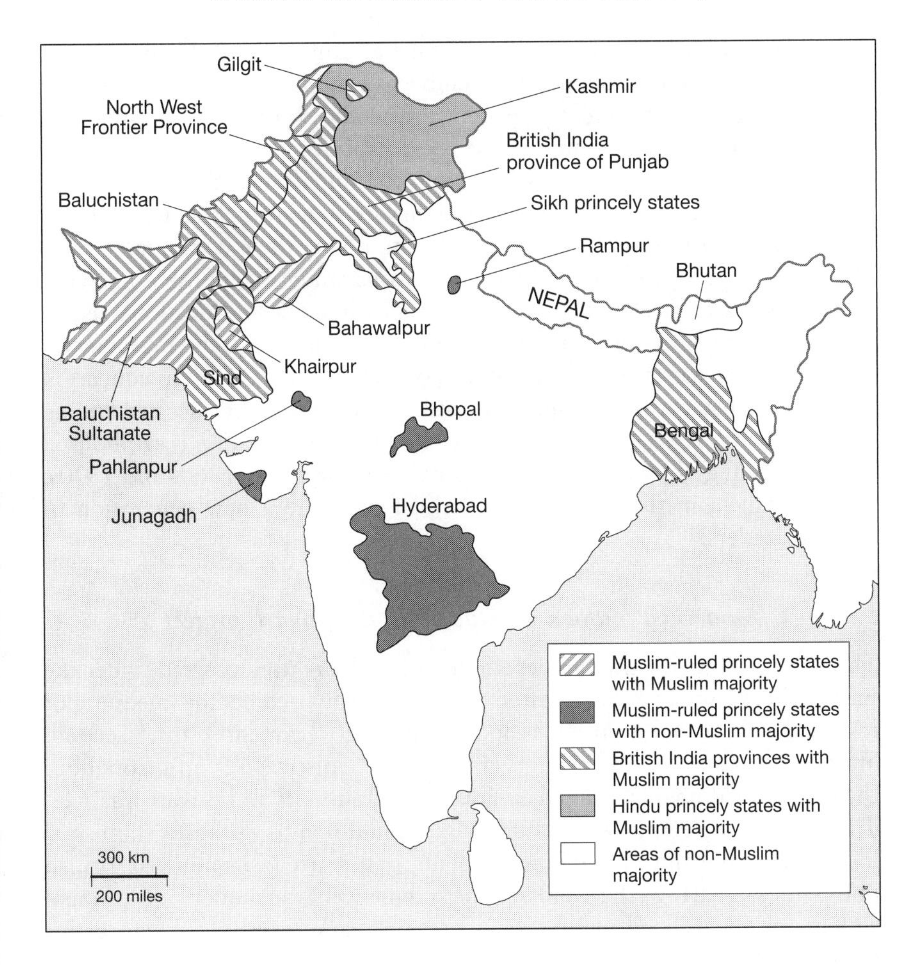

Map 9.1 Muslim South Asia: 1858–1947

(including Indian Muslim) nationalism and Muslim separatism, leading to the fateful separation of Pakistan from the rest of India in 1947. The events of the last two and a half decades of British rule will therefore be examined in the next chapter.

Education and Muslim identity

In her proclamation of 1858 Queen Victoria also willed that her subjects, 'of whatever race or creed, be freely and impartially admitted to offices in our service, the duties of which they may be qualified, by their education, ability and integrity, duly to discharge'.[3] This promise could not come to fruition until the government employed suitable Indians within its ranks.

The aspiring Indians who were able to take advantage of such employment were mostly the English-educated members of the Hindu intelligentsia and the higher castes, particularly in Bengal and the south, and the minority Parsee community of Bombay; but their aspirations were severely constrained by overt and covert institutional discrimination within government service.[4] Aspiring Muslims, too, suffered this discrimination but, additionally, they also had to endure official suspicions about their community after 1857.[5] Although Muslims had left the serious study of English somewhat late, a number of their middle-class leaders saw in modern education a useful tool for their young men to gain economic security through employment, for example, in government service. For others, however, modern education was of little use unless it helped to instil Islamic pride and identity among young Muslims. This tension between two visions of education is symbolized by the founding of two educational institutions during the 1860s and 1870s, both of which, in their own ways, helped to create a new generation of Muslim leaders.[6]

Modernist outlook in education: the Aligarh project

Three important educational developments had already occurred under the East India Company. First, after 1835, English became the medium of instruction in all institutions supported by a government grant.[7] Second, after 1854, a network of primary and high schools was set up throughout the Company domains, funded by grant-in-aid support and private finance.[8] Third, three universities – Calcutta, Madras and Bombay – were established in 1857,[9] while the first Indian students had started enrolling at British universities as early as the 1840s.[10] The rudiments of the modern educational system for the subcontinent were therefore marked out even before this period began.

A common perception of British policy makers was that the Muslims were not only educationally more backward than the Hindus but that they were also not interested in modern education.[11] The reasons for this, it was argued, lay in their religious orthodoxies, vainglorious pride in their past and hatred of Western civilization. The linkage between Muslim educational backwardness and religious conservatism was forcefully argued by a scholar and civil servant, W.W. Hunter, in his book *The Indian Musulmans* (a common term used in place of 'Muslim' in this period), published in 1871.[12] His arguments, based on the case study of Bengal, helped to popularize a misplaced stereotype about Muslim educational aspirations in general. It was misplaced because, first, it was really not religion but poverty that deterred many Muslims from seeking modern education and, second, the attendance and attainment figures for Muslim children were at least as good as those of Hindu children in many parts of British India.[13]

Hunter's arguments, though exaggerated, found support among a section of the Muslim intelligentsia. This was because, although at school level Muslim children performed tolerably well, the achievement of Muslim students at college and university levels was much inferior to that of the Hindu students. Thus, for example, the results of the BA degree examination at Calcutta University in 1870 showed that out of a cohort of 151 Hindu students fifty-six were successful, while there were only two Muslim students, both of whom failed.[14] Between 1858 and 1878, of the 1,373 Bachelors of Arts emerging from India's colleges and universities only thirty were Muslims; and among the 326 Masters of Arts just five were Muslims.[15] Such extreme disparity was bound to have a negative effect on Muslim chances of securing relatively high posts in government service or the professions. That, in turn, could mean that Muslims would be unable to contribute creatively to policy making at various levels of public life. The person who articulated these concerns most clearly in this period was Sir Sayyid Ahmad Khan (1817–98), whose analysis of the Indian revolt of 1857 was referred to in Chapter 8.

Sir Sayyid Ahmad Khan must be considered as the nineteenth century's most distinguished Indian Muslim social, educational and political thinker and reformer. He actively campaigned for the adoption of modern Western education in India, particularly for Muslims.[16] He both started and joined a number of organizations whose purpose was to make European knowledge accessible to young Muslims and other Indians in Urdu vernacular.[17] His appreciation of British and European civilization was greatly enhanced after a stay of nearly a year in Britain where he was much feted by both aristocratic and learned circles (Excerpt 9.1). It is worth bearing in mind that he was descended from an old and genteel Mughal aristocratic family,[18] and that behind his educational fervour lay one central motive, which was to reconcile the post-1857 alienated upper- and middle-class Muslims of the United Provinces and Delhi to British rule. It was to this class of people that he appealed for funds and general support for a Muslim educational institution that he wanted to start at Aligarh. He wanted this institution to become a training school for Muslim administrators, managers, lawyers and doctors. Although he had strong concerns for mass education, his first interest lay in the training of a Muslim professional class. The Muslim community's true place in British India, he argued, could only be secured by its educated elite. Within seven years, between 1870 and 1877, he had secured enough funds to establish a Muhammadan Anglo-Oriental College (MAO College) at Aligarh.[19] The term 'Muhammadan' (implying a worshipper of Prophet Muhammad) is, for Muslims, a pejorative term, originally coined and popularized by some Islam-hating Christian missionaries; but it was widely used in India at this time, even by Muslims. It took another half a century before the college was recognized as Aligarh Muslim University in 1920.[20] Today, it is one of India's premier higher education institutions.

Sir Sayyid had a clear vision for his Muslim College. He expected Muslim students to predominate there, but he was not a sectarian; non-Muslim students were freely admitted. He wanted the college to impart modern scientific education rather than a traditional Islamic syllabus,[21] and placed great emphasis on scientific research methods and rational discussion. He also hoped that the students would develop a strong sense of camaraderie, duty, loyalty and leadership by participating in varied clubs, societies and sporting activities.[22] Since his ideal models for the MAO College were the Victorian English public school and the University of Cambridge,[23] it was crucial for him that Englishmen rather than native educationists were appointed as heads. The first five principals were all Englishmen – H.G.I. Siddons (Principal 1875–83), Theodore Beck (Principal 1883–99), Theodore Morison (Principal 1899–1905), W.A.J. Archbold (Principal 1905–9) and J.H. Towle (Principal 1909–16).[24] Although the college had three departments to begin with – Urdu, English and Persian/Arabic – by the 1880s market forces had ensured that only the English department remained.

The Aligarh MAO College/Muslim University was not the only centre of higher education for Muslims, but the Aligarh graduates became leading figures in Muslim social, educational and political life, and 'it is difficult to trace the history of any Muslim movement in India after the 1870s without some reference to Aligarh' (Plate 12.1).[25] The movement for Muslim women's education, although not part of Sir Sayyid's initial plan, was also later inspired by the Aligarh project (Chapter 10). The entrenchment of Aligarh graduates in key positions in India, and until not too long ago in Pakistan, has symbolized the continuation of that sense of power and confidence which the older Muslim elite exuded for centuries.

Education and Islamic revivalism: the Deoband project

The orthodox Muslim religious leaders had grave reservations about Sir Sayyid Ahmad Khan's unabashed championship of British rule, advocacy of Western education and speculative scepticism in matters of theology.[26] They appreciated the necessity of adapting to British rule, but they did not wish to jettison Islamic traditions. While the Sufis and various groups of *ulama*, recognizing the general decline in their status under the British, were content enough to secure patronage for mosques, shrines, *dargahs*, etc., and to concentrate on prayers and studies, mosque rituals, elementary *madrasa* education and counselling the poor and the needy, more reform-minded religious leaders were keen to promote and propagate Islamic ideas through education and modern means of communication. Two such leaders – Maulana Muhammad Qasim Nanautawi (1833–80) and Rashid Ahmad Gangohi (1829–1905) – had established in 1867, a decade earlier than the MAO College, a religious seminary at Deoband, 90 miles north-east of Delhi.[27] In due course this seminary became the nerve centre of Islamic revivalism in the late nineteenth and early twentieth centuries.

The curriculum at Deoband was radically different from that at Aligarh. Modern scientific subjects were not taught, and English was not the medium of instruction. However, medieval Greco-Arabic science was taught, while English was not completely ignored. The traditional Islamic Studies curriculum at Deoband put greater emphasis on the learning of revealed sciences, the *manqulat* (Chapter 7), which concentrated on Quranic themes. The rational subjects, the *maqulat*, such as logic, philosophy or medieval science were taught, but not too intensively. The entire course of studies lasted between six and ten years; and 106 major Islamic texts had to be mastered.[28] The authorities conceived the Deoband curriculum as complementary to Western education, not against it, which meant that the government did not view the project with suspicion or hostility, at least in the early years. The long term objective was that a young Muslim person had to have his cultural identity secured through an Islamic education before he ventured out into the modern world. The Deoband leaders had no complaint against Western education *per se*.

Apart from teaching, the Deoband seminary aimed to send out to the wider Muslim public a particular normative message, which was to keep to the straight and narrow path of the Quran and the *Shariah* and to eschew various un-Islamic customs which had been absorbed by Muslims unconsciously over a long period of time.[29] Modern methods were adopted for communicating with a mass Muslim audience.[30] Easy-to-read handbooks were printed, published and distributed to many bookshops across the subcontinent. The seminary graduates in due course began opening smaller branches on the Deoband model, thereby beginning a country-wide programme of Islamic public education. The aim was to inculcate in ordinary Muslims a sense of pride and confidence about their religion. This in turn brought valued contributions from appreciative ordinary people, which allowed the Deoband authorities to remain independent of great *zamindars* and other nobles. The abundance of funds helped to make the seminary a model of a well maintained and well run educational institution, with a fine building, a well stocked library, a professional staff, a systematic curriculum, competitive examinations, awards and trophies, affiliations, incorrupt financial management, etc.[31] An Islamic curriculum in a modern setting was what Deoband wanted to provide; and its success in achieving this is testified by the fact that within a few decades of its foundation it was attracting many overseas students.[32]

The success and popularity of the Deoband project help to explain the dynamism of nineteenth-century Islam in India. The *ulama*, who both taught and studied at the institution, were deeply engaged in trying to make sense of their own traditions in the context of the tumultuous secular changes happening all around them. Their dedication to learning and teaching endowed them with a rare authority in society. Their rulings or *fatwas* on social and moral issues carried weight among ordinary Muslims.[33] One such ruling even declared that the learning of English did not pose a danger to Islam.[34]

Muslim political desires and outcomes

From the 1880s onwards, the Aligarh MAO College and the Deoband seminary both in their own ways, along with many other colleges and universities, contributed towards creating a new generation of Muslim leaders. Until the end of the nineteenth century, these leaders remained in the shadow of Sir Sayyid Ahmad Khan, whose political thinking was shaped by his loyalty to the British rulers and an intense mistrust of those educated Indians who challenged British rule. After his death in 1898, the Muslim leaders' political desires came to be expressed in two significant campaigns: one, within British India, for separate communal representation; and the other, outside India, for the preservation of spiritual and temporal authority of the Ottoman sultan as the Caliph of Islam. The first was a success, but the second resulted in abject failure.

Background to political participation

Despite the benevolent nature of Queen Victoria's proclamation of 1858, the mindset of the British elite remained closed against meaningful Indian participation in the political process for another fifty years. There was great unease about the long term consequences for its control of India if Indians had a hand in shaping government policy. The policy that was adopted was to make alliances with groups such as the princes, the *zamindars*, the urban commercial classes etc., offering them varied favours in return for their cooperation;[35] but there was never any question of relinquishing ultimate British control over power. What was to make this policy increasingly unsustainable was the creation of a new class of English-speaking Indians emerging from modern schools and colleges. In the 1830s, Lord Macaulay might have wished to produce educated Indians to oil the wheels of the East India Company's bureaucracy; what he failed to foresee was that Indians would not for ever be content with a limited English education for serving in low paid government jobs. They would continue to seek higher professional education.

Education became the passport to politics for such Indians with professional qualifications, particularly those in law. Dissatisfied with the grace and favour approach that allowed for a few selected Indians to have the ear of the government, they demanded greater transparency and more self-government. Mass political organization was also seen by them as a necessity if the British were to be forced to move at a quicker pace. We thus witness in this period the founding of many Indian political and social associations among the Westernized Indian public residing in cities like Calcutta, Bombay, Madras, Allahabad and Poona.[36] It was through these bodies that Indian critics of the practices of the Raj found a voice. Among several issues of concern, for example, the discrimination within the judiciary was a particularly irksome matter for Indian lawyers. An aspect of this discrimination was the right of Europeans (a term used in the British Empire to describe Britons and other white people) living in country areas not to be tried by an Indian judge. When, under the influence of the liberal-minded Lord Ripon (Viceroy 1880–4), the government attempted to ban this practice by proposing the so-called Ilbert Bill of 1883, a raucous and vicious protest against the legislation was led by the Europeans of Calcutta, who unabashedly displayed all their racial prejudices and hatreds.[37] The withdrawal of the bill disappointed the politically conscious Indians, but it also became the catalyst for the formation, in 1885, of a future umbrella association, known as the Indian National Congress (henceforth, Congress in this volume).

The Congress was the brainchild of a retired British civil servant, A.O. Hume (1829–1912), who was able to bring together a small group of public-spirited and highly placed group of Indians in Bengal and Bombay.[38] Men like Surendranath Banerjee (1848–1925), Romeshchandra Dutt (1848–1909), Manmohan Ghose (1849–1909) and W.C. Bonnerjee (1844–1906) were Hindu Bengalis; while from Bombay enthusiastic support came from two prominent Parsees, Pherozshah Mehta (1845–1915) and Dadabhai Naoroji (1825–1917, who then resided in Britain) and a Muslim leader, Justice Badr ud-Din Tayyabji (1844–1906). They, and the other delegates, started meeting annually at the Congress sessions in one or other of the big cities, and proposed moderate reforms over such issues as a parliamentary inquiry into Indian affairs, the modification of the India Council in London, the reform of the legislative councils, civil service recruitment procedures, and so on.[39] Adopting a moderate tone, they pledged loyalty to the British Raj; but over a passage of time the government began to view with alarm the long term implications of the issues they raised. Most Congress delegates belonged to the legal profession, followed by those from the landed gentry and the commercial classes.[40] Others were doctors, teachers or journalists. Regionally speaking, Bengal, Madras and Bombay sent most delegates, while smaller numbers came from the United Provinces, Central Provinces, Bihar and Punjab. Delegates from the princely states and from Burma too attended during some years.[41] In terms of ethnicity, what was clearly noticeable was that upper-caste Hindus had a dominating hand in the proceedings; and, with the exception of the Lucknow Congress of 1899, most Congresses were attended by only a handful of Muslims.[42]

Sir Sayyid Ahmad Khan's antipathies towards the Congress

From the very beginning the Congress struggled to secure the wholehearted cooperation of Muslim leaders. The one exception, Justice Badr ud-Din Tayyabji, well connected in high circles in Bombay, had little influence on the broad mass of Muslims. Another potential leader, the Bengali aristocrat, Justice Ameer Ali (1849–1928), was more concerned with securing educational and employment opportunities for Bengali Muslims who were at a disadvantage vis-à-vis Hindus; his advocacy of positive discrimination in favour of Muslims in Civil Service recruitment in Bengal, however, went against the Congress policy of a competitive approach to civil service recruitment which, in his view, inevitably favoured upper-class Hindus who had had a head start in Western education, particularly in Bengal.[43] The Muslim leader with the greatest authority, Sir Sayyid Ahmad, who could have helped to secure a strong Muslim voice within the Congress, was in fact vehemently opposed to the organization.[44]

Three major reasons explain Sir Sayyid's antipathy towards Congress. First, the increasingly strident demands for reform by Bengali Hindu intellectuals in Congress were far too radical for him; he put his faith in the essential goodness and benevolence of the British government; and, in his view, Muslims should not be seen to be against the government in case they rekindled the post-1857 period of British concerns about their loyalty.[45] Second, like Justice Ameer Ali, he was against the Congress demand for civil service recruitment through a meritocratic system of competitive examinations, but for a different reason. While in Bengal Muslim numbers in the civil service were abysmally low, in his own area of the United Provinces, Muslims had until then enjoyed a healthy recruitment for varied government positions, owing to the traditional Muslim aristocratic influence since Mughal times that the British seemed to have acknowledged. He, therefore, did not wish for any unnecessary disruption to Muslim advantage through competition from successful Bengali candidates who might become entitled to regional postings outside Bengal.[46] Third, Sir Sayyid rejected all attempts by the Congress progressives to inject an elective principle in the running of public institutions. With his aristocratic background, he thought more in terms of the non-communal Mughal service elite, Persian- and Urdu-speaking, including Hindus as well as Muslims, and both serving the British loyally. His patriarchal background and his political instincts led him to champion the principle of selection rather than election.

Sir Sayyid's reservations about the Congress were a consequence of the fracturing of the idea of the old non-communal elite that he strongly believed in. A symptom of this fracturing was the rise of a zealous form of Hindu revivalism in North India and Maharashtra in the 1880s and 1890s. The early nineteenth-century Hindu social and religious movements were essentially concerned with the reform of traditions, but the founding of the Arya Samaj by Swami Dayananda Saraswati (1824–83) in 1875 marked a critical moment in transition from reform to revivalism.[47] The Hindu revivalists promoted among their followers a certain type of self-righteous fanaticism which expressed itself in resentment and hostility towards both Muslims and Christians. Three separate forms of agitation, not necessarily or always led by the Arya Samaj, alarmed Muslim opinion considerably. One was the campaign to ban the killing of cows, which could directly threaten the livelihoods of hundreds of thousands of Muslims engaged in meat trade. Many riots ensued as a result.[48] The second campaign was over the replacement of Urdu with Hindi as the language of lower courts and subordinate tiers of government administration. This was viewed with great concern particularly in the United Provinces, the home of numerous prominent personalities in Urdu literature. Any tilt by British officials towards Hindi and its Nagri script was considered by the Muslim intelligentsia as a signal of their declining influence; and Sir Sayyid Ahmad, who was attached to the idea of an Urdu-speaking elite made up of Hindus and Muslims, was alarmed about that.[49] The third agitation, led by Hindu Marathas like Bal Gangadhar Tilak (1844–1920) in Bombay and Maharashtra, was over the holding of festivals in the honour of Shivaji, the seventeenth-century Maratha leader who fought Emperor Aurangzeb and the Mughals.[50] From now on, Aurangzeb was to become a highly demonized figure in the imagination of fanatical Hindus.

The sort of campaigns organized by both Hindu and Muslim revivalists exacerbated unnecessary tensions between the two major faith communities, leading eventually to what is called 'communalism'. In the Indian context, communalism is the defining of a community in no other way except in terms of religion.[51] Although every Hindu and every Muslim carries multiple identities, many of which are similar and overlap, for a communalist there can be no synthesis or reconciliation between the followers of Hinduism and Islam. In other words, a monolithic Hindu community confronts a monolithic Muslim community. Sir Sayyid Ahmad remained a non-communal elitist all his life but, by the time he died, communalism was fast imprinting itself on the Indian socio-political landscape. The Muslim demand for separate communal representation was to be the consequence.

Steps towards communal representation

The imperial government was generally willing to grant moderate demands for privileges or concessions from the loyal elite of Hindus and Muslims. The situation was, however, changing rapidly in the late nineteenth century. The government had to deliberate over two sets of demands it was facing. On the one hand, the Hindu-dominated Congress, pressured by its

Bengali wing, was clamouring for radical political reforms which could only lead in the end to the weakening of the Raj; on the other hand, Muslim leaders, while professing undying loyalty to the Raj, were requesting separate political recognition of their community. The headstrong Lord Curzon (Viceroy 1899–1905) had little time for Congress demands; and in a private letter to Lord Hamilton (Secretary of State for India 1894–1903) he vented his feelings that 'the Congress is tottering to its fall, and one of my greatest ambitions while in India is to assist it to a peaceful demise'.[52] It is not unreasonable to speculate that some form of divide and rule strategy lay behind his dramatic decision to partition the province of Bengal. The decision made sense administratively, as the province was too big and unwieldy; but suspicion lingered about an ulterior motive behind the partition.[53] The eastern areas of Bengal, predominantly Muslim, were cut off from western Bengal and joined up with Assam, creating a new Muslim-majority province. In the reduced western part, the Bengali populace became in fact a minority in relation to Biharis and Oriyas, since parts of Bihar and Orissa were also part of the old Bengal province. Politically aware Hindus and Muslims both protested at the Viceroy's plan, but Muslim anger subsided once the advantages of a separate Muslim-dominated province came to be appreciated, such as more jobs being available for the Muslim educated class no longer facing Hindu competition.

A prolonged period of violent agitation by mostly Hindu revolutionary students and dissidents ultimately forced the government to revoke the partition; but even before that, plans were afoot to regain the confidence of the Congress by bringing forward some positive constitutional changes. Curzon's successor, Lord Minto (Viceroy 1905–10), and his superior, John Morley, Secretary of State for India in the British Cabinet, worked together for over three years to introduce an elective element in the choice of members for the Imperial and Provincial Legislative Councils, which was a crucial Congress demand. The franchise was to be very limited, with only the wealthy and the well educated people given the right to vote; but the conceding of the elective principle was seen as a recipe for calming tensions among the Hindu elite. The Muslim leaders did not stand by passively as they digested the political implications for their community of the intended extension of franchise, however limited.[54]

The Muslim campaign was led by both the Aligarh-trained new educated elite and the traditional landlord class, particularly in the United Provinces. It suited the interests of both groups to fight for separate Muslim representation. A fragmented voting system could enable more of the educated young Muslims to be elected as representatives, while rich and powerful landlords in places like the United Provinces could continue to play the traditional role of pulling strings at local level in order to preserve their economic interests. Both groups decided to campaign vigorously in favour of separate electoral lists, whereby the Muslim electorate qualified to vote could choose fellow Muslims for various levels of assemblies, and for a number of reserved seats for Muslims in provincial and imperial legislatures. They prepared their ground carefully under the astute leadership of the Secretary of the Aligarh College, Mohsin ul-Mulk. Through the influence of his principal, W.A.J. Archbold, Mohsin ul-Mulk was able to secure an agreement from the Viceroy to meet a Muslim delegation at his residence in Simla on 1 October 1906. The delegation, led by the Aga Khan (1877–1953), the progressive leader of the Khoja Ismaili community, was promised by the Viceroy that their demands would receive serious consideration in the drafting of the new legislation. The Muslim leaders showed little complacency after that meeting; indeed their campaign continued in a determined fashion during the next three years, and the government was sufficiently unnerved to the extent that, in one of his official letters, Lord Minto even wrote that the people of England needed to be reminded that the Bengali (meaning Hindu agitator) was not everybody in India; in fact 'the Mohammedan community, when roused, would be a much stronger and more dangerous factor to deal with than the Bengalis'.[55] After the formation of the Muslim League on December 30, 1906, the members of its London branch used all political means within the Westminster Parliament to influence the government's thinking and deliberations over the reforms to be introduced.[56]

The Indian Councils Act of 1909 (often known as Morley–Minto Reforms) built upon earlier such acts as those of 1861 and 1892 and extended Indian participation in the legislative process. Although an official European majority was retained in the Imperial Legislative Council, the non-official Indians elected to the council had the right to ask questions on the budget and to propose amendments. The Congress leaders felt satisfied with the principle of election to both the Imperial and Provincial Legislative Councils conceded by the government, but they were clearly unhappy with the concession that the Muslims had secured (Excerpt 9.2). This was that all Muslims who were entitled to vote were now to be put on a special communal list of electors; and for all elections of non-official members for the Imperial and Provincial Legislative Councils, such Muslim voters would have the right to vote for a specifically Muslim candidate of their choice; additionally, the Muslim electors would further enjoy the right to vote for any other candidate on the basis of the general electoral list.[57] Muslim leaders had won a significant victory, particularly for the ambitious and politically aware young people of the Muslim upper and educated middle classes.

Pan-Islamism and the Khilafat Movement

Pan-Islamism is an ideology that emphasizes the solidarity of Muslim peoples worldwide. In its modern form, the ideology arose in the nineteenth-century

Ottoman Empire, at a time when Turkey was facing political and military pressures from varied European nations.[58] The Ottoman Sultan, Abd al-Hamid II, used the ideology to prop up his power within Turkey with the claim that, as the Caliph of Islam, he exercised spiritual authority over Muslims throughout the world. The Turkish sultans had claimed the caliphate on the ground that, following the Mongol sack of Baghdad in 1258, the Abbasid Caliphate had moved to Egypt where, afterwards, the Mamluk sultans had kept members of the Abbasid family as titular caliphs in Cairo until the Ottoman conquest in 1517.[59] With a strong sentimental attachment to the Turks, and with the word 'Turk' being practically synonymous with Muslim since the Middle Ages (Chapter 3), it was no wonder that Indian Muslims with an interest in international politics not only recognized the Ottoman sultan as the caliph but also saw in Turkey a Muslim bulwark which could come to their aid if they felt threatened by the British. This conflation of Pan-Islamism and the Ottoman Caliphate in the Indian Muslim mind was also the result of anti-British rhetoric of an Afghan revolutionary, Jamal al-Din al-Afghani (d. 1897), one of the most prescient of modern Muslim thinkers, who had travelled and preached across British India.[60] The result was that pro-British Indian Muslim intelligentsia felt psychologically secure when Britain and Turkey were on friendly terms. This, however, became increasingly difficult in the late nineteenth century, as Turkey's European subjects looked up to Britain for moral support in their struggle for freedom. With Turkey becoming the 'sick man of Europe' in the nineteenth century, the strategic interest of Britain in the Levant and the eastern Mediterranean came to the fore.

Sir Sayyid Ahmad Khan had ridiculed the notion of a titular Ottoman caliph having any authority in India and had consistently advised the Muslims to stay loyal to the British.[61] Many of his opponents continued to agitate in favour of Turkey. The British occupation of Egypt in 1882, the Italian conquest of Tripoli in 1911 and the Balkan Wars of 1912–13 were seen by them as acts inimical to Turkey's interests.[62] And when, owing to Turkey's alliance with Germany in the First World War, the British declared war on her, religious passions in sympathy for Turkey were inflamed. Many secular and educated Muslims joined the religious lobbies in favour of pan-Islamism. The Muslim League had remained loyal and quiescent since the victory over separate electorates gained by the 1909 reforms; but in 1912 a so-called 'young Party' captured the league's organization and moved towards greater militancy on pan-Islamism. Some of their leaders, particularly Abul Kalam Azad (1887–1958), Muhammad Ali Jinnah (1876–1949) and two brothers, Muhammad Ali (1878–1931) and Shaukat Ali (1873–1938) wished to reach accommodation with nationalist Hindu leaders in the Congress, in order to create a strong anti-British alliance.[63] Their efforts succeeded in 1916 with the signing of an agreement, the Lucknow Pact, by which the Congress and the League sought to resolve differences over separate electorates and representations in the legislative councils. A formula was worked out, by which the Muslim leaders accepted under-representation in Muslim majority areas, in return for over-representation in Muslim minority provinces.[64] This unity between the Congress and the league, however fragile, forced the government to move towards the idea of a new constitutional settlement after the war; but that was overshadowed by great unrest at the end of the war.

General unrest in India had been caused by the war's pauperization of the Indian economy, rising inflation and a lack of preparations in Punjab for receiving returning wounded soldiers.[65] The government's anti-libertarian and repressive Rowlatt Acts and the massacre of innocent people in the Jallianwala Bagh (garden) at Amritsar in April 1919, brought to the fore the leadership of Mahatma Gandhi in the Congress. His reputation, charisma and vision created the Non-Cooperation Movement of 1920–2, which aimed to de-stabilize British rule in India by means of a non-violent protest on a massive scale. When he heard of the humiliation of Turkey by the terms of the Treaty of Sèvres in 1920, he ardently joined in the pan-Islamist campaign of saving the caliphate.[66] Although the Turkish Caliphate had little to do with India and her problems, Gandhi soon appreciated the emotional impact of his support for the so-called Khilafat agitation on Muslims. His instinct proved correct, when influential Muslim leaders, like the Ali brothers, Abul Kalam Azad and Maulana Hasrat Mohani (1875–1921), mobilized millions of Muslims on Gandhi's side. The two movements of Non-cooperation and the Khilafat now coalesced into one great protest against British rule. Both campaigns ultimately failed, despite heroic efforts. The first one failed because Gandhi, the ardent defender of non-violence, abruptly called off the campaign after violence inevitably ensued, while the other failed because a Turkish Republic was founded in 1923 by Kemal Ataturk who also, shortly afterwards, abolished the Caliphate itself.[67] The rationale for the Khilafat agitation had gone.

One positive outcome that emerged from the short-lived period of amity between the two anti-British movements was the establishment in 1921 of an educational institution, the Jamia Millia Islamia, by nationalist and Khilafatist Muslims with strong support from Gandhi. The two earlier models of Muslim education in this period had been the Aligarh Muslim College and the Deoband seminary. The former, while providing modern secular education, was politically very pro-British at first and then, afterwards, pro-Pakistan; the latter promoted strong Islamic education in a modern setting. The Jamia Millia Islamia offered a third model for Muslim education: an education in secularism, nationalism and Hindu–Muslim unity. Through the dedicated work of its early group of scholars like Dr Zakir Hussain (Vice Chancellor 1926–48) and Professor Muhammad

Mujeeb (Vice Chancellor 1948–73), the Jamia came to attain a prestigious reputation among India's universities.[68]

The Muslim poor in three regions

While the educated urban elites were fighting their political battles with the imperial authorities, a more raw kind of struggle faced some of the Muslim communities across India. This was the struggle of poorer rural Muslims of different regions for socio-economic justice. Three particular examples – from Kashmir, Malabar and the United Provinces – will provide us with useful insights into their diverse social conditions and experiences.

Muslim cultivators in Kashmir

In 1846, after their victory in the first Anglo-Sikh War, the British broke up the Sikh kingdom of Ranjit Singh by annexing a part of Punjab and creating a new princely state of Jammu and Kashmir under the leadership of Gulab Singh, the leader of the ancient Hindu dynasty of the Dogra Rajputs.[69] Few restrictions were imposed on the state, and there was no British Resident until 1859. This was because Kashmir was considered as a safe Hindu buffer state in a region where there was a fear of an alignment between the Russians, Afghans and the local Pathans. This meant that the Dogra dynasty was more or less completely autonomous in the way it arranged its internal affairs. After Gulab Singh (r. 1846–57) came Ranbir Singh (r. 1857–85), followed by Pratap Singh (r. 1885–1925) and finally Hari Singh (r. 1925–49). Under these four Dogra rulers a highly discriminatory regime operated in favour of Hindus and, overtly and covertly, against the vast majority of Muslims.[70] An umbilical alliance bound the Dogras to Kashmiri Brahman Pandits who controlled the great Hindu temples and their wealth. Since a traditional Hindu ruler's legitimacy was derived from his association with and patronage of Hindu temples, the Dogra rulers had to assuage the Kashmiri Pandits with many worldly favours. This could only become possible by exploiting the Muslim majority.

Most Kashmiri Muslims were poor peasant cultivators who eked out a miserable living from the land which was continually being alienated from them by powerful Hindu interests.[71] They suffered from a heavy and unjust revenue squeeze from a class of free-booting contractors and revenue farmers, known as *chakladar*s, specially employed by the Pandits to avoid face to face confrontation with the cultivators. A strict food control system forbade the peasant cultivator to freely export foodstuffs; all rice had to be sold at a low price to the state, in order that it could be sold cheaply to the manufacturers of shawls in the towns. Every year, the peasant cultivators had to provide free labour for work on empty lands controlled by the state or the Pandits. Quite often, while away carrying out free labour, the cultivators faced the loss of their own lands which were, with artful legal means, confiscated by the Pandits for their friends and relations. The Dogra rulers themselves reserved large tracts of empty lands, to be sold off at favourable terms for the benefit of their own clan of Dogra Mian Rajputs. The exception to the general rule of Muslim discrimination was the special favour and status granted to the very wealthy Muslim families of shawl manufacturers in Srinagar and to the Naqshbandi Sufis.[72]

The Dogras also maintained a secret fund, known as Dharmath Fund.[73] Into this fund went money collected as tax from the Muslim cultivators; but the money was diverted to the upkeep and maintenance of Hindu temples and charities controlled by the Pandits. Public money was thus being channelled into private institutions; and yet, at the same time, the salaries of many public servants remained in arrears. The Dogra misrule went unchecked for many years both before and after 1858; but from the late 1870s onwards the imperial authorities became gravely alarmed, particularly after a great famine in 1877–9 in which many thousands perished. With the British becoming more sensitive to Muslim concerns in general throughout the country, the Resident therefore began to interfere more actively in the affairs of Maharaja Pratap Singh; and, in fact, the Maharaja was divested of a number of powers between 1889 and 1905.[74] A commission was set up in 1890 to inquire into the finances of the Dharmath Fund; and eventually a compromise settlement was reached, which placed a strict limit on the fund's expenditure on Hindu charities; part of the money had to be spent on public services and utilities. A more wide ranging commission of inquiry into land ownership in Kashmir was held in the years 1889–95 under the chairmanship of Sir Walter Lawrence whose book *The Valley of Kashmir* provided graphic accounts of the life of Kashmiri peasant cultivators. The Lawrence Commission clarified the terms and classes of land occupancy, which helped the peasants to a certain extent.[75] However, most of the Hindu privileges remained undisturbed, owing to the Pandits' intellectual and legal skills at defending their interests (Excerpt 9.3).

Further British commissions of inquiry did little to ease the hardships of Kashmiri Muslims, but they sparked off in their community a visceral hatred of the Dogras and the Pandits. This hatred, later channelled into a political struggle from the early 1930s onwards, was intensified by the events of 1947–8 which we shall examine in the next chapter.

The Mappillas of Malabar

Oppressed by Dogra rule, the Muslim population in Kashmir came to recognize the British as their saviours. In contrast, the equally disadvantaged Mappilla Muslims of Malabar (modern Kerala) blamed the British for many of their ills.[76] Known as Moplahs in the British period, the Mappillas were the descendants of the earliest of Arab traders who came to the shores of Malabar in the first millennium CE (Chapter 2). They had struck deep roots in the soil of Malabar; and some of them had risen to high positions and acquired wealth during the Middle Ages. Their problems began with the domination of Malabar by the Portuguese from the late fifteenth century onwards. They despised the Portuguese for their religious intolerance and commercial greed.[77] They also felt alienated from the Hindu populace because, first, the Hindu rulers had not supported them in their wars with the Portuguese and, second, the system of land tenure was unfairly structured against them.[78] Land ownership in Malabar was monopolized by the Namboodiri Brahmans who were the *jenmis* or the absolute owners. Under them was another group, the Nayars, who were the leaseholders. The actual cultivators, whether Mappilla or lower-caste Hindus, were allowed to work and survive at subsistence level at the pleasure of the *jenmis* and the Nayar leaseholders. Their constant fear was eviction from the land.

In the last quarter of the eighteenth century, the two Muslim rulers of Mysore, Haidar Ali and his son, Tipu Sultan, came to control the Malabar region; and, for a short period of time, the Mapillas experienced a more secure existence under Muslim rule.[79] However, with the first defeat of Tipu by the British in 1792 (Chapter 8), Malabar became a possession of the East India Company. British rule lasted from 1792 to 1947. The British restored the ancient land tenure arrangements and the *jenmis*, who had earlier fled, returned to their lands. Feeling insecure and threatened, the Mappillas normally resorted to ill-thought-out strategies of violence. More than fifty outbreaks of such violence were reported between 1821 and 1921, many of them against Hindu estates;[80] and successive British rulers adopted harsh and cruel methods to put down the rebellions. In official reports the Mappillas were castigated as fanatics; the almost *jihad*-like fervour with which they rebelled over their rights was never quite understood or appreciated.[81]

The most ferocious Mappilla rebellion against the British broke out in August 1921.[82] For a whole year before that, a heady brew of emotions had been aroused among the Mappillas by the combined rhetoric of Non-Cooperation and Khilafat Movements, led by Gandhi and the Muslim leaders like Ali brothers.[83] The call for self-government and the preservation of the Turkish Caliphate caught the imagination of the Mappillas, but unfortunately Gandhi's call for a strictly non-violent campaign was not easily understood. During the six months that the rebellion lasted, the government nearly lost control of southern Malabar, although it eventually won by deploying the army in the countryside. Many thousands of Mappillas were killed, and plans were made to exile others to places like the Andaman Islands.[84] A major tragedy of the rebellion was that what began as an anti-British struggle also became transformed into an anti-Hindu frenzy as well. Most Hindus in positions of local power or influence took fright at Mappilla violence and refused to join them. Anti-landlordism drove the rebels to seek vengeance on vulnerable Hindu landlords and leaseholders. Gandhi ruefully observed that the Mappillas understood something about the caliphate but nothing about non-violence.[85]

Muslim weavers in North India

There was a time, in the eighteenth and early nineteenth centuries, when Muslim weavers in North India lived relatively well in their *qasbah*s. Decades of colonial economy, however, destroyed their livelihood and sense of well-being. The fine cloths they used to manufacture faced stiff competition from British mills; and market forces dominated the economy of numerous large and small towns in which the weavers predominated.[86] Many of them were forced to migrate to large cities like Calcutta and Bombay to find employment in menial jobs. The vast majority who remained rooted felt trapped and alienated. This had an impact on Hindu–Muslim relations in North India. In the age of colonial capitalism, the merchant, the banker and the money lender, who generally happened to be Hindus, enjoyed a far higher status than Muslim weavers and Hindu and Muslim cultivators. Many of the weavers became highly indebted. On the other hand, they saw the wealthy Hindus attempting to improve their status among their co-religionists by ostentatious forms of religious practices, building temples in Muslim *qasbah*s and displaying much piety.[87] Economic envy contributed to rising tensions between communities. Most British observers and government inspectors were quick to characterize North Indian Muslim weavers as extremely fanatical and warlike, but they seemed to be unable to understand the underlying economic reasons for such behaviour.[88]

Victorian policy makers and social thinkers increasingly categorized Indians in terms of castes; and from 1871–2 onwards the decennial censuses of India reflected that attitude.[89] Caste characteristics became an obsession with British sociologists and anthropologists of this period that was witnessing the rise of what came to be known as Scientific Racism, a pseudo-science.[90] Entire castes and groups of people, sometimes in their hundreds and thousands, were often described as 'martial' castes or 'criminal' castes without any proper evidence. In this situation, the Indian response to the census queries was normally to present their caste in a favourable light to the authorities; they took steps to upgrade their status in every way they could; and in our period a new configuration was shaping up the Hindu caste hierarchy. The North Indian Muslim weavers, originating from the Hindu low castes, had generally been known by the term *julaha* (originally Persian *Julah*, meaning weaver); but they, too, now began to re-brand themselves by shedding their ancestral names with Hindu connotations. Increasingly, names like Momin, Ansari, Sheikh, Siddiqi, Madari or Muhammadi became more fashionable among them. These new names suddenly appear in great numbers in the 1911 census, in place of the collective appellation *julaha*.[91] This may be considered as one variant of Muslim revivalism.

Modern Muslim scholarship and literature

Muslim learning and scholarship in this period reflect the cumulative impact of Western intellectual styles that were a part of the package of modernity that Indian middle classes were buying into under the Raj. This process could be seen at work in the works of Hindu Bengali intellectuals before 1857; it solidified further after 1857, affecting and influencing Muslim intellectuals, with varied and interesting results.

Two patrician scholars

Sir Sayyid Ahmad Khan from the United Provinces and Justice Ameer Ali from Bengal, both descended from old aristocratic families, were the most prominent sages of the modern Muslim intellectual movements of the time. They were the first Muslim scholars steeped in the use of Western tools and methods of learning; they used their skills both to attack and defend their religion, and additionally to educate their co-religionists and give them a voice. Sir Sayyid Ahmad was the senior of the two. He articulated, both in English and Urdu, his social, political and educational views which we have touched upon in this chapter. In Chapter 8, we also referred to his incisive account of the causes of the Indian Mutiny/Rebellion of 1857, which greatly impressed the British authorities. His achievement in producing such a level of historical analysis was due to to his own self-study of modern historical methods and the historiographical traditions of the Islamic world and the modern West.[92] His intellectualism and modernism shone through all his works; and there can be no doubt about the profound influence of his thought and methods on an entire generation of Muslim scholars in the early twentieth century.

Justice Ameer Ali had trained as a lawyer, and reached the pinnacle of his career as the first Muslim member of the Privy Council in London.[93] He wrote critical legal works, such as *The Personal Law of the Muhammadans* (1880) and *The Legal Position of Women in Islam* (1912), but his real influence was felt through his historical and religious works. His *History of the Saracens* went through thirteen editions between 1889 and 1961, while his *Spirit of Islam*, published in 1891, went through nine editions.[94] These two books enjoyed a wide readership both in India and Britain, but his target audience was the Western public. He wished to familiarize this public with the history and religion of Islam; and he was successful at that. He accepted that there were issues that made it problematic for his Western contemporaries to appreciate Islam as a religion suited to the needs of the modern world, but he remained unapologetic on central Islamic beliefs.[95]

The Aligarh circle

The Aligarh circle was a group of Urdu scholars and writers who supported Sir Sayyid Ahmad Khan's vision of a modern Muslim society. Chiragh Ali (1844–95), a lawyer from Hyderabad, argued for the reform of Muslim civil law and the establishment of a humane Islamic law based on the Quran rather than on later accretions and interpretations.[96] Zaka Allah (1832–1910), who had studied at the Delhi College, showed his support for Sir Sayyid's views and the Aligarh project by his work in education.[97] He rose to be an inspector of schools and a principal of a college, wrote a monumental history of India in Urdu and provided an interesting account of the changes wrought in the country under the British.[98] Professor Khuda Baksh, a middle-class scholar from Calcutta, unreservedly accepted Western civilization and, like Sir Sayyid, saw no reason why Islam could not co-exist with modern industrial capitalism. Mohsin al-Mulk (1837–1907), closely associated with Sir Sayyid, was for many years the Secretary of the Aligarh College; and in that capacity he helped to arrange a key meeting of Muslim leaders with the Viceroy in 1906, in the promotion of separate Muslim electoral privileges (referred to earlier in the chapter).

A historian and a poet were two other close members of the Aligarh circle. Shibli Numani (1857–1914), a diligent Urdu historian, specialized in writing biographies.[99] Having travelled much and observed the general decadence of life in the wider Muslim world, he had a keener appreciation of Sir Sayyid's wish to promote modernization among his people. Another influential figure was the poet Altaf Husain Hali (1837–1914), a scholar of Arabic and Persian, who took his first steps in writing Urdu poetry with help from Ghalib, the great master. He came under the influence of Sir Sayyid from 1875 onwards; and the latter, greatly struck by the majesty of his verses, asked him to compose a poem on the fallen condition of Muslims. Four years later, the result was the *Madd o jazr-i Islam*, or 'The Ebb and Flood of Islam', a national poetic treasure, usually called *The Musaddas* after its six-line stanza form. Using powerful imagery, Hali described the drama of Islam, its glories and its tragedies, in simple, sensuous and passionate poetry; and, at the same time, he highlighted the need for the reform of Muslim society.[100] *The Musaddas* became a best-seller in Urdu-speaking India; and it is still read and enjoyed by thousands even today.

We can appreciate the enduring influence of Sir Sayyid Ahmad Khan through some common themes that emerge from the writings of the foregoing six intellectuals of the Aligarh circle. First, all six fully accepted British rule and the reality of both Western material and intellectual superiority. Second, they were firmly of the opinion that, without the use of English language and science, Muslims could not really participate in the affairs of the modern world. Third, they believed that Muslims had to re-interpret Islam in the light of new developments and, while defending

Islam, should attempt to bring Islamic law in line with modern Western law. Fourth, they wanted to secure for Muslims political and civic rights, one of which was the maintenance of Urdu as a key language of government and public discourse. Naturally enough, their successors and followers, in later years, also became the most ardent and idealistic protagonists of the Pakistan Movement.

Urdu literature beyond the Aligarh circle

The scope and spirit of Urdu literature greatly widened during this period. In poetry, Muhammad Husain Azad (1829–1910), a Delhi writer, produced new poems of nature, patriotism, seasons of the year and other subjects, often imitating the best of imageries and descriptions found in English poetry. Akbar Allahabadi (1846–1921), famed for his biting satire, criticized Sir Sayyid Ahmad Khan's zest for Western culture and compared Muslims aping Western lifestyles as 'performing monkeys' for their British trainers (Excerpt 9.4).[101] Muhammad Ismail Meeruthi (1844–1917) gave pleasure to schoolchildren by his descriptive poems in the so-called Urdu readers. His poetic description of Agra fort is considered as the first attempt in Urdu to create a love for architecture through the medium of poetry.[102] A number of Urdu poets specialized in rendering translations from English Literature, particularly the romantic poetry. The poet Ali Haider Tabatabai (1852–1933), for example, translated the English poet Thomas Gray's much loved 'Elegy written in a country churchyard'.[103] The translations of nineteenth-century English poets' works were in great demand at the turn of the century; today we might find their contents over-sentimental.

The works of most of these poets have been overshadowed by those of another major poet, philosopher and political thinker, Muhammad Iqbal (1877–1938). His political philosophy and his support for autonomous Muslim states will be referred to in the next chapter; but essentially he remained a poet, and his poetry began in this period. One of his earliest poems, the much loved *Tarana-i Hind*, or the 'Song of India', is still widely sung (Excerpt 9.5).[104] Sensing a wide cultural and social gap between the Western and Islamic civilizations, he supported Sir Sayyid's ideas on modernity. His poems, *Shikwa*, 'Grievance' and *Jawab-i Shikwa*, 'Answer to the grievance', written just before the First World War, depict the problems faced by Muslims and how God chastises them for their defects.[105] He went on to produce many more profound works in the 1920s and 1930s; but already in 1922, by the end of our period, the British government had recognized his greatness and bestowed a knighthood upon him.[106]

Much good Urdu prose was being written in this period, but three particular developments greatly enriched its range and quality. The first one was the growth in journalism. Although the Urdu press goes back to the late eighteenth century, it began truly to flourish and prosper from the

1870s onwards. Papers like *Akhbar-e Am* in Lahore (1870), *Oudh Punch* in Lucknow (1877–1936) and *Paisa Akhbar* in Delhi (1888) revolutionized journalism by adopting new trends, such as eye-catching headlines, advertisements, cheap price, tabloid lay-out, along with the editorial skills of satirical writing, extemporaneous improvisation and forceful polemic.[107] A literary magazine, *Makhzan*, started in 1901 in Lahore by the distinguished judge and educationist, Sir Abdul Qadir, set a high standard of literary articles and became a nursery of training for budding writers.[108] Anti-British sentiments were commonly expressed in such political papers like the *Zamindar* of Lahore, the *Al-Hilal* and *Al-Balagh*, started by Maulana Abul Kalam Azad in 1912 and 1915, and the *Hamdard* edited by Muhammad Ali.[109]

The second important development in Urdu prose was the birth of the novel. There were three early pioneers in this field. Nazir Ahmad (1836–1912), generally considered to be the first Urdu novelist, concentrated on family themes.[110] Abdul Halim Sharar (1860–1926), from Lucknow, wrote twenty-five novels in all, including many historical romances.[111] He also composed a scintillating account of Lucknow life before and after the overthrow of the Nawab Wajid Ali Shah in 1856 (Chapter 8).[112] Mirza Hasan Mohammed Ruswa (1858–1931), also from Lucknow, wrote highly realistic novels, with the help of systematic plots and characters boldly brought out rather than used as types. His most famous work, *Umrao Jan Ada* (1899), the life story of a Lucknow courtesan, broke new ground in realism. Its didacticism, however, reminds us of the limitations under which these early novelists worked.[113] The popularity of the novel further increased with the cheapness of printing, speedier postal communication and the availability of numerous translations of classic Western works.[114]

Finally, we must record the rise of Urdu drama in this period. The first Urdu play, *Inder Sabha*, 'Indra's Court', was performed in 1853, at the court of Nawab Wajid Ali Shah in Lucknow. Written by Aga Hasan Musawi Amanat (1816–59), it was basically a musical comedy modelled on the European opera, with many fairies, jinns and gods, drawn from such sources as classical Sanskrit drama, Hindu mythologies and Persian poetry and legends.[115] It set the pattern of many early amateurish productions that contained excessive pathos, passion and clowning. The real change in quality came with the works of Aga Hasan Kashmiri (1879–1931) whose plays bore the imprint of a thorough mastery of Shakespeare and his art. In fact, Kashmiri came to be known as Shakespeare-i Hind.[116] Further developments in scene painting, mechanical devices, structure of the stage and stage music, all helped Urdu drama to improve in quality and gain larger audiences. Nevertheless, certain negative factors, such as the extravagances of language, excessive rhyming prose, melodramatic plots, low repute in which actors were held, and Muslim sexual conventions and restrictions, all tended to limit the artistic creativity of the stage.[117]

Muslim 'bhadramahila' *writing from Bengal*

Thus far, we have concentrated on works in Urdu, because it was for Muslims the prime language of literary culture; but we should be mindful of the fact that Islamic literature was at the same time being composed in many other regional languages. One such language was Bengali. The Muslim literary class in Bengal was able to write in both languages, although a certain amount of upper-class snobbery gave Urdu an edge over Bengali. The preference for Bengali arose partly out of admiration for what was being achieved in that language by the educated and upwardly mobile bourgeois Hindu elite, generally called *bhadralok* (for men) and *bhadramahila* (for women).[118] The members of the smaller and traditionally established Urdu-speaking Muslim elite had called themselves the *ashraf* class of people; but, increasingly, the newly educated Bengali Muslims very much wanted to be counted as part of the *bhadralok* society. Like their Hindu sisters, the Muslim *bhadramahila* were also very keen on writing and expressing themselves in Bengali. They were not necessarily Westernized, but they were able to articulate some of the issues of modernity that Muslims faced: issues concerned with family life, patriarchy, colonial identity, *purdah*, polygamy, etc.[119] One of the earliest of these writers, Faizunnessa Chaudhurani (1834–1903), whose work *Rupjalal* was published from Dhaka in 1876, discussed family life and relationships in a sensitive manner.[120] Another, Rokeya Sakhawat Hossein (1880–1932), explored the complexity of the Bengali household and gender roles in her novel *Padmarag* published in 1925, while in her 1922 satirical poem *Appeal* she poked fun at those members of the *bhadralok* who were reluctant to take up the nationalist cause in case they lost their cherished titles granted to them by the British.[121] A third figure, Akhtar Mahal Syeda Khatun (1901–29), explored romance outside marriage in her novel *Niyantrita*.[122] Some of the *bhadramahila* women also went into journalism.

Select excerpts

9.1 Sir Sayyid Ahmad's injudicious remarks on British and Indian character

Sir Sayyid was so carried away by the marvels of British civilization that, for once, he lost all sense of balance in his comments. In view of his rude and insulting remarks about the Indians, as expressed below, it was no wonder that many of his fellow countrymen and co-religionists considered him a sycophant of the British. This reputation stuck to him until the end despite the great respect and deference shown to him from the highest to the lowest in nineteenth-century India.

> In India we used to say that the British behaved extremely rudely to Indians, and that they look upon Indians as no more than animals. (But) this is what we really are. I tell you without exaggeration and with complete sincerity that Indians . . . rich and poor alike . . . from the most learned and accomplished to the most ignorant, when one compares them with the education and training and cultured ways of the British, bear the same resemblance to them as a filthy wild animal does to an extremely talented and handsome man. And do you think any animal deserves to be treated with honor and respect? So we have no right (even if we have reason) to object if the British in India look upon us as wild animals . . . The British behave extremely badly in their dealings with Indians, and they ought not to. This is not because I think that the Indians are so cultured that they deserve better treatment, but because when the British, with all their culture, behave like this they discredit their own culture and education . . .
>
> (Russell 1995: 188–9)

9.2 Mahatma Gandhi's feelings on Morley–Minto Reforms

In her Journal of October 1932 Lord Minto's wife, Mary, has left us an interesting little account of her conversation with Gandhi on the Morley–Minto Reforms.

> When Mr. Gandhi attended the Round Table Conference in London, the Maharajah of Bikanir brought him up to me at a party to be introduced.
>
> 'Do you remember my name?' I asked.
>
> 'Remember your name!' exclaimed Mr. Gandhi. 'The Minto–Morley Reforms have been our undoing. Had it not been for the Separate Electorates then established we should have settled our differences by now'.
>
> 'You forget, Mr. Gandhi', I replied, 'that the Separate Electorates were proposed by your leader and predecessor, Mr. Gokhale'.
>
> 'Ah', said Mr. Gandhi with a smile, 'Gokhale was a good man, but even good men may make mistakes'.
>
> 'Yes', said a Punjab landowner now on the Council at the India Office, who was standing near, 'and if Lord Minto had not insisted on Separate Electorates we Mohammedans should not be in existence'.
>
> (Minto 1934: 20, n.1)

9.3 Kashmiri Pandits condemned

Walter Lawrence was a fair and impartial surveyor of the Kashmir scene. While appreciating the useful role that the Pandits had played in the affairs of a state where standards of official morality were generally speaking non-existent, he could not but censure their behaviour.

> In a country where education has not yet made much progress it is only natural that the State should employ the Pandits, who at any rate can read and write with ease. They are a local agency, and as they have depended on office as a means of existence for many generations, it is just and expedient to employ them. Still it is to be regretted that the interests of the State and of the people should have been entrusted to one class of men, and still more to be regretted that these men, the Pandits, should have systematically combined to defraud the State and rob the People . . . In recent times there were few Pandits who were not in receipt of pay from the State, and the number of offices was legion. But though this generosity in the matter of official establishments was an enormous boon to the Pandit class, it was a curse and misfortune to the Musulmans of Kashmir; for the Pandit does not value a post for its pay, but rather for its perquisites, and every post in the valley was quickly made a source of perquisites.
>
> (Lawrence 1895: 401)

9.4 Biting satire of Akbar Allahabadi

Akbar Allahabadi's irreverent and caustic verses provide us with the measure of the man: an independent, critical thinker who was not afraid to speak his mind about the negative effects of colonialism on the Indian mind and character.

On Sir Sayyid Ahmad Khan's fawning of the British

> Sir Sayyid had an intellect that radiated learning
> And strength enough to vanquish any foe you care to mention
> And I for one would readily have counted him a prophet
> But that there never was a prophet yet who drew a pension.

On the British impudence to accuse Muslims of spreading Islam by sword

> You never ceased proclaiming that Islam spread by the sword:
> You have not designed to tell us what it is the gun has spread.

On the Westernized elite's strong attachment to European civilization

Today when my petition was rejected
I asked the Sahib, feeling much dejected,
'Where shall I go to now Sir? Kindly tell'.
He growled at me and answered 'Go to Hell!'
I left him, and my heart was really sinking;
But soon I started feeling better, thinking,
'A European said so! In that case
At any rate there must be such a place!'

On self-important Indians

Look at the owl! What airs and graces! What a way to talk!
Because the British told him he is an honorary hawk!
(Russell 1995: 202, 204–6)

9.5 *Tarana-i Hind*

Among the many patriotic songs of this period three have become most well known. The first one, 'Bande Mataram' (Salutation to Mother (India)), was composed by the Bengali writer, Bankim Chandra Chattopadhyay, in 1882. Some of the Muslim leaders objected to the song on the grounds that only God can be the object of veneration, not a country. Hindu nationalists have been its strongest protagonists and, were it not for opposition from Pandit Nehru, the song might have been adopted as the national anthem of India in 1947. In the event, the song chosen for the national anthem was the 'Jana Gana Mana', written in 1911 by Rabindranath Tagore. If there had been no campaign for Pakistan, it is more than likely that the song adopted for the national anthem of united India might have been the 'Tarana-i Hind', written by Mohammed Iqbal, in 1911. The haunting two lines of its first verse in Urdu have a magical effect on the listener:

One of the numerous English translations of the song is as follows:

Better than the entire world, is our Hindustan,
we are its nightingales, and it is our garden abode.
If we are in an alien place, the heart remains in the homeland,
know us to be only there where our heart is.
That tallest mountain, that shade-sharer of the sky,
it is our sentry, it is our protector.
In its lap frolic thousands of rivers,
whose vitality makes our garden the envy of Paradise.
Oh, waters of the Ganges, do you remember that day,
when our caravan first disembarked on your waterfront?

Religion does not teach us to bear ill-will among ourselves
We are of Hind, or Homeland is Hindustan.
In a world in which ancient Greece, Egypt, Rome have all
vanished
our own attributes (name and sign) live on today.
Such is our existence that it cannot be erased
Even though, for centuries, the cycle of time has been our enemy.
Iqbal! We have no confidence in this world
What does anyone know of our hidden pain?

(Iqbal 1911: www.enotes.com/topic/Saare_Jahan_Se_Achcha)

10

UNITY OR SEPARATISM?

Muslim dilemma at the end of the Raj (1924–1947/8)

The struggle for independence led by Gandhi and the Indian National Congress dominated the history of the last twenty-five years of British rule in India. Parallel to this struggle, there took place a gradual shift of opinion among Muslims towards separatism. Yet a closer study of the period tells us that separatism was not inevitable. Until the end of the 1930s, it was almost inconceivable that there could be a separate Muslim nation; most political efforts were directed at accommodating Muslim interests within the wider body politic of India. The Second World War, however, changed the situation. What appeared in 1939 a possibility became inevitable by 1945 and a certainty by 1947. Then, out of the maddening confusion of the years 1945 to 1947, there came about a double partition of the subcontinent with its human tragedies. This momentous story of the twentieth century, told in the first two sections, is the centrepiece of this chapter. It is a depressing story, the origins of which lie in historical circumstances, human follies and lack of empathy on the part of principal players. The mood will be lightened in the last two sections that tell a positive story of Muslim women and literary personalities.

From Khilafat to Pakistan: a political narrative

At the end of the First World War, Gandhi's Non-Cooperation Movement and the pan-Islamic Khilafat Movement had forged a coalition against British rule (Chapter 9). At the end of the Second World War, the Indian National Congress and the Muslim League were set on a path of complete collision. The circumstances that led to this transformed Indian political scene were shaped by a variety of prominent individuals from all sections of Indian society: here we introduce a selection of Muslim leaders and groups.

Muslim actors in the drama

The highly sophisticated, rational and secular lawyer and politician, Mohammad Ali Jinnah (1876–1948) was, after Gandhi, perhaps India's second most influential political figure. Instinctively a nationalist, he had first joined the Congress through his circle of legal friends. He had been inspired by men like Dadabhai Naoroji (1825–1917) and Gopal Krishna Gokhale (1866–1915) who had formulated the Congress's moderate anti-imperial programme.[1] He remained a Congress member until well after he joined the Muslim League in 1913. As president of the league's session in Lucknow in 1916 he helped to draft the Lucknow Pact of that year with the Congress (Chapter 9). This agreement doused communal passions during and, for a brief period, after the war; and Gokhale hailed him as the ambassador of Hindu–Muslim unity.[2] Jinnah's disenchantment with the Congress began with the rise of Gandhi in that organization from 1920 onwards. His coolness towards Gandhi might have stemmed from his sceptical and detached attitude to what he considered to be Gandhi's generally pietistic and patronizing tone of language; Gandhi's use of Hindu religious symbolisms had no place in Jinnah's secular vocabulary.[3] It was, however, not Gandhi who made Jinnah truly alienated from the Congress; it was a whole group of right-wing Hindu communalists who drove him towards defending Muslim interests. He did not abandon his nationalistic fervour, but increasingly demanded the recognition of Muslim political concerns. It was not a call for Pakistan, a term that first appeared in a shortened form, 'Pakstan', in a pamphlet *Now or Never: Are We to Live or Perish for Ever?* written by a young Cambridge-based Punjabi, Choudhary Rahmat Ali (1897–1951) and three associates in early 1933. Jinnah showed indifference to their radical plan for Muslim-majority provinces to combine into one federal state whenever the British left India.[4]

It took three decades for Jinnah's position to evolve from that of a Congress member to a Muslim separatist. Indeed, most of the Muslim leaders of this period were inclusivists rather than separatists. Some, deeply attached to secular nationalism and liberalism, had joined the Congress even before the Khilafat Movement.[5] Drawn from different professions, they were united by a common goal of resisting British colonial rule. The most prominent early leaders of this group – the two brothers Mohamed Ali (1878–1931) and Shaukat Ali (1873–1938), along with Zafar Ali Khan (1873–1956), Hasrat Mohani (1877–1951) and Choudhury Khaliquzzaman (1889–1973) – left the Congress only after a long period of disenchantment over the unwillingness of official Congress leaders to countenance further electoral concessions to Muslims in Muslim-majority provinces.[6] Yet others, for whom any idea of Muslims being a separate national entity was complete anathema, remained loyal to Congress throughout this period and beyond. The most distinguished among them were Hakim Ajmal Khan (1863–1928),

Dr Mukhtar Ahmad Ansari (1880–1936), Gandhi's devoted physician, the educationist Dr Zakir Hussein (1897–1969), Justice M.C. Chagla (1900–81) and the writer, thinker and educationist Abul Kalam Azad (1888–1958). The last three went on to achieve high office in the post-1947 Indian Republic.

Apart from individuals, large Muslim organizations and parties were also affiliated to the Congress. The most dynamic of these groups, Khudai Khidmatgar (Servants of God), included many Pathans of the North West Frontier Province (NWFP), led by Khan Abdul Ghaffar Khan (1890–1988), Gandhi's earnest disciple.[7] The Pathans, stereotypically famed for revengeful blood-feuds and inter-tribal warfare, were inspired by Abdul Ghaffar's personality and force of character. His insistence on non-violence, self-discipline and principled resistance to exploitative landlords and colonial authorities created a new image of the 'Pathan Unarmed'.[8] He and his party, dressed in their red shirts, and believing in pristine Islamic ideals of simplicity and honesty, successfully helped to implement the Congress-led 'No Rent' campaign of 1930 in the NWFP. Other not so well known organizations, such as the Ahrar Party, the Momin Group, the Indian Majlis, etc., formally supported the Congress but did not necessarily agree with its national-secular approach. Their religious views were in fact quite reactionary.[9] Their support for the Congress fell away in the late 1930s and the 1940s, owing to both the persistent propaganda of the revived Muslim League and the neglectful attitude of the Congress towards its Muslim constituency.

There were many provincial Muslim leaders who, secure in their regional power bases, wished to maintain their own independent line of thinking and course of action. Although both Punjab and Bengal were two of the largest Muslim-majority provinces, neither the system of separate electorates secured in 1909 nor any gains made by Muslims through the Lucknow Pact of 1916 could help to produce an overall Muslim majority in their legislatures. It was, therefore, incumbent on the provincial Muslim leaders to reach out to other communities in order to run their administrations and, at the same time, to keep on the right side of imperial authorities. Thus in Punjab, the leaders of the Unionist Party, men like Fazl-i Husain (1877–1936), Sir Sikander Hayat Khan (1892–1942) and Khizr Hayat Khan Tiwana (1900–75) looked for support from Hindu and Sikh legislators; and, in Bengal, Fazl ul-Haq (1873–1962), Khwaja Nazimuddin (1894–1964) and Husain Suhrawardi (1893–1963) could not form successful provincial governments without non-Muslim support. In any case, some of the strong-minded provincial leaders, like Sir Sikander Hayat Khan in Punjab and Khan Bahadur Allah Baksh Soomro (1900–43) in Sind, sincerely believed in the correctness of forming creative coalitions rather than keeping out minority representatives from their provincial cabinets.[10] Such leaders had to play a subtle game of their own, alternately opposing and favouring the British, the Congress and the Muslim League at different times, with a view to securing and preserving their own power.

From the Islamic religious standpoint in general, there was little consensus about creating Pakistan.[11] Two groups initially argued strongly against separatism. One, led by the Principal of the Deoband seminary, Hussain Ahmad Madani (1879–1957), who belonged to Jamaat al-Ulama, an organization linked to the Congress, believed that the creation of Pakistan would destroy national solidarity. Being closely allied to the Congress Muslim leader, Abul Kalam Azad, Madani strongly favoured inclusion rather than separatism.[12] He failed in his efforts, because Jinnah was able to persuade other influential *ulama* at Deoband, such as Ashraf Ali Thanwi (1863–1943) and Shabir Ahmad Usmani (1886–1949), to provide theological legitimacy to the Pakistan project. As a result, a large number of *ulama* eventually migrated to Pakistan; many others, however, including Madani, stayed on in India. Another group, not trained as *ulama*, opposed Pakistan for a different reason altogether. Their main concern was with the nature of the future state of Pakistan. They wanted it to be an Islamic state, not a mere state for Muslims. They were not convinced of the Islamic credentials of highly Westernized, secular leaders like Jinnah or his great acolyte, Liaquat Ali Khan (1895–1951). Their doubts were eloquently articulated by Abu Ala Mawdudi (1903–79), one of the twentieth century's greatest exponents of Islamist ideas.[13] Through a monolithic religio-political organization, the Jamaat-i Islami, started in 1941, Mawdudi fought against both Madani's vision of Muslim–Hindu solidarity and Jinnah's vision of a Muslim state of Pakistan. He, too, did not succeed; and ultimately he and other radical Islamists left for Pakistan.

The Second World War saw a decisive shift in favour of separatism, with pro-Congress Muslim nationalists going on the defensive. The idea of a separate Muslim nation looked attractive to many different groups. The professional Muslim salaried class, consisting of high government officials, could see the possibilities of gaining advantageous offices and posts in a new state.[14] The landlords of Punjab and the United Provinces, anxious to preserve their feudal privileges and taking fright at the left-wing language of some of the Congress socialists, could see themselves as natural leaders of a new Muslim state. On the other hand, Muslim peasants were being promised by the Muslim League that they would be liberated from the tutelage of Hindu money lenders and *zamindars*.[15] Muslim commercial and industrial interests, represented by the Ispahani, Haroon and Adamjee families, could appreciate the monopolistic advantages for themselves in a new state, without the fierce competition from the well established non-Muslim conglomerates of Tatas, Birlas and Dalmias.[16] Increasingly, during the war, commercial and financial men muscled in and secured key positions within the league hierarchy.[17] Progressive women, too, came to be affected by separatism. For many decades they had fought both for freedom from

British rule and justice for their gender with their non-Muslim sisters; yet by the 1940s some were making Muslim separatism their first priority. Jahanara Shah Nawaz (1896–1979), the daughter of the influential Punjabi politician, Sir Mian Mohammed Shafi (1869–1932), and a woman who had been at the forefront of meetings and progressive resolutions at the All India Women's Conferences during the 1920s and 1930s, suddenly lost interest in Indian issues and started persuading Muslim women to support the idea of Pakistan.[18]

Jinnah's evolving strategy: 1924 to 1945

It is most unlikely that, without Jinnah, Pakistan could have come into existence. His shrewdness, patience and political skills enabled him to create single-handedly a new Muslim nation in 1947. This did not mean that he had a fixed political goal which he pursued unrelentingly. He was flexible in his thinking and was prepared to make compromises; but circumstances played a great part in shaping his eventual hard-line position.[19] In order to appreciate the changes in his strategy, we need to divide the two decades of his career between 1924 and 1945 into three stages. The first stage covered the years 1924–9. In this post-Khilafat period, when the relations between the Congress and the more radical Muslim leaders were rapidly deteriorating, Jinnah was nevertheless working for solutions within the spirit of Khilafat. Thus, for example, he supported the Congress boycott of the Simon Commission of 1927 sent by the imperial government to report on the constitutional progress since the Montagu-Chelmsford reforms of 1919 (Chapter 9).[20] He also persuaded the Muslim League to offer the Congress an olive branch in order to reconcile the two parties. The offer was to replace the separate electorates with joint electorates, in return principally for an increase in the number of Muslim seats in the legislative assemblies of Muslim-majority provinces, and with one-third of the seats in the Central Legislative Council to be reserved for Muslims.[21] This was a generous offer, and the proposal to scrap separate electorates was indeed a major Muslim concession; unfortunately, the offer remained unappreciated by the increasingly Hindu-revivalist dominated Congress of that period; and the Nehru report of 1927–8 rejected Jinnah's offer.[22] A 'parting of the ways' now took place between Jinnah and the Congress and also between him and those Muslim nationalists who continued to support the Congress.[23]

The second stage covered the years between 1929 and 1940, when Jinnah moved away from fruitless negotiations with the Congress over separate electorates and reserved seats and argued for a form of federalism. In what are known as Fourteen Points of March 1929, he envisaged the five Muslim majority provinces of British India – Punjab, Sind, Bengal, Baluchistan and the North West Frontier Province – becoming autonomous parts of an Indian federation.[24] Relations with the Congress rapidly cooled, and severe disagreements between Jinnah and Gandhi were exposed before and during the British-sponsored Second Round Table Conference of 1931.[25] An opportunity arose for both Indian parties to come together when, following the Round Table Conferences, the British government came forward with a comprehensive Government of India Act of 1935.[26] The Act addressed the concerns of both the Congress and the Muslims, but it also envisaged the possibility of a future Indian federation. It was a bold and measured plan of action that revealed the best of British statesmanship. Unfortunately, the Congress paid scant attention to the Act; and Gandhi on his own admission later regretted that he had not even bothered to read it.[27] The surprising thing was that Jinnah also remained equivocal, perhaps for fear of being branded a British stooge. If he and the Congress leaders had worked together to make a success of the federal proposals, Indian politics would have been set on a more hopeful trajectory.

Jinnah's decisive leadership and political skills became more noticeable after the provincial elections of 1937 had ended disastrously for the Muslim League. He vigorously resisted the so-called Congress Mass Contact campaign to woo back those Muslims who had withheld their support to it in the elections, and started his own campaign to mobilize Muslim masses behind the Muslim League.[28] His unequivocal support for Britain at the beginning of the war also stood out in marked contrast to the negative attitudes of Gandhi and the Congress leaders. In appreciation of this gesture, Viceroy Lord Linlithgow declared on 1 November 1939 that all institutional progress in India from then on would need the cooperation of both the Congress and the Muslim League.[29] By one master stroke of diplomacy, Jinnah had secured almost a veto on future constitutional plans for India. Encouraged, Jinnah next persuaded the Muslim League to adopt the so-called Lahore Resolution on 24 March 1940.[30] While leaving vague any particular territorial definition or any mention of the term 'Pakistan', the Lahore Resolution explicitly demanded the near-independence of Muslim-majority provinces in the event of British departure from India.

The third stage covered the years 1940 to 1945. Along with declaring support for the British war effort, Jinnah showed political adroitness in his dealings with the government throughout these war years. When the ambivalent offers of the Cripps Mission of March 1942, sent by the British government to assuage Indian national feelings and keep India quiescent at a time of low British war morale, were rejected by Gandhi as a 'post-dated cheque on a failing bank', Jinnah adopted the strategy of using negotiations as a means of impressing upon the British the inevitability of separatism, before finally rejecting Cripps's offers.[31] He acted astutely by standing aside when the Congress organized the massive civil disobedience campaign called by Gandhi under the name of 'Quit India Movement' with a catchy slogan of 'Do or Die'; in doing so it handed Jinnah an advantage that he used effectively until the end of the war. With Gandhi and all the leading Congress leaders languishing in jail, Jinnah re-positioned himself as the leader of all Muslims, able to put pressure on the non-League provincial Muslim leaders to accept him as the 'sole spokesman'[32] on behalf of Muslim interests. The war years proved most useful to Jinnah, as it was then that he secured the support of many different groups of Muslims, mentioned earlier, for his policies.

By the end of the war, the British, hard pressed militarily and financially, were faced with an agonizing dilemma of whether to leave India as one integral nation or divide it up, basically for the benefit of Muslims but, in the long run, also for the benefit of the princely states that wished to become independent. Their instinct was to leave India united; and they had devised an ideal federal formula in the 1935 Government of India Act. Since the elections of 1937, however, the acrimony between the Congress and the Muslim League had put paid to the success of the formula. The failed Cripps Mission was followed up with three further attempts to bridge the gap: the so-called Rajagopalachari Formula of 1943, the Gandhi-Jinnah meetings of 1944 and the Simla Conference called by Viceroy Lord Wavell in June 1945.[33] Each attempt failed, partly because of the lack of imaginative thinking on the part of the Congress and partly because of Jinnah's increasing obstinacy. The one thing that did happen was the calling of the provincial elections in December 1945. The results were excellent for Jinnah, because the Muslim League won all the reserved Muslim seats, in contrast to the 1937 elections. It still could not muster sufficient majority in the two provinces that mattered: Bengal and Punjab, where coalitions had to be formed;[34] but there was no doubt that Jinnah's star was now in the ascendant, and his hard work during the war had paid off. Both the Congress and the British government could not be under any illusion about his final goal of establishing separate Muslim states (cf. the Lahore Resolution) amounting to Pakistan within the federal system of India.

From adroitness to obduracy: Jinnah's bitter-sweet triumphs

For twenty years Jinnah had navigated through the rough and tumble of politics with both flexibility and patience. By 1945, he had behind him a vast mass of Muslim population ready to follow him wherever he wished to lead them. He had also pressured the Congress leaders into giving serious consideration to the idea of a federation that he had been championing since 1929. This was now the moment when he needed to pause and show a degree of generosity and empathy. He needed to make it clear to Gandhi, Nehru and all other Congress leaders that his demand for federation did not mean complete separation or the break-up of the subcontinent. He failed to do that because he had become far too obdurate and obsessive. His vision of a federation was one in which a central government with limited powers would simply oversee the provinces, and Muslim-majority provinces had to have the right to secede. No Congress politician could countenance such a scenario. That was why a special Cabinet Mission of May 1946 under Lord Pethick Lawrence, which merely hinted at the provincial right to secede, was rejected by the Congress.[35] Jinnah's obduracy reached dangerous proportions when he called on his supporters to observe a so-called Direct Action Day on 16 August 1946. This was a truly unwise decision, because on that day thousands of people engaged themselves in inter-communal disharmony. Initially, unthinking Muslims in Calcutta began attacking Hindus but, ultimately, many more innocent Muslims became victims of Hindu violence during the four days of massacres (16–20 August 1946).[36] Jinnah had not only acted out of character; he was also out of his depth since, unlike Gandhi, he had never truly led a mass action campaign which he could properly control. Until this point in time, despite all the disagreements, there was still hope that a compromise might be found between the Congress and the Muslim League; but something more sinister would now prevent that. Although the Viceroy was able to get both organizations to work together in an interim government in September 1946, a new factor had entered into the equation: the spectre of communal violence on a mass scale.[37]

Upon his arrival in India in March 1947, the last Viceroy, Lord Mountbatten (1900–79), was confronted with two clearly opposed positions among Indians. The Congress wanted power to be transferred to a united federal government in Delhi which, after the new constitution had been framed, might devolve certain powers to the provinces. Jinnah and the Muslim League now demanded the most extreme form of separatism: the creation of a new Muslim state. They wanted a Pakistan to be carved out of Muslim-majority provinces, and to be treated as an equal partner of what he termed 'Hindustan' (Hindu-majority provinces), within what would be a federation in name only. The Congress position was more appealing to Mountbatten, in view of the fact that the British strategy for the future had been predicated on friendship with a united India (Excerpt 10.1). At the same time, he could not ignore Jinnah without triggering off a massive civil conflict that might destroy Britain's reputation on the eve of her departure. However, as he progressed with his interviews with Indian leaders during April and May, 1947, two particular facts helped him come to a final resolution. First, despite the official clarion calls by the Congress for a united India, it was clear that some of its key leaders (although not Gandhi) were no longer averse to a division of India. Increasingly, they felt that the turbulence caused by a forced integration of intransigent Muslim-majority areas into India would hinder the peaceful progress of the country after independence.[38] Pakistan was therefore the price that they were prepared to pay. But not the Pakistan that Jinnah was proposing, as Mountbatten also came to understand the second vital fact: that Muslim majority in the provinces of Punjab and Bengal was not overwhelming and that there were large concentrations of Sikhs and Hindus in the former and of Hindus in the latter. In Punjab, particularly, the Sikhs showed a threatening militancy that alarmed Mountbatten.[39]

Mountbatten's first proposal for independence, commonly known as Plan Balkan, was rejected by Nehru, the principal Congress leader, on account of lack of clarity over the issue of provincial secession.[40] Mountbatten was a man in a hurry, and acted with undue haste;[41] he simply jettisoned the Plan Balkan and, on the advice of a senior Indian civil servant, V.P. Menon (1894–1966), opted for a simple solution based on the creation of two dominions: India and Pakistan. This was the real Mountbatten Plan of 3 June 1947 which brought forward independence from 30 June 1948, the date originally proposed by the British government, to 14/15 August 1947.[42] The reality of an independent state of Pakistan was accepted under this plan which required the partition of British India. However, the two great

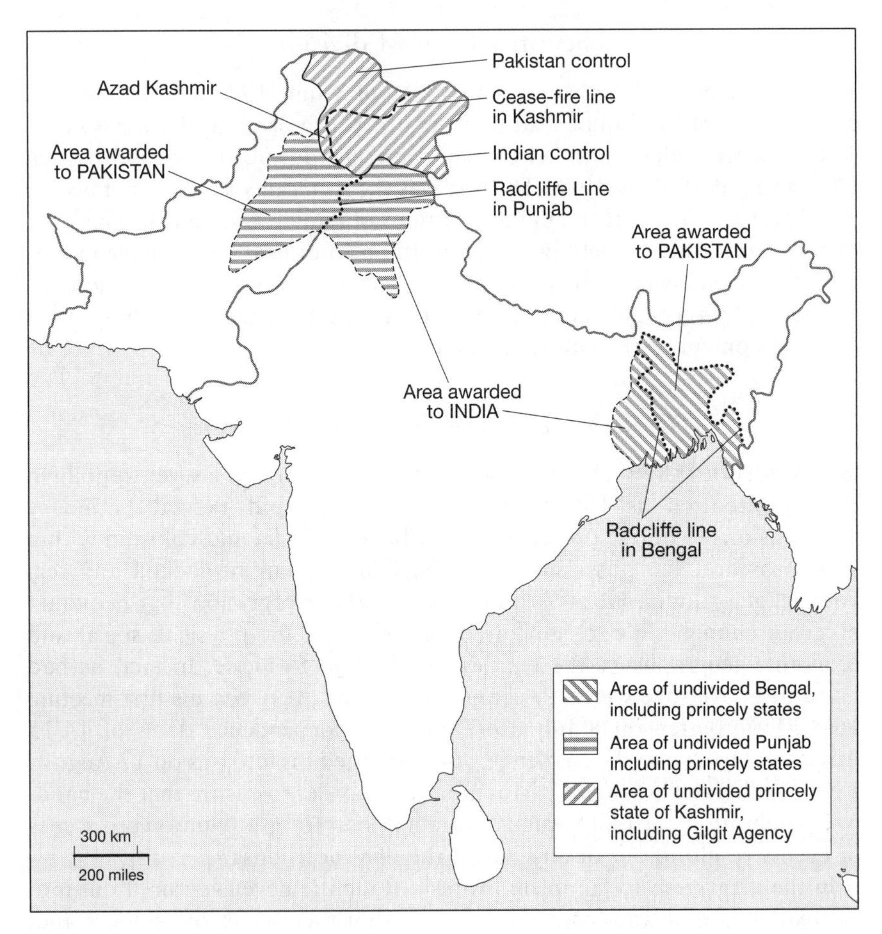

Map 10.1 Dividing lines of 1947–8

Muslim-majority provinces of Punjab and Bengal had also to be partitioned in order to satisfy their large non-Muslim populations (Map 10.1). It was difficult for Jinnah to reject the Mountbatten proposal, first, because he could not deny to large concentrations of non-Muslims in Punjab and Bengal what he had been asking for the Muslims generally: communal separation; and, second, his Muslim League had at that time only a tenuous authority in the provincial assemblies of the two provinces. Reluctantly, therefore, he had to accept a 'truncated' Pakistan.[43] In doing so, he of course abandoned millions of Muslims living in Hindu-majority areas to their fate in the new India (Excerpt 10.2). The Congress leaders had reason to feel satisfied to the extent that they secured British India minus Pakistan in one whole piece at least; but in their impatience they betrayed the hopes of those Muslims who had fought for a unitary state.

The bitter fruits of division

The endorsement of the Mountbatten plan of 3 June 1947 and the creation of two new states did not lead to an era of peace or amity between them. Two great tragedies engulfed the subcontinent during the second half of 1947 and the whole of 1948. The first was the suffering of millions of people who became refugees through the partitions of Punjab and Bengal. The story of their fate, today indelibly etched in the memories of their descendants, has been movingly described in numerous accounts, historical and fictional.[44] The second tragedy, affecting the Kashmiri people, arose out of the conflict over the control of their princely state.

The Radcliffe Awards and the refugee crisis

Cyril Radcliffe (1899–1977) was a highly regarded British lawyer, appointed by Mountbatten as Chairman of the Punjab and Bengal Boundary Commissions to decide on the boundary between India and Pakistan within each province. He possessed a fine legal mind, but he lacked any real knowledge of India. He took on the task in the expectation that he would be given enough time to familiarize himself with the physical, social and economic geography of the regions that he had to incise. In fact, he had just a little over five weeks to complete the work, between his first meeting with Mountbatten on 8 July 1947 and the independence dates of 14/15 August.[45] He rose to the challenge, and published his findings on 17 August, a date deliberately chosen by Mountbatten in order to ensure that the handover celebrations of 14/15 August were not marred by any untoward scenes or events resulting out of anger over the final decisions.[46]

In the great rush to complete his task, Radcliffe never ventured out into the real world; he and his commissioners shut themselves off indoors and, poring over innumerable maps of varied scales and out of date census figures, began to draw the line section by section, for both Punjab and Bengal. The main criterion in determining the line was the population ratio between Muslims and non-Muslims; but there was inconsistency in its application. In most cases, entire districts were handed over to Pakistan on the basis of an overall Muslim majority, yet in some other cases a particular *tehsil* or a sub-district was itself partitioned for 'other factors'. In the case of Lahore, for example, part of its *tehsil* of Kasur was handed to Indian East Punjab, possibly to minimize disruption to railway communications and water systems. On the other hand, the whole district of Amritsar, with only a bare majority of non-Muslims (53.5 per cent) was allotted entirely to East Punjab, although its northern *tehsil* of Ajnala had a Muslim-majority population. The religious-demographic principle retained its primacy in the Radcliffe team's calculations and considerations. With a few exceptions, the geographical, environmental, historical, industrial, agricultural or emotional considerations were ignored. Lahore had 600 *gurdwara*s and was a bastion of Sikh culture, but it had a small Muslim majority; and so it went to Pakistani West Punjab.[47] Calcutta's jute industry secured raw jute from the Muslim-majority hinterland, but it went to Indian West Bengal because of the Hindu majority in the city. Radcliffe received numerous deputations from different communities pleading their case for inclusion in one of the two parts in each region; but all exchanges were deliberately kept secret from the public. There has been much speculation as to whether Radcliffe was at all under discreet pressure from Mountbatten or Nehru to favour new India's geographical and strategic interests concerning Kashmir, by incorporating certain key salients, like Firozpur and Gurdaspur, in East rather than West Punjab;[48] this alleged collusion has been difficult to prove, however.

While Radcliffe and his team fulfilled their remit in the most difficult of circumstances (Excerpt 10.3), no one could be in any doubt as to the human consequences of their judgements. The massive refugee crisis unleashed by them was the culmination of a great tragedy in the making, as soon as Mountbatten's Partition Plan of 3 June had been accepted. Also, a great scramble for the division of assets of British India began in earnest. Everything, from the most trivial to most essential, was to be divided on the ratio of 80 to 20 for India and Pakistan.[49] Muslim soldiers were being transferred to Pakistani territory at a short notice of just four hours.[50] Thousands of Muslims working for the Indian Civil Service were asked to make a precipitous and momentous decision to opt for India or Pakistan; and once they had opted for Pakistan, it became well nigh difficult to reverse their decision.[51] The top and the upper-middle echelons of civil servants, particularly in the United Provinces and North India generally, had for some years been supporters of Jinnah and the Pakistan Movement; and they had little physical or psychological difficulties in making the decisive break. It was their humbler colleagues who experienced the greatest pain in making the transfer.

Even before the secretive Radcliffe decisions were finally published on 17 August 1947, a great movement of refugees had begun across the new international boundaries. One hundred thousand people had already moved internally within Punjab for safety amidst their relations, unsure of where the frontier would be drawn.[52] In both Punjab and Bengal, desperate people were making their own moves. The rush of people turned into a mighty torrent after 17 August; for example, just between 18 September and 29 October 1947, 849,000 refugees entered India from Pakistan.[53] By November 1947, 8 million people had crossed the border in both directions, while one in ten in Pakistan was a refugee. Unimaginable cruelty and barbarity of some of the people on both sides have been narrated in graphic detail in a variety of historical and literary works. The looting of abandoned property, the arson attacks on trains, the raping and hijacking of women, ethnic cleansing of villages and parts of urban areas were some of the ugly manifestations of people – Hindu, Muslim, Sikh – devoid of humanity and sick in mind. On the other hand, there were countless examples of compassion, personal heroism, charity and mercy shown by people from all communities. Both India and Pakistan lost valuable human resources.[54] Pakistan suffered by the departure of Hindu and Sikh merchants, bankers, entrepreneurs and diligent clerks; India lost its stalwart Muslim railwaymen, weavers, craftsmen, agriculturists, scholars, professionals and administrators. Inter-ethnic tensions arose out of refugee presence. Karachi was in complete turmoil. On the one hand, the native Sindhis resented the Urdu-speaking *muhajirs*; on the other hand, there was a massive exodus of Hindus from the city.[55] In Gujarat the arrival of thousands of Hindu Sindhis generated tensions over properties etc.; while, in Delhi, the Sikh refugees did not enjoy the warmest of welcomes. Communal-minded organizations, like the Jamaat-i Islami in Pakistan and the RSS and the Mahasabha in India, also took advantage of the refugee crisis and created a toxic situation in local politics.

The dispute over Kashmir

The intensity of the refugee crisis was felt severely in the short term; but, while countless families and their descendants among Muslims and non-Muslims suffered long term psychological damage, both India and Pakistan at least managed to absorb and settle in some manner, however inadequate, those who had arrived from across the borders. The immediate crisis mercifully passed by the early 1950s. In contrast, the dispute between the two countries over the former princely state of Kashmir has dragged on until today; and no solution seems to be in sight. It has been painfully difficult for policy makers of both India and Pakistan to shake off the ghosts of 1947 and 1948 and begin anew.

Countless volumes and millions of words have been written about the Kashmir issue,[56] but here we shall confine ourselves to sketching the outline of the story before reflecting on some attitudes of the main protagonists in the dispute. The Mountbatten Plan of 3 June 1947 (referred to earlier) was essentially concerned with the partition of British India on the basis of religion. It was also generally expected that the 584 princely states of India, enjoying autonomy under British 'Paramountcy' since the great rebellion of 1857/8, would sooner or later have to join one of the two new states that were to be created on 14/15 August 1947. For some years, the princes of the largest states had been engaged, with discreet help from some senior British civil servants, in a rear-guard action to become independent themselves once British rule ended;[57] but on 25 July 1947, appearing before the Chamber of Princes, Mountbatten disabused the princes of their illusions by clearly stating that while technically and legally they could become independent on 14/15 August they had to be mindful of 'certain geographical compulsions that cannot be avoided'.[58] That was Mountbatten's polite way of saying that all princely states would have to attach themselves to either India or Pakistan, depending upon the contiguity of their borders with those of either state. In order to ensure the continuity of basic services like electricity, supply and communications, Mountbatten advised each of the princes to sign a Standstill Agreement with whichever state they wished to join before finally signing an Instrument of Accession to that state.

By 14/15 August 1947, the rulers of only three princely states – Junagadh, Hyderabad and Kashmir – had refused to be committed to accession to India or Pakistan. Tiny Junagadh in Gujarat and massive Hyderabad in south-central India both contained Hindu majority populations, but were ruled by a Muslim *nawab* and a *nizam*. The two rulers' dreams of independence were quashed by the new Indian state. Kashmir had a Muslim majority population, but its Hindu Dogra ruler, Hari Singh (1895–1961), and his Prime Minister, Pandit Kak (1893–1983), both played long and hard to stay independent.[59] What made it difficult for them to achieve their goal was that neither India nor Pakistan was prepared to acknowledge Kashmir's independence. It was Pakistan, however, with which Kashmir had a truly contiguous border, and not India; indeed, if it were not for the fact that the Radcliff Award had granted a section of a Muslim majority area in East Punjab to India, the latter would have lost an accessible route into Kashmir.[60] Jinnah and the new Pakistan government were determined to secure the accession of Kashmir to Pakistan, both on the basis of its contiguous borders and its majority Muslim population. India, particularly its Prime Minister Nehru, felt that the inclusion of Kashmir within India would strengthen its secular credentials and that, in any case, one of the most influential of local Muslim leaders, Sheikh Abdullah (1905–82), had been a fervent champion of Indian nationalism.[61] Through strong pressure on the Kashmiri ruler, India had managed to secure the release of Sheikh Abdullah from prison.[62]

By October 1947, despite having signed a Standstill Agreement with Pakistan, the Kashmiri ruler's options were narrowing; he faced, on the one hand, an internal Muslim revolt in the hardy district of Poonch and, on the other hand, the massive infiltration of the formidable Afridi and Mahsud tribesmen from the North West Frontier Province.[63] His Hindu sentiments, and the knowledge that Sheikh Abdullah did not care for Pakistan, persuaded him to sign the Instrument of Accession to India in return for Indian military help. Mountbatten, as the new governor-general of India, and the Indian Cabinet sent sufficient help to save the capital Srinagar, the district of Jammu and the Vale of Kashmir. They promised to hold a plebiscite after a while to determine the will of the Kashmiri people; and, indeed, on 1 November 1947, Mountbatten offered to resolve the issue through the verdict of the people, which Jinnah rejected.[64] Although clearly upset by the fact that Kashmir had legally acceded to India, Jinnah was dissuaded by the supreme British commander, Field Marshal General Auchinleck (1884–1981), from sending the regular units of the Pakistani army to fight the Indians. Irregular troops and guerrillas continued their struggle against the Indian army. The fighting raged throughout 1948, although India took the case to the United Nations on 1 January 1948. The typical dilatoriness of the proceedings at the United Nations meant that an entire year had passed before a ceasefire line was established between Indian forces and the irregulars from Pakistan on 1 January 1949 (Map 10.1).[65] The military situation has remained frozen since then, adversely affecting all Indo-Pakistani relations.

On reflection, we may conclude that the difficulty in amicably resolving the Kashmir dispute owed to the fact that the principal parties lacked trust in each other and behaved less than honourably. First, the ruler of Kashmir hankered for his own independence when he knew perfectly well that the vast majority of Muslim people, under the leadership of Sheikh Abdullah, would only be satisfied with genuine popular democracy, which he had no intention of granting. He could have acted swiftly and boldly on Lord Mountbatten's correct legal advice and signed an Instrument of Accession to Pakistan before 14/15 August; and neither Sheikh Abdullah nor India could have raised any legal objections to that move. He could then have negotiated favourable terms with a gratified Pakistan. Instead, his hesitations created great suspicions in Pakistani minds and confused the Indians. Second, Jinnah made the arrogant presumption that a Muslim majority was always bound to favour Pakistan; he had reportedly asserted that Kashmir was 'in my pocket'.[66] He could not accept the fact that most Kashmiri Muslims at that time favoured Sheikh Abdullah's anti-Pakistan stance and, perhaps for that reason, he rejected Mountbatten's offer of 1 November 1947, to resolve the dispute through the people's verdict. His government also quietly encouraged the infiltration of Muslim guerrillas into Kashmir, while officially denying any responsibility.[67] Third, in their rush to secure Kashmir as a useful strategic territory in the geo-political interests of India, Mountbatten and the new Indian government sent troops to help the Kashmiri ruler without any discussion or consultation with Pakistan which had an obvious interest in the matter. Although India took the issue to the United Nations, it refused to accept the various resolutions of the Security Council which tended to favour the Pakistani position, and thus stalled any progress. It basically relied on its military superiority along the Ceasefire Line. India was also less than honest in the matter of the plebiscite it had promised. Apart from the Mountbatten offer of 1 November 1947, the Indians have continually thwarted any move towards holding a plebiscite in Kashmir under proper international supervision. The tragedy of Kashmir, we can therefore conclude, arose from the fact that the principal parties were concerned less with the feelings of Kashmiri people and more with the land of Kashmir as a desirable territorial acquisition.

Fighting illiteracy and *purdah* seclusion: Muslim women's advances

For countless centuries, the vast majority of the subcontinent's women, from all faiths and ethnic groups, suffered a great variety of restrictions and disabilities at the hands of the patriarchy, in the name of religion, hallowed traditions and obscurantist customs. Lack of education and degrees of seclusion shaped their lives. With the arrival of new ideas and attitudes from the West, however, there began in the early nineteenth century a movement for the reform of Hindu society, led by members of the Hindu intelligentsia in Bengal. This reform movement, first pioneered by Raja Ram Mohan Roy and, after him, carried forward by campaigners in other parts of India, envisaged social emancipation and liberation of Hindu women from such practices as *sati* and child marriage, seclusion at home and illiteracy.

In some respects, Muslim women were more fortunate than their Hindu peers. They did not, for example, experience the horror of *sati* immolation; and they had more liberal rules on divorce, remarriage and inheritance. On the other hand, they suffered from illiteracy, *purdah* seclusion and other disabilities more severely than their Hindu counterparts. Illiteracy, even in their own vernaculars, and general lack of education had to be addressed most urgently in any movement of Muslim social reform. Unfortunately, the Muslim reformers of the nineteenth century were at first quite hesitant about taking up this cause. Although Sir Sayyid Ahmad Khan, for example, strongly believed in modern education and had insisted on a Western-style curriculum to be followed at his Aligarh MAO College (Chapter 9), he was primarily concerned with the education of males. His instincts were basically liberal but, being brought up in the Mughal-style *ashraf* family within the sexist milieu of north India, he found it difficult at first to envisage a role for women outside the safe and secure sanctuary of the family home. He was not a fanatical supporter of the *purdah* system of seclusion, but he was prepared to accept its usage.[68] Neither was he against female education, but he thought it was best to promote it within the home.

The change in attitudes, paradoxically, first came through those young men who, on graduating from the Aligarh MAO College, moved away from Sir Sayyid's ideas; it was then that the movement for Muslim women's education began to be more successful. The new graduates' horizons had been widened by their academic success resulting in greater economic security; many of them aspired to marry educated women who were at the same time good Muslims. Although most were conventional enough to keep their wives in *purdah*, others wanted to go further and wished to be seen in public with their wives without a *burqa*. In this process of their 'embourgeoisement', they were prepared to challenge the traditional orthodoxy.[69]

By the early twentieth century, small town middle-class Muslim men of modern outlook – government officers, teachers, merchants, doctors, writers, even some of the *ulama* – were taking their daughters' education with greater seriousness than ever before. Addressing these concerns, an early pioneer, Sayyid Karamat Husain (1854–1917), founded the women's section of the Muhammadan Educational Conference (Chapter 9) in 1896 and went on to establish a Muslim Girls' College in Lucknow in 1912 with a Canadian headmistress in charge. He did not believe in equality of sexes, but he wanted girls to be able to cope in the modern world.[70] A couple, Sheikh Muhammad Abdullah (1874–1965) and his wife Wahid Jahan Begum (1886–1939), having been inspired by the Aligarh MAO College, campaigned for the establishment of a sister institution at the same site for girls. The then principal of the existing college, W.A.J. Archbold (1865–1929), revealed all his male prejudices in his determination to keep out girls from the college; but Sheikh Abdullah stood his ground and the girls' college was established in 1914. It is now one of India's premier women's educational establishments.[71] The most outstanding example of a Muslim woman establishing her own school was set by the writer, Rokeya Sakhavat Hosein (1880–1932), a prominent member of the *bhadramahila* circle of Bengali literary women (Chapter 9), who established the Sakhavat Memorial Girls' School in Calcutta in 1911.[72] These examples are evidence of growing interest in the promotion of Muslim women's education, particularly through vernacular literacy, during this period.

The spread of such literacy among Muslim women is evidenced by the increasing number of new ladies' journals published in Urdu or Bengali. One such journal, *Tahzib un-Niswan* (Cultivation of Women), was started by a strident defender of women's rights, Mumtaz Ali (1860–1935), in partnership with his wife, Muhammadi Begum (1878–1908).[73] A Delhi based publication, *Ismat* (Honour), founded by the novelist Rashidul Khairi

(1868–1936), encouraged women to contribute creative work in its pages, while eschewing political articles.[74] Many of the articles in such journals popularized the contents of specific books written for women. These books contained both pietistic advice on Islamic teachings and practical information concerning such subjects as women's and children's illnesses, *rites de passage*, decorating, cooking, sewing, etc. A revolution in Muslim women's thinking was therefore the result of what they read and absorbed from such journals and books that were now being published. While these works did not advocate complete openness in discourse or abandonment of *purdah* seclusion, they nonetheless educated and motivated Muslim women into modern ways.

With literacy levels rising and greater social awareness being fostered by books and magazines, it was but natural that some educated women began to challenge the lifestyle of *purdah*.[75] The custom of veiling was, and is, its most conspicuous feature, but in its wider context the *purdah* legitimized elaborate codes of seclusion and female modesty which have been used to protect and control the lives of women. The focus in Muslim culture was 'the seclusion of girls who had reached puberty, whether they were married or not'.[76] Muslim social reformers, like Hali and Mumtaz Ali, condemned this seclusion as a breeding ground for ignorance, superstition and mental distress, but were nevertheless confronted with the reality of the long established tradition that had provided women with some form of security and mental comfort within a Muslim household.[77] A small group of reformers were just lone voices among the mass of Muslim men who, however much they admired female education, kept the tradition going, partly out of respect for their elders and partly out of the fear of being ostracized by the wider community. For the coming out from *purdah* to be successful, it required a band of very courageous Muslim women to break the taboo and resolutely face the onslaught from the orthodox and the conventional. Since the *purdah* restrictions were far more stringently enforced within middle-class families, in contrast to the relatively liberal regime that prevailed among rural women who worked alongside men in the fields, it was necessary for educated urban women to make a stand against the custom. And indeed, a number of them, predominantly well educated and well established, emerged in the early twentieth century.[78] Two of the earliest were Atiya and Zohra Fayzee from the very rich Bohra Tayabjee family of Bombay.[79] In Punjab, Jahanara Shah Nawaz (Plate 12.2) led the trend, with much encouragement from her father, Mian Muhammad Shafi, the Education Member of the Viceroy's Cabinet.[80] Again, in Punjab, there was Begum Qudsia Aizaz Rasul (1908–2001), whose father Sir Ali Khan Zulfikar (1875–1933) sent her to the best convent school in Lahore but insisted on her wearing a *burqa*. Once married, however, she discarded her veil.[81] In Lucknow Rashid Jahan (1905–52), the daughter of Sheikh Abdullah and Wahid Jahan Begum, the founders of the Aligarh Girls' school, parted with the veil during her medical studies; and later wrote short stories about its deleterious effects.[82] The last Begum of Bhopal, Sultan Jahan Begum, having worn a *burqa* all her life, told the All-India Women's Conference in 1928 that *purdah* did not sit easily with modern life (Excerpt 10.4).[83] The cumulative efforts of these brave women ultimately resulted in the release from the veil of literally millions of women in the subcontinent. It is, of course, the right of every Muslim woman to wear a veil if she wishes to, and large numbers still do: but blind adherence to the restrictive features of *purdah* is now confined either to the most orthodox of families in the subcontinent or some women in the diaspora who wish to assert their Islamic identity in the face of increasingly intolerant attitudes of Western governments and media.

Higher levels of literacy and freedom from *purdah* encouraged Muslim women to advocate further social reforms, in connection with issues like polygamy, child marriage and divorce. A radical resolution against polygamy was proposed by Jahanara Shah Nawaz in 1918 at the All-India Muslim Ladies Conference (Anjuman-i-Khavatin-i-Islam) in Lahore, but it was difficult to get legislation enacted on this issue owing to the strident opposition from orthodox men.[84] Muslim men's apathy did not, however, stop Muslim women supporting the all-India campaign for raising the age of child marriage, thereby helping to get the Child Marriage Restraint or Sarda Act of 1929 passed.[85] With the cooperation of some of the more liberal *ulama* and politicians like Jinnah, Muslim women also won a major victory over divorce liberalization, with the passing of the Muslim Dissolution of Marriages Act of 1939.[86] The demand for other social reforms began to grow.

Eminent literati before independence

Educational advances and social reforms of the late nineteenth and early twentieth centuries led to greater self-confidence within the varied ranks of Muslim society. One particular result of this was the flourishing of Muslim cultural life in the subcontinent during the inter-war years. Muslim participation enriched both old and new forms of entertainment. The *mushairas* and the all-night *qawwali* sessions continued to attract iconic poets and singers, and Hindustani North Indian classical music reached new heights under great masters like Fateh Ali Khan (1901–64), Mubarak Ali Khan (d. 1971), the instrumentalist Alauddin Khan (1881–1972) and the young *shehnai* player, Bismillah Khan (1916–2008). While Muslim actors and actresses have played a noted role in the success of the Hindi/Urdu cinema since independence, it is worth remembering that one of the earliest cinema heroines was a Muslim woman, the now forgotten Mumtaz Shanti, who acted in so many of the pre-Second World War films. Some literary figures and academics continued to produce outstanding works of merit.

Underneath this exuberance, however, there lay the spectre of separatism that inevitably led to the splintering of the hitherto subcontinental Muslim cultural and intellectual life.

The most intimate link between religious thought and political vision can be traced in the writings of two of the most important religious thinkers of this age: Muhammad Iqbal and Abul Kalam Azad. We have referred to Muhammad Iqbal's early poetry in Chapter 9. Most of it was written in Urdu; but from the First World War onwards he increasingly wrote in Persian, partly because he was addressing a wider Muslim community than that of India. He articulated the necessity for a synthesis of Islamic tradition and the spirit of modern reform. He also laid stress on social justice the roots of which, he believed, lay in the moral perfection of man. In his 1928 lectures, *The Reconstruction of Religious Thought in Islam*, he aimed to reconstruct Muslim religious philosophy with due regard to the philosophical traditions of Islam and scientific knowledge;[87] and he emphasized the principle of *ijtihad* or enlightened judgement.[88]

Iqbal's majestic poetry and religious works created a profound impression on Muslims all over the world, but for many Indian Muslims it was his passionate desire for a Muslim homeland within the subcontinent that provided them with a spark of inspiration. Jinnah's later conversion to separatism owed much to Iqbal's vision of a Muslim state based on Islamic ideals of social justice.[89] Iqbal stated his position in 1930, when he argued for such a state based in the north-west of India and incorporating the Muslim majority regions of Punjab, Kashmir, North West Frontier Province, Sind and Baluchistan.[90] A central concept in his complex thinking involved identifying what he called a 'cognizable centre' for Indian Muslims, in the same way that the House of God in Mecca, the *Kaaba*, served as such a centre for the entire Islamic world.[91] Such a centre, for him, was to be the ultimate safe sanctuary for them. He appreciated that there were Muslims living in other parts of India, but he did not see any contradiction between the notions of an Indian Muslim 'cognizable centre' and the Indian Muslim diaspora. They could both co-exist in Iqbal's vision, because the Muslim base would continue to be a unit within an all-embracing Indian federation of many units. By proposing a geographical base within which a Muslim identity could develop and flourish, Iqbal went further than Sayyid Ahmad Khan.

Abul Kalam Azad first gained public fame in 1912 with the publication of a pan-Islamic and anti-British journal, *Al-Hilal* (Chapter 9).[92] For the next forty-six years he remained at the centre of Indian nationalist politics. Yet he was not just a politician; he was a man of profound religiosity, and his Islamic passion was as brightly burnished as that of Iqbal. After engaging over many years with a deep study of his faith, he arrived at definite conclusions on the fundamental message of Islam.[93] He believed that Islam enjoined upon all the duty to do good, to promote justice and to relieve suffering. Islam also forbade the narrow-mindedness based on racial and religious prejudice. It was the Islam of diversity, not sectarianism or belligerence that Azad searched for; and it was in the course of that search that Azad produced his most mature and comprehensive religious work, *Tarjuman al-Quran*, a translation and commentary on Islam's holiest book, written between 1930 and 1936. His scholarship was wide ranging, and he wrote with sincere conviction.[94] The same Islamic religious thought that had led Iqbal to yearn for a Muslim homeland of Pakistan was conversely interpreted by Azad as justification for his deep conviction in the idea of Hindu–Muslim unity. He remained an Indian nationalist until the end and wrote convincingly about the damage that the Pakistan Movement would inflict upon the subcontinent's Muslims.[95] He cherished the ideals of secular liberalism and, as Minister of Education in independent India, emphasized the inculcation of humane civic values in the education of young people.[96]

The academic progress of the Muslim middle classes continued through the three Muslim-inspired institutions of higher learning: the Aligarh Muslim University, the Jamia Millia Islamia in Delhi, and the Osmania University at Hyderabad; but separatist debate was to lead their academics towards different directions. The Aligarh Muslim University, originally very pro-British and favoured by the imperial government, came to be strongly influenced by those who supported Jinnah and the Muslim League.[97] After partition, a great number of Aligarh-trained academics left for Pakistan. The position was markedly different at the Jamia Millia Islamia, an institution originally founded by nationalist Muslims of the Khilafat period and supported by Congress luminaries like Gandhi and Nehru. Academics there stayed loyal to the nationalist-secular ideals of Indian unity.[98] At the Osmania University in the princely state of Hyderabad, feelings among students and academics veered between the desire for an independent Hyderabad state and Pakistan; many fine scholars and intellectuals left for Pakistan when it became clear that Hyderabadi independence was unattainable.[99] Support for Pakistan was, however, much more nuanced at religious seminaries like Deoband, owing to theological differences over the nature of a future Pakistan.[100]

Both fiction and poetry of this period display sophistication of styles and a variety of themes. In Bengali literature, the influence of Rabindranath Tagore was overwhelming, and it was difficult for most Bengali literary figures to break loose from the patterns set by him; but one Muslim writer, Kazi Nazrul Islam (born 1899), did just that by producing works of an original nature. After serving in the army during the First World War, he began his literary career in the early 1920s, producing a number of novels, poems and short stories with underlying political and social themes.[101] Although he did not take an active part in the nationalist movement, he had an instinctive feel for the contemporary struggle for freedom that was being waged. Some of his writing was indeed revolutionary; his readers, enthused by his spirit of revolt and patriotic zeal, called him a *bidrohi* (rebel) which was also the title of one of his poems. Similarly, while not being ideological, he nevertheless wrote much against economic and social injustices of his times.

A more stridently left-wing approach was, however, adopted by a group of Urdu writers who formed the Progressive Writers' Association.[102] They wanted to change the tone of Urdu literature, from romanticism to realism; and they therefore took on challenging themes such as women's rights, dysfunctional family, social taboos, political and religious freedom, etc., in their early collection of short stories entitled *Angaray* (Live Coals or metaphorically a raging anger), published in 1932. One such story, written by Sajjad Zaheer (1905–1973), attacked the incongruence of private thoughts and public hypocrisy of the men of religion;[103] and the violent storm it provoked among many Muslims at that time was a foretaste of the storm that broke over Salman Rushdie's *Satanic Verses* in the late 1980s. Others wrote less sensationally but had powerful messages to relay. Dr Rashid Jahan (1905–52), for example, wrote intimately about the lives of Indian Muslim women in *purdah* seclusion; it was the women who came to her medical practice who were the source of her knowledge. Mumtaz Shah Nawaz (1912–48), the daughter of Jahanara Shah Nawaz, the feminist politician who switched her support from Congress to Pakistan, left behind in her posthumously published work *The Heart Divided* a moving account of the agonies caused by communal differences in the five years before independence.[104] In the end, the 1947 partition also splintered the Progressive Writers' Association; and new circumstances and political realities adversely affected its fortunes in both India and Pakistan.

Select excerpts

10.1 Sir Claude Auchinleck on future British strategy

In a top-secret note on the strategic implications of the inclusion of future Pakistan in the British Commonwealth, written from his GHQ in Delhi on 11 May 1946, the supreme British commander came to the following conclusion.

> a) The inclusion of Pakistan in the British Commonwealth of Nations . . . and the assumption by Britain of the consequent responsibility for its defence could be justified on the following grounds:
>
> (i) That it would enable us so to dominate and control an independent Hindustan as to prevent her or her potential allies

> from disrupting our sea and air communications in the Indian Ocean area.
> (ii) That it would aid us in maintaining our influence over the Muslim countries of the Near and Middle East and so assist us to prevent the advance of Russia towards the Indian Ocean and the Mediterranean.
>
> b) If the arguments contained in this note are being based on correct surmises, it seems perfectly clear that the first of these objects is unattainable, because of the large forces which its achievement would require, relative to the resources likely to be available to the British Commonwealth, at the outbreak of a major war.
>
> If the first object cannot be achieved, it would be useless to attempt to achieve the second, because it would be quite obvious to all the Muslim countries that Britain had ceased to be a power in Asia.
>
> c) If we desire to maintain our power to move freely by sea and air in the Indian Ocean area . . . we can do so only by keeping in being a United India which will be a willing member of the Commonwealth, ready to share in its defence to the limit of her resources.
> (Singh 2010: 476)

10.2 Abul Kalam Azad on Muslim delusions

In the following passage, Abul Kalam Azad, a staunch Indian nationalist, describes the pathetic confusion in which certain Muslim leaders found themselves after 1947.

> After partition, the most ridiculous position was that of the Muslim League leaders who remained in India. Jinnah left for Karachi with a message to his followers that now the country was divided and they should be loyal citizens of India. This parting message created in them a strange sense of weakness and disillusion. Many of these leaders came to see me after 14 August . . . I could not at first understand what they meant by saying that Jinnah had deceived them . . . (But) as I talked to them I realized that these men had formed a picture of partition which had no relevance to the real situation. They had failed to realize the real implications of Pakistan. If the Muslim majority provinces formed a separate State, it was clear that the provinces in which the Muslims were in a minority would form part of India . . . It is strange but the fact is that these Muslim Leaguers had been foolishly persuaded that once Pakistan was formed, Muslims, whether they came from a majority or a minority province would be regarded as a separate nation and

> would enjoy the right of determining their own future. Now when the Muslim majority provinces went out of India and even Bengal and Punjab were divided and Mr. Jinnah left for Karachi, these fools realized that they had gained nothing but in fact lost everything by the partition of India . . . It was now clear to them that the only result of partition was that their position as a minority was much weaker than before. In addition, they had through their foolish action created anger and resentment in the minds of the Hindus . . . I reminded them of what I had said during the Cabinet Mission Plan. In my statement of 15 April 1946 I had warned the Indian Muslims in unambiguous words. I had said that if partition ever became a reality, they would one day wake up to find that after the majority of Muslims went away to Pakistan, they would remain in India as a small and insignificant minority.
>
> (Azad 1988: 226–7)

10.3 Partition*: a poem about Radcliffe*

In the following verses the poet W.H. Auden brilliantly captures the multiple agonies of Cyril Radcliffe.

> Shut up in a lonely mansion, with police night and day
> Patrolling the gardens to keep assassins away.
> He got down to work, to the task of settling the fate
> Of millions. The maps at his disposal were out of date
> And the Census Returns almost certainly incorrect.
> But there was no time to check them, no time to inspect
> Contested areas. The weather was frightfully hot,
> And a bout of dysentery kept him constantly on the trot.
> But in seven weeks it was done, the frontiers decided.
> A continent for better or worse divided.
>
> The next day he sailed for England, where he quickly forgot
> The case, as a good lawyer must. Return he would not.
> Afraid, as he told his Club, that he might get shot.
>
> (Khilnani 1997: *The Observer*
> (Review Section), 22 June 1997)

10.4 Purdah *in the Bhopal royal family*

When Abida Sultan, the eldest granddaughter of Sultan Jahan Begum, returned from her European trip in 1926, the dowager queen attempted to force *purdah* on her. She resisted successfully. In the following passage she expresses her feelings on the subject.

> Purdah was a cruel blow . . . No effort had been made to mentally prepare an athletic, outdoor girl like me for a life long incarceration behind the veil. I wondered what the meaning was of all the emphasis and pride in making us ride, drive cars, play rough games, shoot and be constantly reminded that we had to prove ourselves better than our 'backward' male cousins, if it was all to end abruptly at the age of 12 behind a burqa?

Shortly afterwards, Sultan Jahan Begum herself changed a lifelong habit and made the following sensational remarks in her address to the All India Women's Conference of February 1928.

> I have no hesitation to own that the purdah system as it is observed among Muslims of India is not exactly Islamic and is indeed very harmful to the progress of education among our girls. It is a hindrance moreover in the way of their physical and mental development. The Musalmans should coolly and calmly reflect and decide whether by respecting a mere custom they would keep their women in a state of suspended animation, whether they would sacrifice the prospect of their future generations at the altar of blind prejudice.
>
> (Lambert-Hurley 2007: 121–2)

11

EPILOGUE

New challenges in a fractured subcontinent (1947–2011)

The political history of the South Asian subcontinent dramatically changed on 14/15 August 1947, with the emergence of two new independent states of India and Pakistan. Direct British authority and British paramountcy over the Indian princes all lapsed. Although the new nations of India and Pakistan began life as dominions, vital decisions now lay entirely in the hands of Indians and Pakistanis. They were the masters of their own destinies. Unfortunately their leaders have since 1947 failed to prevent continuing political and military tension in the subcontinent (Map 11.1).

The idea of Pakistan had emerged out of the fear of Muslim separatists about Hindu domination in independent India, expressed in the credo of the so-called Two Nations theory. A safe and separate homeland for Muslims in the Muslim-majority provinces of British India was Jinnah's ultimate goal from 1939 onwards; and, as we saw in Chapter 10, he achieved it, except for the fact that Bengal and Punjab had to be partitioned. The creation of Pakistan, however, left behind many other millions of Muslims as a much reduced minority within the new Indian state. The first section of this chapter will examine some of the problems that this minority has faced in India.

Jinnah almost single-handedly created a new Muslim nation in Asia; and Pakistanis have rightly felt gratitude and admiration for their founding father. He did not, however, appreciate sufficiently that a Muslim homeland was no guarantee against Muslims perpetrating injustice or brutality against fellow Muslims; and that faith alone may not reconcile ethnic or cultural differences amongst the populace. Within just twenty-five years the tension between Western Pakistanis and the Bengali-speaking Eastern Pakistanis eclipsed the apparent unity of the two people conforming to a single faith. The break-up of Pakistan in 1971–2 resulted in the emergence of a new state: Bangladesh. Both Pakistan and Bangladesh, somewhat less so, are overwhelmingly Muslim nations. Muslims there are citizens of countries whose constitutions mention, in varying degrees, the role that Islam and

Islamic codes of behaviour are meant to play in their governance. Their countries have also been subject to extensive periods of military rule which has thwarted their democratic maturing. The relevant issues facing the two nations will be highlighted in the second and third sections. The final section will deal briefly with developments in literature and arts of the subcontinent since independence.

The Muslim minority in independent India

Secularism and populist politics

In most parts of India, partition made little difference to Muslims leading their daily lives. It was mostly in North India, particularly around Delhi and Agra and in the old United Provinces (now Uttar Pradesh), that they encountered the greatest hostility and suspicion of communally minded Hindus who unfairly vented their anger about Pakistan on them. The refugee crisis of the late 1940s helped to ignite communal passions; and the Muslims of Delhi felt quite isolated and vulnerable, given the fact that while in 1941 they had constituted nearly 33 per cent of the city's population they formed just under 6 per cent in 1951.[1] Sectarian Hindu politicians within the ruling Congress Party and from the Hindu extremist parties, such as the Jana Sangh, clamoured for Hindi to replace Urdu in every public institution and for a ban on cow slaughter. Even the most eminent of Hindu politicians, like Vallabhbhai Patel (1875–1950) and Dr Rajendra Prasad (1884–1963), found it difficult to restrain their anti-Muslim prejudices.[2] The fact that these prejudices were not translated into hostile parliamentary legislation was due to the overwhelming authority of the Prime Minister, Jawaharlal Nehru (1889–1964), who remained the principal torch-bearer of the ideals of democracy, equality and secularism. Nehru, along with his mentor Gandhi, insisted that Muslims who stayed behind were an integral part of India, not second-class citizens. The revulsion felt by most Indians at Gandhi's assassination in January 1948 by a Hindu extremist made it easier for Nehru to propagate the Gandhian message of tolerance to the nation at large.[3] The Muslim minority consequently began to feel at ease with itself in the new India.

Another factor in the Muslims' sense of security was the promulgation of a new constitution in 1950, under the direction of Dr Bhimrao Ambedkar (1891–1956), the leader of the Dalit community of the co-called 'untouchables', the most deprived section of people within the Hindu world. Ambedkar's strong sense of justice, fairness and egalitarianism is manifest throughout the constitution; and every group and community in India can seek redress for injustice or inequality from this secular constitution. In the Western world the term 'secular' normally implies a non-religious or

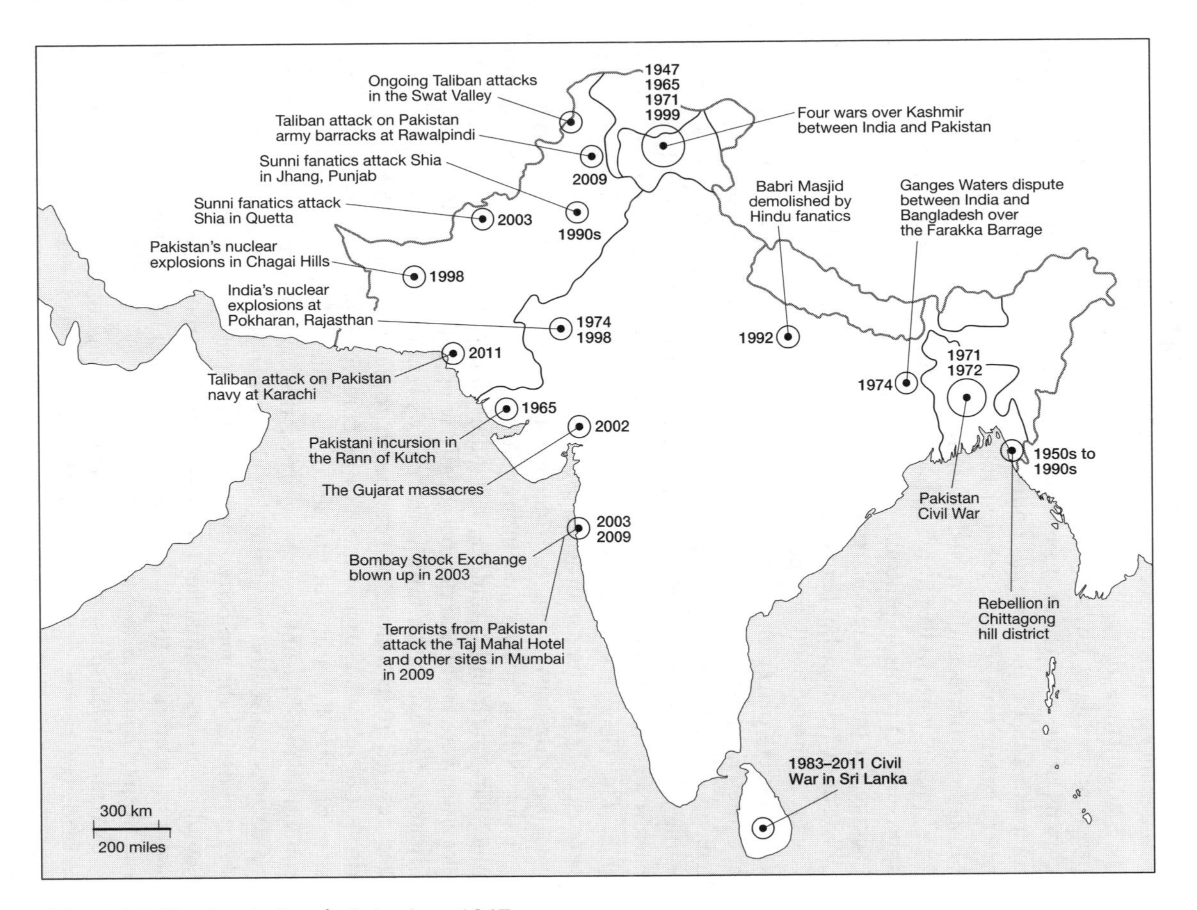

Map 11.1 Tension in South Asia since 1947

an anti-religious outlook. In India, it has a different connotation. The Indian constitution is not non-religious or anti-religious; it aims to maintain a neutral position within the arena of diverse religions.[4] The Indian state actually often provides help and encouragement to its citizens in the maintenance of their respective faiths, but no one faith can claim precedence over another as of right. The constitution, while conceding no special privilege for Islam, gives Muslims the right to be treated as free and equal citizens of the country in the eyes of the law.[5] Muslims are also permitted to make use of the *Shariah* courts to settle personal issues concerning marriage, divorce, death, etc. On the other hand, such ideas as separate Muslim electorates and reserved Muslim seats have no place within the post-1947 Indian state. Indian Muslims have been loyal supporters of this secular approach.[6] In terms of practical politics, this translated into strong Muslim backing for the Congress Party over many decades.

The close relationship between Muslim leaders and the Nehru-dominated Congress carried risks for both parties. The former were often perceived as the latter's placemen committed to the political status quo.[7] Their apathy and their inability to engage themselves effectively in debates over challenging issues led in time to greater receptivity for Islamic religious parties within the Muslim community.[8] The exclusivist political agenda of such parties was, in the end, bound to arouse suspicion across the wider population. The Congress Party, too, came to be perceived as the party that appeased the Muslims in order to use them as a mere vote bank for winning elections. Majority Congress rule first lasted for thirty years (1947–77) – under Nehru, Shastri and Indira Gandhi – and then again for another nine years under Indira and Rajiv Gandhi, from 1980 until 1989. Although elections were fairly and democratically won, corruption and complacency in public life had much tainted the party's reputation. Gradually, over the decades, the party's share of vote began to decline, and a variety of both right- and left-wing parties ate into its core vote, particularly in the large state of Uttar Pradesh. The parties of the Hindu Right, with their barely concealed anti-Muslim feelings, worried the Congress most. Their propaganda was centred on the claim that the secular state favoured Muslims at the expense of the majority Hindus.[9] To counter this trend, it was Indira Gandhi who first shifted her position from her father's (Nehru's) principled stand on minorities to that of playing off Muslims against Hindus.[10] Her son, Rajiv Gandhi, also used similar tactics.

Two fateful decisions taken by Rajiv Gandhi's government (1984–9) were to lead to a period of profound crisis for secularism and inter-communal relations within India. The first one was over the case of Shah Bano, a Muslim woman divorced by her husband in 1978 by the traditional method of repeating thrice '*talaq, talaq, talaq*' and left without maintenance after three months. When the case reached the Supreme Court in 1985, the husband argued that he had fulfilled his maintenance obligations in accordance with the *Shariah* conventions in matters of divorce. The court disagreed and ordered him to pay her a modest alimony every month, in line with the general law that applied to non-Muslim women. Instead of respecting the court's decision, Rajiv Gandhi's government was forced into passing a new law, the Muslim Women (Protection of Rights on Divorce) Act of 1986, by noisy protests from the conservatives in the Muslim Personal Law Board. This Act, while introducing some flexibility over maintenance payments, essentially re-confirmed the traditional Islamic position on divorce and maintenance; it was a regressive measure that hindered the modernization of Islamic law on women's issues.[11] Many felt that the government's motive in passing this legislation was the fear of losing the Muslim vote bank. Appeasement was the charge levelled at the Congress by the Hindu parties.

Rajiv Gandhi's government compounded the problem further by taking another inept decision over what is known as the Ayodhya Babri Masjid Controversy. The Babri Masjid, a sixteenth-century mosque named after the Mughal Emperor Babur (Chapter 4), was situated in the city of Ayodhya that is renowned for its association with the Hindu god, Lord Rama, and the great epic of *Ramayana*. The mosque was on a site that Hindu tradition has claimed to be the birthplace of Rama; and it was possibly built upon the ruins of an earlier Hindu temple. For nearly three centuries, both Hindus and Muslims had been able to use the site jointly for worship; and, although from the 1850s onwards Hindu claims to the site had become more assertive, the mosque had remained a Muslim centre of worship. During his premiership, however, Rajiv Gandhi shied away from confronting the Hindutva politicians of the 1980s who displayed strong anti-Muslim feelings by using Ayodhya as a symbol of their resurgence. After what had seemed like appeasing reactionary Muslims by passing the Muslim Women Act, he attempted to please the reactionary Hindus by disturbing the status quo over the Babri Masjid: first, in 1986, by endorsing a local court's decision to open the gates of the inner compound and then in 1989, just before the national election, supporting the laying of the foundation stones of a new Rama temple at the site. The die was cast from that moment onwards. With their appetites whetted by Rajiv's actions, the Hindu fundamentalists put out a strident call to the Muslims to surrender the mosque in order to build a grand Rama temple. Meanwhile, the secular consensus began to collapse during a succession of weak governments after Rajiv's defeat in 1989. On 6 December 1992, thousands of fanatical Hindus demolished the Babri Masjid brick by brick and set India on a path of Hindu–Muslim riots and disturbances for over a decade, culminating in the shameful Gujarat massacres of 2002.[12]

Definitive judgments by the Supreme Court of India are still pending over numerous misdeeds by both Hindu and Muslim zealots; and a stalemate prevails over the future of the Babri Masjid site. The politics of Indian democracy provide continuing opportunities for inciting communal disharmony; however, the fact that the attacks on Mumbai in 2009 by Pakistani terrorists brought the Hindus and Muslims of the city together in grief may be a sign of hope for Indian secularism. Some Muslims have also advocated a more radical approach by their co-religionists to integration within Indian life (Excerpt 11.1).

Socio-economic marginalization

Congress has ruled India for fifty-two out of the last sixty-four years and, by and large, it has received overwhelming support from Muslims. However, owing to its obsession with winning elections by balancing the vote banks, it has refrained from adopting positive measures to tackle Muslim socio-economic disadvantages.[13] Since most of the upper- and middle-class Muslims of North India left for Pakistan, the disadvantages mostly affect an economically depressed Muslim minority. These disadvantages have been exhaustively chronicled in the reports of two important commissions – the Gopal Singh Commission of 1983 and the Sachar Commission of 2005–6 – that were in fact set up by Congress governments. The indices for Muslim share in education, health, employment, infrastructure facilities, credit and finance all show negative trends. Let us take one example, that of employment: although the Muslim population of India was 13 per cent in the 2001 census (138 million out of 1,029 million), Muslim employment was just 3 per cent in the administrative service, 4.4 per cent in health services, 4.5 per cent in railways, and so on.[14] One or two positive indices balance the picture a little. Thus, for example, the sex-ratio among Muslims (986 girls to 1,000 boys) is more positive than among Hindus (914 girls to 1,000 boys), implying less bias against new-born females. A small Muslim middle class has also emerged out of the artisan communities which have benefited from globalization and greater international commerce. Several proactive steps have been suggested by the Sachar Committee, such as passing the equal opportunities legislation, establishing a diversity index in all jobs, spending more money on improving the *madrasa*s, and so on; but all such measures are generally more talked about than acted upon. Some Muslim radicals have suggested as a solution a link-up with the 'untouchable' Dalit community. Dalits and many backward groups enjoy the benefits of positive discrimination in public-sector jobs; but only Hindu, Buddhist and Christian Dalits are eligible. Since the origins of most Muslims of India lie in the lower castes and classes of the social pyramid, it makes sense to press for legislation approving the reservation of jobs and educational places for Muslim Dalits who may number as many as 100 million.[15]

The treatment of Kashmiri Muslims

India faces two conflicts in Kashmir. One is a long drawn out and well known conflict with Pakistan; the other is that between the Muslim civilian population and the Indian administration, the police and the army within the Indian sector. The state of Jammu and Kashmir is the only state in the Indian Union with a Muslim majority; and the relationship between this population and the Indian state has markedly deteriorated since the 1980s. Among the factors causing this, three are of special significance: the persistently high and endemic level of corruption within the political establishment of Kashmir, the active Islamist militancy encouraged from across the border, and brutal suppression of peaceful demonstrations. Consequently, over a period of more than two decades, the Muslim civilian population has become highly radicalized and restless. Kashmiri Hindus and Sikhs have withdrawn into the relative safety of the district of Jammu. Notwithstanding the rights and wrongs on all sides, it is increasingly clear that continuing interference by the central government into the political life of the state and the ferocious response of both the police and the army have created an atmosphere of uncertainty, suspicion and fear. Renowned human rights bodies, such as Amnesty International and the New York-based Human Rights Watch have provided graphic evidence of the Indian state's brutality towards its own people in Kashmir, which does not bode well for the future of democracy in India as a whole.[16]

The travails of Pakistan

Pakistan has endured a turbulent history over the last six decades. The human and material wreckage from partition, and the ensuing psychological distress, took years to mend and heal. Political instability arose out of frequent changes of government, oscillating periodically from rule by civilian leaders to rule by military dictators. The country suffered defeat in all four wars with India (1948, 1965, 1971 and 1999). The involvement in Afghanistan since 1980 and the continuing War on Terror at the United States's behest for the last ten years have brought the nation's finances to near bankruptcy that is avoided only by American military aid and international loans. Natural disasters, like the great Indus floods of 2010, have every few years adversely affected the agricultural economy that is supported by a relatively small industrial base. The complex history of Pakistan can be studied from different angles from the works of a number of scholars and researchers;[17] in this section we shall restrict ourselves to highlighting three critical issues that have bedevilled the Pakistani state: the democratic deficit of the political system, the influence of Islamic orthodoxies on the legal system, and a dangerous entanglement in the affairs of Afghanistan.

Democracy on a short leash

It is understandable that a European model of democracy would have been hard to establish in Pakistan, a nation with a distinct social and political culture and identity. Nevertheless, the failure to establish democratic accountability as a key principle in the governance of the country remains a major shortcoming of its political system. Three crucial factors help to explain this. First, within the deeply rooted feudal social structure of Pakistan that is overlaid with a veneer of elitist Western modernity, power is narrowly confined to a coalition of the politically ambitious members of the *salariat* middle class, the landlords, the merchant elites and a sophisticated military-officer class. This coalition, although tiny in numbers, has usurped all the tools of authority in the state. Second, the leadership has from the beginning invested in the armed forces complete confidence and trust to defend the nation against its perceived enemy, India. The burning desire to possess and perfect an arsenal of nuclear weapons also arose out of the rivalry with India. This has meant that Pakistan has been turned into a bureaucratic-military state or, to put it less generously, a garrison state. Such a state is maintained by military power and structured to secure primarily its own need for military security. Third, the Pakistani leaders have for long willingly allowed themselves to be used as a pawn in the struggle for global hegemony by the great powers. The securing of large scale financial and material aid from both the United States and China over many decades has permitted the state to be run without any need for true dialogue with the people at large. The furtherance of genuine democracy has not been permitted to enter into the equation of power.

Perhaps the very first blow to Pakistani democracy was delivered by Jinnah himself when he insisted on becoming the first governor-general of the country after independence.[18] Both India and Pakistan enjoyed dominion status, and the convention was that a governor-general should command respect but little political authority. Mountbatten stayed on in India for some time as governor-general, leaving Nehru and his cabinet with the daily task of governing the country in a democratic manner. Jinnah's power and influence, as the founder of Pakistan, were of such an overwhelming dimension that they could not but unconsciously strangle development of a democratic structure from its conception. After his death in September 1948, his close colleague, Liaquat Ali Khan, was prime minister for three years; but, owing to his instincts for authoritarian centralism and reliance on bureaucracy inherited from the Raj, democracy was left in the cold.[19] In the absence of a set of transparent rules about democratic accountability, Pakistan was plunged into political chaos and confusion after 1951.[20]

A disturbing factor that sabotaged effective democracy in the 1950s was the manner in which civilian politicians managed the friction that was emerging between the two divisions of Pakistan. East Pakistan contained more people than West Pakistan, and some Bengalis were indeed heads of state and prime ministers during the decade. The real power, however, lay in West Pakistan where most of the civilian and military bureaucracy was based. Those who manned this bureaucracy – mostly the Punjabis, the Pathan elite in the armed forces and the ablest of the *muhajir*s, the refugees from India – did not only resent the aggregate Bengali numerical majority; they also wished to thwart its democratic implications. The constant attempts to deny the East Pakistanis their rightful share of resources were strongly evident during this period, and came to be reflected in the first Pakistani constitution of 1956 which allowed for rule by decree.[21] The party system, the life blood of representative democracies, was allowed to languish in a moribund state, with the historic Muslim League making no impact. Although regional political parties like the Awami League in East Pakistan were vibrant, they were just considered to be too dangerous for the state's unity. The ground was thus prepared for the first military coup by Field Marshal Ayub Khan in October 1958.

The military regime of Ayub Khan (October 1958 to March 1969), introduced populist and cosmetic changes to political infrastructure, but the vast majority of Pakistanis remained essentially disenfranchised.[22] Democracy was dead for over a decade. An expensive war with India in 1965 depleted the nation's resources; and it was only when the mounting popular anger against the armed forces reached its crescendo in the late 1960s, particularly in East Pakistan, that a dynamic politician, Zulfikar Ali Bhutto (1928–79), left Ayub Khan's cabinet and formed his own party, the Pakistan Peoples' Party (the PPP).[23] The fall of Ayub did not mean the end of the military regime; power simply passed to his senior commander, General Yahya Khan (March 1969 to December 1971). The latter agreed to hold genuine multi-party elections, but the rift between the eastern and western divisions of the country was past healing. The first true genuine election, held in 1970, reflected the numerical majority of the East. Neither party made an impression across the whole country, but the Awami League gained overall ascendancy, while Bhutto's PPP was confined solely to the West.[24] Since neither Bhutto nor Yahya Khan was prepared to accept the consequences of the democratic verdict, the irretrievable political chasm between the two Pakistans resulted in a civil war, which then turned into a war between India and Pakistan.[25] The much delayed resurrection of democracy could not prevent the break-up of Pakistan.

In the diminished post-war Pakistan (confined to the former West Pakistan) Zulfikar Ali Bhutto's authority reigned supreme from December 1971 to July 1977. His election victory secured him the democratic credentials and, outwardly at least, Pakistan could claim to be a democracy. Bhutto also introduced a new constitution in 1973 which, despite various amendments, is still in force.[26] Through his strong mandate and charisma[27] he was able to carry out a number of socio-economic reforms, but democracy itself suffered because he had little time or inclination to strengthen the institutions of his party. More dangerous for him, the party was accused of fraud in the elections of 1977, which in the end brought about a military coup by his army commander, General Zia ul-Haq (1922–88).[28] Bhutto's execution on specious charges in 1979 and Zia's populist authoritarianism meant the crushing of democracy for another decade. A civilian administration that Zia set up after a sham referendum was merely a façade behind which lay strong presidential powers.

After Zia's assassination caused by a bomb exploding in an aeroplane in 1988, a form of democratic restoration took place under two civilian leaders, Benazir Bhutto (1953–2007) and Nawaz Sharif (born 1949), whose rule alternated during the next eleven years. Without effective democratic institutions they were trapped in the quagmire of difficulties abroad and allegations of corruption at home.[29] Since no clear boundaries of power were delineated between the great offices of the state, it was no wonder that the army chief, Pervez Musharraf, unilaterally embarked in May 1999 on a disastrous military adventure in the Kargil district of Kashmir and, a few months later, launched a military coup in order to prevent his own dismissal. A further nine years of unrepresentative military rule followed, notwithstanding the fact that Musharraf's declared wish was to rule in the spirit of 'enlightened moderation'.[30] He copied all the methods of Ayub Khan and Zia ul-Haq, such as holding manipulated referendums, establishing nebulous local structures and continually changing the legal framework in order to strengthen his own powers. It was only after a great amount of public dissatisfaction and the pressures from the United States after the assassination of Benazir Bhutto that he finally left office in 2008. The democratic structure has, in theory, been restored since then; but, in the absence of tested democratic institutions, it is proving difficult for the present government to manage the multiple crises which engulf it.

The break-up of Pakistan in 1971 was only the most dramatic consequence of the lack of a proper representative democracy in the country; two other negative consequences should be noted too. Power structures were somewhat diffused in the earlier decades, with Karachi being the capital and East Pakistan having the majority population. Over time, however, for various reasons, power has become increasingly concentrated in Punjab, particularly across the Lahore-Rawalpindi-Islamabad axis. The *crème de la crème* of the Pakistani elite is based there, which means that the interests of diverse ethnic groups are often ignored and marginalized.[31] Until the rise of the Bhutto dynasty, the politicians of Sind felt such marginalization very strongly, particularly after Karachi ceased being the capital.[32] With little attention paid to the concerns of hundreds of thousands of desperately poor refugees, the *muhajirs*, crammed into the city from the partition years and the Bangladesh war of 1971, extreme violence and gang warfare have blighted Karachi.[33] This situation stands in marked contrast to the greater attention that was paid to demands of Punjabi Muslim refugees who had moved over from East Punjab to West Punjab. In Baluchistan, too, there has been over many years the growth of a Baluchi sense of nationalism that is resentful of the dominance of the Punjab and the Pashto military.[34] The roots of such ethnic problems, which do not necessarily pose a threat to the survival of Pakistan, lie in the absence of a genuine democracy.

The technical competence of the military machine has been the *raison d'être* for showering on it a great amount of public finance. The exaggerated trust that ordinary Pakistanis held for decades in the army's ability to deliver a knock-out blow against India permitted the disproportionate allocation of resources by the state for the defence establishment. The recent War on Terror has further bloated the military budgets, at the expense of such items as health, education and social welfare. These general remarks can be substantiated and vindicated by figures and statistics from official budgets of Pakistan, along with various international budgetary comparisons, etc. But a more sinister development has been the deliberate camouflaging of information and statistics concerned with the military's own control of the country's economy for private gain. This secret economy, described by one writer as *Milbus*, refers to a form of predatory control of military capital by the officer class, which is neither recorded nor is mentioned in the defence budget.[35] It works through secret foundations and charities whose tentacles are all-enveloping in the control of land, banks, fishing right, industrial enterprises etc., all meant for the enhancement of profits for the military establishment.[36] Its pervasiveness is the clearest proof of the lack of democracy in Pakistan.

Religious extremism

It was mentioned in Chapter 10 that, before independence, there was a significant split among the influential *ulama* and the religious ideologues about the desirability of creating Pakistan; but Jinnah's passionate oratory and appeal to Islamic sentiments, particularly between 1942 and 1947, influenced many who felt insecure about their mission in a secular independent India. They eventually committed themselves to Pakistan, but they expected their voice to command respect within the ruling circles of the new state. From the outset, this was to prove challenging for the Pakistani government. First, while Jinnah and the majority of the middle-class intelligentsia who formed the government acknowledged Islam as the fountain head of the new state, they had little interest in creating an Islamic state (Excerpt 11.2). Second, the religious establishment did not speak with a united voice. While in theory they were all agreed on such fundamentals as the centrality of the *Shariah* law and some form of hierarchical separation of non-Muslims from Muslims, in practice there were many differences of opinion and interpretation among them.

It took nine years of wrangling within the Constituent Assembly, the body charged with drafting a constitution, before a compromise document was finally agreed in 1956 between the politicians and the religious establishment. The latter gained a number of concessions: Pakistan was proclaimed an 'Islamic Republic'; only a Muslim could be president; the sovereignty of Allah was affirmed in the Declaration of Objectives; Islamic education and morality were to be promoted; and a Consultative Commission on Islamic Ideology was to ensure that the laws passed by parliament were in conformity with the Quran and the *Sunnah*. On the other hand, no mention was made of the term *Shariah*; the Consultative Commission remained just that, possessing only a consultative role; and the members of the Commission were to be chosen by the government rather than by their fellow *ulama*. The result was a reasonably balanced compromise.[37]

If those in power in Pakistan had kept to the letter and spirit of this aspect of the 1956 constitution, the country might have developed into a moderate Muslim secular state. This was, however, not to be, owing to the policies of two Presidents – Zulfikar Ali Bhutto and General Zia ul-Haq – who between them ruled through the 1970s and most of the 1980s. Both of them, in their own ways, were populist opportunists, and they used religious ideologues and religious parties to score short term gains. In their eagerness to portray themselves as good and true Muslims in front of their own people and Muslims worldwide, both promoted Islamization by their rhetoric about such nebulous concepts as Islamic economics, Islamic television, Islamic clothing and even Islamic bomb.[38] They also made Pakistan a proactive member of international Islamic organizations and blocs. After the humiliating defeat at the hands of India in 1971, Bhutto looked towards the Middle East and signed agreements with Saudi Arabia, Iran and the small but rich Gulf kingdoms.[39] Large scale Pakistani migration to those lands, resulting in substantial remittances, was thus secured, along with financial aid. Zia continued with Bhutto's policy, but with one big difference. Iran was no longer a favoured country with the coming to power of Ayatollah Khomeini in the revolution of 1979. Zia's Sunni prejudices drew him towards Saudi Arabia and Iraq. Saudi Arabian money and patronage helped Zia in his desire to Islamize the country through a network of *madrasas* run by Sunni *ulama*.[40] These *madrasas* were more than educational institutions; they served the political purpose of popularizing the military state.

One of the results of Zia ul-Haq's close relationship with the Wahabi-influenced Sunni establishment of Saudi Arabia has been the growth of a vicious form of sectarianism which has torn apart the generally amicable relations between the Sunni and the Shia. The latter, following their specific traditions, and encouraged by the 1979 revolution in Iran, had resisted Zia when he introduced an element of compulsion over the collection of *zakat*. Other Pakistanis who wished to pass on their inheritance according to Shia law and avoid paying the compulsory *zakat* also declared themselves to be Shia. The apparent rise in numbers of Shia within the population and the granting of exemptions to them by the government provided the Sunni militants with the excuse to begin an anti-Shia campaign. When many of the mainstream Islamic parties, such as Jamaat-i Islami, shied away from inciting anti-Shia rage, the Wahabi patrons encouraged other more extremist groups to carry out their hate campaigns.[41] A highly sectarian Sunni group that emerged from the Jhang district in Punjab, the Sipah-i Sahabah Pakistan, embarked on a murderous anti-Shia campaign for many years.[42] The Shia too, covertly supported by Iran, engaged in violence but, being a minority, bore the brunt of the sectarian brutality.

In their desire to gain legitimacy from the religious establishment, Bhutto and Zia had encouraged the Islamization of Pakistani law by proclaiming the *Shariah* as its basis. The specific *Shariah* regulations put into place by President Zia included stoning or whipping for adultery, amputation for theft, whipping for the use of intoxicants and imprisonment or the death penalty for blasphemy.[43] Although these severe punishments were meted out only very occasionally to the tiniest fraction of the populace, their very existence sullied the reputation of the country in the eyes of eminent lawyers and judicial authorities across the democratic world. More specifically, three groups of people have suffered grievously. The educated and motivated Ahmadiyya community was declared non-Muslim on the grounds that their nineteenth-century leader, Mirza Ghulam Ahmad (1838–1908), had considered himself to be endowed with the spiritual qualities of a prophet and thereby disobeyed the traditional Muslim credo that the Prophet Muhammad was the seal of all prophets.[44] Yet the members of this community have been a leading force in promoting Islamic missions worldwide. Vicious riots have been incited against the community. The Christians of Pakistan, a poor and marginalized community living in rural areas, face severe punishment if they are unfortunate enough to live alongside prejudiced and ignorant Muslim neighbours who are ready to provide anecdotal evidence of blasphemy against them in the *Shariah* courts. The blasphemy laws of Pakistan are among the most severe in the world; and Christians are the main victims. Third, it is the poorer women living in rural areas or in the city ghettoes who endure the hardships of the so-called Hudood Ordinances of 1979 (slightly modified by the Women's Protection Bill of 2006) that regulate the law relating to adultery and rape. Thus a woman alleging rape has to provide four male witnesses; and the failure to find such proof places her at risk of prosecution for adultery. If the offender is acquitted, then the woman invariably faces prosecution for adultery or fornication. All these varied violations of human rights, arising out of extreme Islamization measures, have upset many progressive people in Pakistan, but the dead weight of tradition still weighs heavy in Pakistani society at large. Two brave liberal politicians – Salman Taseer, the Governor of Punjab, and Shabaz Bhatti, the Federal Minister for Minorities – were both killed by fanatics in 2011.

Into the Afghan imbroglio

On 28 December 1979, the Soviet Union invaded Afghanistan in order to support the Communist government that was battling against its various enemies. From that time onwards Pakistan too became profoundly drawn into the Afghan crisis. However, while the Soviets have long departed from that country, Pakistan has for one reason or the other been unwilling or unable to extricate itself safely from its Afghan commitments. This intense involvement has taken place in three successive periods. During the first period (1979–89), Pakistan, a close Cold War ally of the United States, responded positively to American appeals to help to fight the Soviets. The United States was prepared to supply weapons to anti-Soviet militias through Pakistan and give Pakistan financial aid for her support. President Zia showed great readiness to get involved, as he was keen to upgrade his own military forces with American finance; but he also had another objective. He viewed his intervention ideologically too, from the perspective of an Islamic *jihad* against godless communism; and that brought him closer to Saudi Arabia. Since Saudi Arabia's very existence has been linked to US patronage and military support, it was clear that a Washington–Riyadh–Islamabad axis came to be pitted against Moscow. Pakistan became the conduit through which American military hardware was supplied to fanatical *mujahid* fighters trained in Pakistani *madrasas*, set up with Saudi money, and sent across the border to fight the Russians. Pakistan was able to accommodate nearly 2 million Afghan refugees without any financial cost to her; that burden was paid for by the United States and the Saudis.[45] This situation lasted for over ten years. The Soviets were not defeated, but the Soviet leader, Mikhail Gorbachev, alarmed by the unnecessary drain of resources that was bankrupting the USSR, finally withdrew his forces from Afghanistan in January 1989.

With the Soviet withdrawal began the second period. American enthusiasm for further involvement in Afghanistan dissipated, even though the Afghan communists continued to rule until March 1992. Pakistan, despite the fact that she was burdened by the presence of millions of refugees, along with the growth of trade in arms and drugs, particularly along the Afghan frontier, nevertheless carried on supporting Islamic fighters who, along with other Afghan nationalists, were attempting to oust the communists from power. Pakistan's plan, as conceived by her military and intelligence officers, was to maintain a strong influence in Afghanistan and to use that country for strategic depth in any future war with India.[46] The financial help from Saudi Arabia enabled her to continue supporting varied networks of Islamic fighters, particularly those led by a warlord, Gulbuddin Hekmatyar. With the fall of the communists in March 1992 there began a most vicious and violent civil war inside Afghanistan for the next four years. The country was totally ruined, and the beautiful city of Kabul destroyed beyond recognition. To their shame, the civilian governments of Pakistan did not attempt to restore peace or to help reconcile the various Afghan factions. They instead deepened their involvement in the country after switching support from Hekmatyar's forces to an even more extremist group, the Taliban. The word 'Talib' means a student of religion; and the Taliban consisted of those students who were trained in fundamentalist Wahabi ideology and militarism in the *madrasas* of Pakistan and Afghanistan sponsored by Saudi money. The Pakistani *madrasas* produced militant students in prolific numbers, but their Afghan counterparts were not idle either. It was a group of Afghan-based Taliban in Kandahar, led by Mullah Omar, who brought an end to the civil war by successfully assaulting Kabul in 1996 and taking control of government there. Pakistan was one of only three countries that supported the new government. From 1996 until November 2001, the Taliban enforced upon the hapless Afghan population a vengeful form of Islamic puritanism with cruel barbarity. The people accepted their rule simply because they brought a semblance of order to a ravaged land. The Pakistani leaders, having had the benefit of the legacy of modernity inherited from British India, could have helped to moderate the behaviour of the Taliban, but did nothing. They were to regret their inaction once the Taliban permitted Osama bin Laden and his al-Qaeda Islamist terrorists to operate training camps to attack the United States and the West.

The third period may be said to have begun on 11 September 2001. Outraged by the massacre of its citizens from air by al-Qaeda suicide bombers that day, the United States gathered a coalition of willing nations to attack Afghanistan, oust the Taliban and begin a so-called War on Terror. It expected Pakistan to be a member of the coalition and left her with no choice but to join it. General Musharraf, who was in power, obliged, to the chagrin of both the Taliban and many in Pakistan who, like Osama bin Laden, had hated the United States and Western arrogance and dominance since the First Gulf War of 1990. Musharraf closely allied himself to the CIA and the American military machine; he opened up all the facilities in his country for the movement of huge quantities of military hardware across the border. However, in order to maintain his country's influence over events in Afghanistan, Musharraf and the army turned a blind eye to Taliban leaders and their supporters fleeing into Pakistan on the eve of the American onslaught. The latter found refuge among the fellow Pathans on the frontier and in parts of Baluchistan, where many of them are still based. Pakistan has continued to balance the substantial help given to the Americans with the pursuit of its own *realpolitik*. On the one hand, in order to assuage and re-assure the Americans, sections of the Pakistani military and intelligence community have attacked and destroyed the local Pakistani Taliban cells in certain regions of the frontier; on the other hand, they have hesitated to take truly decisive action against Afghan Taliban leaders like Mullah Omar. This is primarily because of their strong dislike of the Afghan government's close ties with India and their desire to keep the Afghan Taliban as insurance for retaining a sphere of influence in Afghanistan when the Americans finally depart. Despairing of this Pakistani ambivalence, the Americans have themselves increasingly entered Pakistan in order to destroy as many of the Taliban that they can target through their drone attacks.[47] In 2011, their superior technology and intelligence allowed them to reach Osama bin Laden's apartment and kill him. This has further exacerbated anti-American feelings in Pakistan and made the government deeply unpopular. The Pakistani Taliban too have become incensed by their own government attacking them in order to satisfy the Americans; their revenge has come in the form of multiple suicide bombings against the police and other institutions. Pakistani society has thus become fractured, and the country finds itself between a rock and a hard place. Further, United States–Pakistani relations worsened dramatically in November 2011 when twenty-four Pakistani soldiers were killed by American drones. We can conclude that Pakistan is reaping the nemesis of her misguided venture into the Afghan imbroglio: and all because of her leaders' obsessive desire to compete with India in South Asian geo-politics.

From East Pakistan to Bangladesh

The Pakistan that Jinnah achieved was, from the outset, beset by one critical geographical problem. This was that the West was separated from the East by a distance of over a thousand miles. The complex difficulties of logistics and communications, arising out of the physical disconnection between the two divisions, became clear in 1971/2 when the government attempted to suppress the independence movement in the East by military means. The geographical separation might have mattered less if other forms of separation and alienation had not developed so virulently since 1947. In reality, the roots of the conflict of 1971 lay, on the one hand, in the cultural and social distance separating East Pakistan from West Pakistan and, on the other, in the political and economic domination of the West over the East.

The cultural gap between the easterners and the westerners was essentially symbolized by the dispute over language.[48] The former felt strongly that there was no official parity of esteem between the Bengali and Urdu languages. The founding fathers of Pakistan envisaged Urdu as the common *lingua franca*; but of the six major ethnic languages – Punjabi, Sindhi, Pashto, Baluchi, Kashmiri and Bengali – it was Bengali that seemed to be furthest from Urdu; and Urdu was spoken in Bengal by mostly the *ashraf* middle-class families of Dacca and Calcutta, a tiny proportion of the population. The vast majority of East Pakistanis spoke Bengali, Sylheti or related dialects;

and their demand for the primacy of their mother tongues became the rallying cry of the Bengali Language Movement, led by a radical politician, Sheikh Mujibur Rahman (1920–75). Apart from the issue of language, the Bengalis resented the lack of adequate democratic representation for their province at the centre and the economic imbalance between the two Pakistans.[49]

Throughout the 1960s, Mujibur Rahman had campaigned against the inequality that East Pakistanis felt strongly about; and his party, the Awami League, finally won an overall numerical majority in the national elections of 1970. The chicanery and subterfuge then employed by both General Yahya Khan and Zulfikar Ali Bhutto in denying Mujibur Rahman the rightful gains of his election victory inevitably led to the start of a major rebellion in the East.[50] The civil war that resulted was fought with great brutality on all sides. In the first instance, the Bengalis terrified the local non-Bengalis, particularly the many thousands of Bihari refugees who had settled in East Pakistan since 1947, by various forms of vengeful attacks. In response, the 100,000 strong West Pakistani led army, in association with the Urdu-speaking and Islamic pro-Pakistani militias, unleashed on 25 March 1971 a reign of terror that lasted for the next nine months. Hundreds of Bengali students and activists were executed; tens of thousands of Bengali guerrilla fighters lost their lives; and millions of refugees, particularly Hindu Bengalis, crossed the border into India. Latest research must caution us against placing all the blame for cruelty on the part of West Pakistanis, as has been conventionally reported. Excessively savage and vicious reprisals were also carried out pitilessly by the Bengali freedom fighters.[51] The fate of the West Pakistanis was ultimately sealed once the Indian army entered the country. The surrender of 93,000 of their elite troops on 16 December 1971, led to complete humiliation of Pakistan. The new nation of Bangladesh came into existence in 1972.

A brief overview of Bangladeshi politics since 1972

The years since 1972 may be divided into three periods: 1972 to 1975 when Mujibur Rahman ruled supreme; 1975 to 1991 when military rule prevailed under two dictators, Ziaur Rahman and Hussain Mohammed Ershad; and 1991 to the present day, with power fluctuating between two women rivals, Begum Khaleda Zia and Sheikh Hasina Wajid.

Mujibur Rahman came to power in a blaze of glory and fame, but very soon his ideas and ideals met with resistance from different quarters. He believed in a secular democracy and banned religious parties; the new Bangladesh constitution referred only slightly to Islam. This did not endear him to the orthodox Muslims. His populist socialism was disliked by the business community. His pro-India policy made him appear too subservient to that country; and, in any case, India did not respond enthusiastically and hastily returned the 3 million refugees back to their impoverished homeland. Within a few years, Mujibur was faced with mounting economic and social problems, some being endemic and others caused by the international oil price rise of 1973 and a huge local famine of 1974.[52] His response merely compounded his difficulties. After declaring a national emergency, he established a personal praetorian guard known as the National Security Force.[53] The violence unleashed by many irregulars within this force alarmed the professional army which had been deprived of resources by him. In August 1975 he and his wife, along with their sons and servants, were assassinated by a group of hostile officers. His two daughters, who were in Germany, escaped the cruel fate of their family.

Apart from differences in styles of leadership, the strategic policies of the two military dictators, Ziaur Rahman (1936–81) and Hussain Mohammed Ershad (b. 1930) were essentially very similar in that they reversed Mujibur Rahman's course of action. They jettisoned socialism and adopted a more robust right-wing economic philosophy that helped to draw in greater foreign investment. The military budget soared to nearly 40 per cent of the revenue,[54] which provided the armed forces with more effective weaponry; and the army structure was rationalized by easing out the irregular forces and remnants of guerrilla fighters from the civil war period. They took personal interest in the administrative re-organization of the country and established a variety of village-based projects and schemes.[55] Agriculture revived, and ordinary people felt a sense of order and security. Both dictators believed that military rule was a short term necessity in order to stabilize the country and attempted to legitimize their rule through elections and referenda. They even formed their own parties: the Bangladesh National Party (BNP) and the Jatiyo Party which, along with the older Awami League, became active participants in democratic revival. The dictators also re-shaped two particularly sensitive policies, in opposition to Mujibur Rahman's ideology. One was adopting a less pliant attitude towards India, while building bridges with Pakistan and other Muslim countries. Friction with India became more common from then onwards, particularly over the share of Ganges waters.[56] The other was the Islamic re-framing of the country's constitution and the abandonment of a secular approach. This, along with the permissive attitude towards Islamic religious parties, oriented the country towards a new trajectory.

Two strong-minded women, Begum Khaleda Zia (b. 1945), the wife of Ziaur Rahman, and Sheikh Hasina Wajid (b. 1947), the daughter of Mujibur Rahman, together built a temporary alliance against General Ershad in the late 1980s; and the rowdy street politics of their supporters and student unions finally brought about his downfall. Democratic politics were restored in 1991. Through elections, some free and fair and others fraudulent, power has oscillated since then between the two women in a regular pattern: Khaleda Zia's ascendancy from 1991 to 1996 and 2001 to 2005; and

Sheikh Hasina's ascendancy from 1996 to 2001 and from 2007 to the present day. The genuine freedom to choose their politicians even amidst widespread corruption has given the people a sense of hope and confidence. The economy grew at a steady rate of 4 per cent between 1991 and 2006, and Bangladesh shifted from 'a basket-case to a middle-ranked developed country'.[57] Although gross poverty has remained persistent, Bangladesh may emerge as the only South Asian country to reduce poverty by the standard of the UN Millennium Development Goal. It also leads both India and Pakistan in female literacy, while impecunious women are actively engaged in micro-credit financing and lending schemes.[58]

The increasing aggregate wealth resulting from investment, modernization, world trade and globalization has, however, eluded one particular group of people in Bangladesh. They are the tribal hill people, such as the Chakmas, who live in the Chittagong Hill Tracts. Neither Bengali nor Muslim, the hill people and their lands had been protected from intrusion and exploitation in British India by the efforts of Christian missionaries. This protection was lost as soon as East Pakistan was created. A massive hydro-electric project started by the Pakistani government with the US Agency of International Development submerged 40 per cent of the arable land and destroyed the livelihoods of nearly 100,000 Chakmas.[59] Even though the Bengalis fought for their independence from Pakistan on the grounds of their distinct ethnic and cultural identity, the Bangladeshi governments, both democratic and military, have shown no particular empathy for the ethnic and cultural characteristics of their hill peoples.[60] Bengali settlers have steadily encroached on their lands over the years, and the natural resources of their region have been exploited by 'Bengalis for Bengalis'.[61]

The challenge from religious forces

For centuries, the dominant Islamic tradition among Bengali Muslims was that of Sufism. During the last fifty years of British rule, however, a more assertive form of Islamic consciousness developed, as a result of struggles for Muslim education and political representation. Jinnah's campaign for Pakistan as a homeland for Muslims also came to be enthusiastically received over the course of time; and the creation of East Pakistan was welcomed by varied Islamic groups and parties that wanted to turn their state into an Islamic state. Splits within the religious establishment arose once the ethnic and cultural differences began to separate the East Pakistanis from West Pakistanis. While most of the *ulama* of mosques with Bengali-speaking congregations wished for an autonomous Bengali Islam based on Bengal's native traditions, the Urdu-speaking *ulama* of mosques frequented by Biharis and other Urdu speakers wanted to preserve the stridently purist strand of Islam that was advocated by the highly organized religious party of the Jamaat-i Islami.[62] Their members fought for Pakistan during the civil war.

The defeat of Pakistan and, following that, Mujibur Rahman's secular policies were a setback for them.

Islamic fortunes revived under military rule between 1975 and 1991, when the ban on religious parties was removed and an Islamic-oriented constitution was adopted. Once democratic politics were restored in 1991, there was a scramble by all the main parties for the support of Jamaat-i Islami and the religious parties at election times. Begum Khaleda Zia showed greatest sympathy towards Islamic involvement in politics; and her election victories of 1991 and 2001 owed much to the support of conservative religious organizations.[63] In order to appease and placate them, she encouraged a phenomenal growth in the number of *madrasas* for mass education during her first period of office (1991–6). Her successor, Sheikh Hasina Wajid, although more circumspect and more secular-minded, did not take any steps to check the proliferation of these institutions, again partly because of electoral calculations and partly because most of their finance came from Saudi Arabia. Thus, from a total of 1830 *madrasas* in 1975–6, the number had jumped to 64,000 by 2002.[64]

The participation of religious parties in Khaleda Zia's coalition government of 2001 created a more sympathetic environment within which more fundamental or extremist organizations were able to flourish. Nearly thirty-three such organizations, some advocating extreme forms of violence such as suicide bombing, were identified by 2005.[65] On 17 August of that year, terror blasts shook sixty-three out of sixty-four Bangladeshi districts.[66] Khaleda Zia could not act decisively without alienating her Islamic coalition partners, but the threat of isolation by the international community finally made her aware of the dangers of giving in too easily to religious forces. Some stability and a more secular spirit have been partially restored since 2007, with the return to power in that year of Sheikh Hasina Wajid and the Awami League. Nevertheless the danger of religious extremism will continue to be a real threat, particularly from millions of disaffected and ill-trained young people graduating from the ideological *madrasas*, owing to general poverty and lack of sustainable employment opportunities.

While terrorism by religious extremists has generally become the prominent news story since 9/11, another form of silent but systematic breach of human rights has been widely prevalent in East Pakistan/Bangladesh for many decades: the persecution of Hindus. Before 1971 most of the Hindu middle class had felt insecure and migrated to India. During the civil war of 1971, when Pakistani soldiers and their collaborators made it a mission to target the rural Hindus, nearly 3 million refugees crossed the border into India. After the war, Mujibur Rahman eloquently defended the returning Hindus, but his government did not help them to recover their properties. A hostile atmosphere was continually generated by economic jealousies and sermons by fanatical Sunni militants against them. Communal hostilities within India during the 1980s and the 1990s did not help; and the destruction of the Babri Masjid at Ayodhya in 1992 further ignited anti-Hindu feelings. Systematic discrimination, sequestration of properties, the hostility of a major political party (the BNP), and the fear of attacks and rapes have all steadily forced the Hindus to continue to migrate to India. Profoundly angered by her country's treatment of Hindu minority, the Bangladeshi writer Taslima Nasrin published in 1993 a powerful novel, *Lajja* (Shame), in which she graphically portrayed the sufferings of a Hindu family. Her life, thereafter, was made intolerable by Muslim zealots, and she has lived in a state of permanent exile for the last two decades.

Literature and the arts

The educational infrastructure throughout the subcontinent has, generally speaking, been poorly resourced in this period; and national literacy figures remain low, in comparison with many other Third World countries. This remains one of the great obstacles to the universal spread of high literary and artistic culture among the vast mass of more than 1.5 billion peoples of South Asia. On the other hand, as we have observed in Chapters 8, 9 and 10, there has been a steady rise of the educated middle class since the early nineteenth century; and today, with Asia providing the principal stimulus to world economic activity and growth, the subcontinent's middle class is burgeoning ever more. Most of its members are, as usual, preoccupied with acquiring monetary affluence and displaying the symbols and products of consumer culture amidst their families and neighbours; but a small group of dedicated South Asians continues to produce cultural works of originality and creativity.

Muslim writers, or writers from Muslim backgrounds, have greatly contributed to modern South Asian literature of this period, both in South Asian languages and in English. Urdu and Bengali remain the principal languages of communication for Muslim authors who do not use English. Rich literary traditions also exist within other vernaculars of India, Pakistan and Bangladesh. English is widely used by authors living in the subcontinent or in the diaspora. Every writer has his or her particular strengths, and our judgements about the quality of their work can at best be subjective. Two prominent names, however, stand out for recognition: Faiz Ahmad Faiz (1911–1984) and Salman Rushdie (born 1947). Faiz, perhaps the greatest of Pakistani poets, wrote the best of his works in Urdu or Punjabi; they range from love poems written in the *ghazal* style to political verses. Romantic imageries and nationalist sentiments are both conveyed in his poems. He had great hopes for Pakistan but soon became disappointed by her rulers who twice in his life imprisoned him on spurious political charges. His prison poems are political in the sense that they encompass national and international concerns (Excerpt 11.3).[67] Salman Rushdie's finest works are in English, and his profound mastery of English words, phrases, idioms and metaphors must place him among the top five or six fiction writers in the English-speaking world. While Faiz suffered persecution at the hands of Pakistani governments for his communist views, Salman Rushdie has had to fear the wrath of Muslim religious fanatics over his book *Satanic Verses* which certainly offended the religious sensibilities of many Muslims who actually read the book.[68] In one sense, Faiz, Rushdie and Taslima Nasrin have all experienced what it is like to be persecuted in the cause of freedom of expression.

The political fragmentation of the subcontinent and the rise of separate forms of nationalisms in three different countries are a reality that cannot be ignored; but contemporary South Asian literature attempts to transcend these differences by exploring some common underlying themes that cross national boundaries. The three such common themes are: the 1947 partition, migration and diaspora life, and women's role in the family. The Muslim authors' contributions have been substantial.

The experience and memory of partition is one of the great trans-national themes of our period. It is the subject of many novels, poems and short stories. It is depicted from a variety of angles. Thus, for example, in her posthumously published novel, *The Heart Divided* (Chapter 10), Mumtaz Shah Nawaz deals with the close ties between Hindus and Muslims before communal differences set them apart during the five years before independence. Although a supporter of Pakistan, she shows empathy for the predicaments of both Hindus and Muslims.[69] Another way of approaching the subject came from the pen of Saadat Hasan Manto (1912–1955). He was a well known short story writer in Urdu belonging to the Progressive Writers' Association, but migrated in 1947 to Pakistan where he lived as a drunken recluse haunted by the dark shadows of partition. The short story that was his literary forte gave him the flexibility to describe graphically, with black humour and brutal imagery, the murderous scenes on, for example, trains. In Manto's world there were no Hindus or Muslims: there were just debauched and wicked people doing terrible things to fellow humans.[70] A popular angle from which a number of Bengali writers have approached partition is that of displacement and its consequences in the form of refugees.[71] The iconic novel, *Midnight's Children*, by Salman Rushdie, is also in a sense a partition novel, in the way Rushdie uses the historical moment of midnight on 15 August 1947 as a symbol to tie together in vivid words and imaginative ways the fortunes of children born then and the larger drama of Indian independence, the socio-cultural milieu of a Muslim family and post-colonial India.[72]

The experience of migration is another important trans-national theme. Hundreds and thousands of post-1947 Indians, Pakistanis and Bangladeshis travelled across the world as migrants. The reasons for leaving their homelands, the difficulties they initially encountered in their 'host' countries, their hard work and sacrifices for their families, the dual culture and the identity crisis of their children internalizing the values and anxieties of the adult world, the 'myth' of return etc. – all these related issues have of course been explored well by academics and sociologists; but they have also formed the core of some of the most imaginative writing by diaspora authors of South Asian origin. Some of the best early writing of Rushdie consisted of essays on the diaspora. Other Muslim writers or writers of Muslim background, like Hanif Kureishi and Kamila Shamsie, have also made distinct contributions to diasporic English literature.[73]

The challenges that women face in South Asia also transcend national, ethnic or religious differences. The simple fact is that, despite the progress made by their gender in the last hundred years (Chapter 10), great swathes of South Asian women are either caught in the trap of customs and conventions or deprived of legal rights. Their predicaments have been accurately described in the works of writers of both sexes, but a Muslim woman writer who broke all bounds of conventionality and was unafraid to handle even the most challenging of feminist issues was Ismat Chughtai (1911–91) from India, who wrote mostly in Urdu. It was her 1942 story, *Lihaaf* (The Quilt), relating the companionship of a frustrated upper-class Muslim woman with her female servant, with implications of lesbianism, etc., that brought her much controversy and a court case (Excerpt 11.4).[74] She greatly admired Dr Rashid Jahan (Chapter 10) and, in turn, she inspired many generations of South Asian women writers.[75]

With a long cultural heritage behind them, Muslim artists across the subcontinent have been very creative in the post-independence period. Their works display a variety of artistic styles: figurative, abstract, calligraphic, post-modern, etc. The most outstanding Indian artist of the second half of the twentieth century was M.F. Husain (1915–2011), a member of the so-called Bombay Progressives, a group of artists experimenting with abstract paintings. Husain's products have given modern Indian art a global reach. Yet, like Taslima Nasrin, Salman Rushdie and Faiz Ahmad Faiz, Husain too faced persecution. While the other three suffered at the hands of Muslim politicians and religious fanatics, Husain's ordeal was caused by the right-wing Hindu politicians and religious extremists who found some of his paintings, completed many years previously, to be insulting to Hindu faith. The works of many outstanding artists of Pakistan also do not simply conform to Islamic conventions or decorations; they take on old and new styles from across the subcontinent and abroad. Their modernist approach has not fitted in easily within the conservative milieu of Pakistan, and artists like Sadequain Naqvi (1930–87) faced hostilities over some of their works.[76] The dominant themes portrayed by many Bangladesh artists, represented by such figures as Zainul Abedin (1918–76) and S.M. Sultan (1923–94), are those of village life, folk culture, scenes from the civil war of 1971, oppressive patriarchy, etc.[77] As with literature, migrant artists from the subcontinent have also established themselves in the diaspora and are helping to fuse Western and oriental traditions.

Select excerpts

11.1 Advice to Indian Muslims

Hamid Dalwai (1932–77) was a radical Muslim social thinker and activist who, during his short life, fought hard to persuade his fellow Muslims that the most appropriate way of fighting Hindu communalism was for them to become more modern and integrated in the post-1947 Indian life. However, some of the suggestions he put forward in his 1968 publication, *Muslim Politics in India*, may be a step too far for both Muslims and non-Muslims.

> We have to support Muslim modernism in India. We have to insist on a common personal law for all citizens of India. All marriages in India must be registered under a common Civil Code. Religious conversion should not be allowed, except when the intending convert is adult and the conversion takes place before a magistrate. Children born of inter-religious marriages should be free to practice any religion but only after they reach legal adulthood. If either a Muslim or a Hindu temple obstructs the passage of traffic on a thoroughfare, it ought to be removed. Government should have control over the income of all religious property. This income should be spent on education and public welfare alone.
>
> (Guha 2011: 456)

11.2 Jinnah's vision of secularism

On 11 August 1947, on his election as first President of the Constituent Assembly of Pakistan, Jinnah addressed the issue of religion and citizenship in secular terms. Unfortunately, his hopes have been unrealized in Pakistan.

> You are free; you are free to go to your temples, you are free to go to your mosques or to any other place of worship in this state of Pakistan. You may belong to any religion or caste or creed – that has nothing to do with the business of the state . . . Today (in England) we might say with justice that Roman Catholics and Protestants do not exist; what exists now is that every man is a

> citizen, an equal citizen of Great Britain and they are all members of the nation . . . Now, I think that we should keep that in front of us as our ideal and you will find that in course of time Hindus would cease to be Hindus and Muslims would cease to be Muslims, not in the religious sense, because that is the personal faith of each individual, but in the political sense as citizens of the State.
>
> (Pakistan Government 1989: 46–7)

11.3 Speak up!

Faiz's poem *Speak*, a product of his prison experience, exhorts people not to be intimidated by forces of oppression and despotism.

> Speak, your lips are free,
> Speak, it is your own tongue,
> Speak, it is your own body,
> Speak, your life is still yours.
> See how in the blacksmith's shop
> The flame burns wild, the iron glows red;
> The locks open their jaws,
> And every chain begins to break.
> Speak, this brief hour is long enough
> Before the death of body and tongue;
> Speak, 'cause the truth is not dead yet,
> Speak, speak, whatever you must speak.
>
> (Hussain 1999: Poemhunter: www.poemhunter.com/poem/speak-4/)

11.4 A section from The Quilt *by Ismat Chughtai*

The nine-year-old niece describes the frustrations of her aunt, Begum Jan, married to Nawab Sahib, before her 'rescue' by Rabbo, the female servant.

> All (Nawab Sahib) liked to do was keep an open house for students; young, fair and slim-waisted boys, whose expenses were borne entirely by him. After marrying Begum Jan . . . he promptly forgot about her! The young, delicate Begum began to wilt with loneliness . . . Despite renewing the cotton filling in her quilt each year, Begum Jan continued to shiver, night after night. Each time she turned over, the quilt assumed ferocious shapes which appeared like shadowy monsters on the wall. She lay in terror; not one of the shadows carried any promise of life. What the hell was life worth anyway? Why live? But Begum Jan was destined to live, and once she started living, did she ever! . . . Rabbo came to her rescue just as she was

starting to go under . . . Rabbo had no other household duties. Perched on the four poster-bed, she was always massaging Begum Jan's head, feet or some other part of her anatomy. Someone other than Begum Jan receiving such a quantity of human touching, what would the consequences be? . . . Rabbo! She was as black as Begum Jan was white, like burnt iron ore! Those puffy hands were as quick as lightning, now at her waist, now her lips, now kneading her thighs and dashing towards her ankles. Whenever I sat down with Begum Jan, my eyes were riveted to those roving hands. . .

(Chughtai 1991: 7–19)

CONCLUDING THOUGHTS

It is legitimate to pursue the study of Muslim history in South Asia as a discreet subject in its own right. It is also legitimate to celebrate the rich legacies of that history. What needs to be kept in mind, however, is that the Muslims share the region with the more numerous non-Muslims. Both have to learn to work together if the true potential of South Asia is to be realized.

At present, despite its riches and resources, and the talents of its people, South Asia remains essentially a region of economic and social backwardness. The per capita GDP and the UN Human Development Index for each of the three countries – India, Pakistan, Bangladesh – remain abysmally low. The situation will not be retrieved without strong and sustained efforts by the intellectual and political leaders of the three countries to explore new pathways that challenge the existing trajectories and thought processes.

A South Asian leadership, drawn from the many diverse communities and faiths of the three nations, can make a start by engaging with three critical issues. First, there is an urgent need to re-formulate fresh arguments for egalitarianism in society. A fatalistic acceptance of gross inequality as a natural human condition needs to be challenged. The economic and social dignity of the poor and the dispossessed cannot be sacrificed at the altar of global capitalism. Second, it needs to be recognized that comprehensive human progress will only be possible when men and women both work creatively in an atmosphere of freedom and mutual respect. Ignorance blights the lives of far too may South Asian women, especially Muslim women. Women's education therefore needs to be prioritized over many other social projects. Third, the educated class in all communities needs to challenge the intolerant bigotry and fanaticism of their co-religionists that is an ever present threat to the peace and stability of the entire subcontinent. Listening to others and seeking to understand what they are saying are the most effective ways of countering our own prejudices. This was how, in the sixteenth century, the great Emperor Akbar endeavoured to change the mindset of the Islamocentric *ulama* at his court (Chapter 6).

SELECT LANDMARK DATES

CE

570–632	Life and times of Prophet Muhammad
622	The *hijrah* marking the start of the Muslim calendar
711	Arab invasion of Sind
1000	Turco-Afghan invasions of India started by Mahmud of Ghazna
1190	Sufi *khanqua* set up in Ajmer by Muin ud-Din Chishti
1193	Muhammad Ghuri defeats Prithviraj Chauhan and takes Delhi
1202	The first stage of Qutb Minar completed
1206	Sultanate of Delhi begins with the Slave dynasty
1280	Nizam ud-Din Auliya settles in Delhi
1333–41	The Moroccan traveller Ibn Battuta resides at the court of Sultan Muhammad Tughluq
1398	Sack of Delhi by Timur
1526	Mughal Empire founded by Babur
1571	Akbar builds a new capital at Fatehpur Sikri
1600–1700	Sufi missionaries undertake mass conversion efforts in rural Bengal
1653	The building of the Taj Mahal completed
1689	Aurangzeb's Deccan campaign marks the start of Mughal decline
1703–62	Life and times of Shah Wali Allah
1724	Nizam ul-Mulk establishes the Asaf Jahi dynasty of Hyderabad
1739	Sack of Delhi by Nadir Shah
1757	Siraj ud Daulah defeated by Robert Clive at the Battle of Plassey

1765	Diwani of Bengal ceded by Shah Alam II to the East India Company
1797–1869	Life and times of Ghalib
1799	Defeat of Tipu Sultan at the Battle of Seringapatam
1825	Delhi College founded
1856	Wajid Ali Shah loses his kingdom of Awadh to the East India Company
1817–98	Life and times of Sir Sayyid Ahmad Khan
1835	Persian replaced by English as official language
1857–8	The great Indian rebellion against the East India Company
1858	The last Mughal Emperor, Bahadur Shah Zafar, exiled
1867	Deoband Seminary established
1876–1948	Life and times of Mohammad Ali Jinnah
1877	The Aligarh Muhammadan Anglo-Oriental College founded
1877–1938	Life and times of Sir Muhammad Iqbal
1906	The founding of the Muslim League
1909	Morley–Minto Reforms
1916	Hindu–Muslim unity at Lucknow Pact
1928	Jinnah's split from the Congress
1933	The term 'Pakistan' coined by Rehmat Ali and colleagues at Cambridge
1940	The Lahore Resolution
1946	Jinnah's call for the Direct Action Day on 16 August
1947	Independence of India and Pakistan
1948	Mahatma Gandhi assassinated by Godse, a Muslim-hating Hindu fanatic
1948	War between India and Pakistan over Kashmir
1958	Ayub Khan establishes the first military dictatorship in Pakistan
1965	Second India–Pakistan war
1971–2	Civil War in Pakistan and the creation of Bangladesh
1975	Mujibur Rahman assassinated in Bangladesh
1977	Zulfikar Bhutto executed in Pakistan
1979	The start of Pakistan's active involvement in Afghanistan

1986	The landmark Indian legal case of Shah Bano
1988	The novel *Satanic Verses* published by Salman Rushdie, to the dismay of Muslims worldwide
1991	Restoration of civilian rule in Bangladesh
1992	The Babri Masjid at Ayodhya destroyed by Hindu extremists
1993	Bombay Stock Exchange bombed by Islamist terrorists
1998	Pakistan explodes nuclear devices
2002	The Gujarat massacres
2007	Benazir Bhutto assassinated in Pakistan
2009	The start of suicide bombing campaign by Pakistani Taliban; terrorists from Pakistan infiltrate Mumbai and carry out shooting outrages.
2011	USA–Pakistan relations worsen with the killing of both Osama bin Laden and Pakistani soldiers by American drones.

MUSLIM ROYAL DYNASTIES

Numerous Muslim dynasties and monarchs ruled over different parts of South Asia between the eleventh century CE and the mid-twentieth century. Muslim dynasties are here classed as imperial or regional. They are chronologically arranged. While a full list of all imperial rulers is given, that for regional rulers is restricted to the kingdoms of Awadh and Hyderabad (regional dynasties numbers 19 and 20). For other regional kingdoms only those rulers mentioned in the book are listed. Complete lists can be found in *The New Islamic Dynasties* (Bosworth 1996).

Imperial dynasties I: dynasties of the Delhi Sultanate (1206–1526) – (Chapters 4 and 5)

1 Slave dynasty (1206–1290)

1206–10	Qutb ud-Din Aybeg
1210–11	Aram Shah
1211–36	Shams ud-Din Iltutmish
1236	Rukn ud-Din Firoze
1236–40	Raziyah
1240–2	Muiz ud-Din Bahram
1242–6	Ala ud-Din Masud
1246–66	Nasir ud-Din Muhammad
1266–86	Ghiyas ud-din Balban
1286–90	Muiz ud-Din Kaiqubad
1290	Kayumarz

2 Khaljis of Delhi (1290–1320)

1290–6	Jalal ud-Din Khalji
1296–1316	Ala ud-Din Khalji
1316–20	Mubarak Shah

3 Tughluq dynasty (1321–1413)

1321–5	Ghiyas ud-Din Tughluq
1325–51	Muhammad Shah Tughluq
1351–88	Firoze Shah Tughluq
1388–9	Ghiyas ud-Din Tughluq II
1389–90	Abu Bakr Shah
1390–3	Nasir ud-Din Muhammad Shah
1393	Ala ud-Din Sikander Shah
1393–4	Mahmud Nasir ud-Din
1394–9	Nasir ud-Din Nusrat Shah
1399–1413	Nasir ud-Din Muhammad Shah

4 Sayyid dynasty (1414–51)

1414–21	Khizr Khan
1421–34	Mubarak Shah
1434–45	Muhammad Shah
1445–51	Alam Shah

5 Lodi dynasty (1451–1526)

1451–89	Bahlul Lodi
1489–1517	Sikander Lodi
1517–26	Ibrahim Lodi

Imperial dynasties II: the Mughals (1526–1858) – (Chapters 5 to 8)

1526–30	Babur
1530–40	Humayun
1540–55	Interregnum: Suri dynasty
1555–6	Humayun
1556–1605	Akbar
1605–27	Jahangir
1627–58	Shah Jahan
1658–1707	Aurangzeb Alamgir
1707–12	Bahadur Shah I

1712–13	Jahandar Shah
1713–19	Farrukhsiyar
1719	Rafi ud-Darajat
1719	Rafi ud-Daulat (Shah Jahan II)
1719–48	Muhammad Shah
1748–54	Ahmad Shah
1754–9	Alamgir II
1759–1806	Shah Alam II
1806–37	Akbar II
1837–58	Bahadur Shah II

Regional dynasties

1 Kingdom of Sind (711–843) (Chapters 2, 5 and 8)

737–1005	Arab Amirs
1005–58	Ghaznavid governors
1026–1351	Sumra dynasty
1351–1522	Samma dynasty
1522–91	Arghun dynasty
1591–1718	Mughal control
1718–83	Kalhora rule
1783–1843	Talpur Amirs

2 The Ghaznavids (977–1186) (Chapter 3)

977–97	Sabuktigin
998–1030	Mahmud (of Ghazna)
1031–40	Masud I
1059–99	Ibrahim
1099–1115	Masud III
1117–57	Bahram Shah
1160–86	Khusrav Malik

3 Shansabanis of Ghur (1011–1215) (Chapter 3)

1173–1203	Muhammad Ghuri

4 Sultanate of Kashmir (1339–1561) (Chapter 5)

1339–49	Shah Mirza Shams ud-Din
1389–1413	Sikander
1420–70	Zainul Abidin

5 Ilyas Shahis of Bengal (1342–1415 and 1433–86) (Chapter 5)

1342–58	Shams ud-Din Ilyas Shah

6 Bahmanis of the Deccan (1347–c. 1518) (Chapter 5)

1347–58	Ala ud-Din Bahman Shah
1422–36	Ahmad Shah I

7 Faruqis of Khandesh (1382–1601) (Chapter 5)

1382–99	Malik Raja

8 Sharqis of Jaunpur (1394–1479) (Chapter 5)

1394–99	Malik Sarwar
1402–40	Ibrahim Shah Sharqi
1440–57	Mahmud Sharqi

9 Sultanate of Gujarat (1396–1573) (Chapter 5)

1407–11	Zafar Khan (Muzaffar Shah)
1411–42	Ahmad Shah I
1459–1511	Mahmud I Beghara
1526–36/7	Bahadur Shah

10 Khaljis of Malwa (1401–1570) (Chapter 5)

1435–69	Mahmud Khalji
1469–1501	Ghiyas ud-Din Khalji

11 *Baridis of Bidar (1489–1619) (Chapters 5–6)*

12 *Adil Shahis of Bijapur (1490–1686) (Chapters 5–6)*

1580–1627	Ibrahim Adil Shah II
1627–56	Muhammad Adil Shah

13 *Sultanate of Berar (1490–1596) (Chapter 5–6)*

14 *Husain Shahis of Bengal (1493–1539) (Chapter 5)*

1493–1519	Ala ud-Din
1519–32	Nasir ud-Din Nusrat

15 *Nizam Shahis of Ahmadnagar (1496–1666) (Chapters 5–6)*

16 *Qutb Shahis of Golconda and Hyderabad (1512–1687) (Chapters 5–6)*

1580–1611	Mumammad Quli
1626–72	Sultan Abdullah Qutb Shah

17 *Suri dynasty (1540–55) (Chapter 5)*

1540–5	Sher Shah Suri

18 *Nawabs of Bengal (1717–1880) (Chapter 7)*

1717–40	Murshid Quli Khan
1740–56	Ali Vardi Khan
1756–7	Siraj ud-Daula
1757–60	Mir Jaffar
1760–3	Mir Qasim

19 *Nawab-Wazirs of Awadh (1722–1856) (Chapters 7–8)*

1722–39	Saadat Khan I
1739–54	Safdar Jang
1754–75	Shuja ud-Daula

1775–98	Asaf ud-Daula
1798–1814	Saadat Khan II
1814–27	Haydar I
1827–37	Haydar II Sulayman Jah
1837–42	Muhammad Ali
1842–7	Amjad Ali Thurayyah Jah
1847–56	Wajid Ali Shah

20 *Asaf Jahis of Hyderabad (1724–1950) (Chapters 6, 7 and 9)*

1724–48	Nizam ul-Mulk, Asaf Jah I
1748–51	Nasir Jang
1751–2	Muzaffar Jang
1752–62	Salabat Jang
1762–1803	Nizam Ali Khan
1803–29	Sikander Jha
1829–57	Farkhanda Ali Khan
1857–69	Mir Mahbub Ali I
1869–1911	Mir Mahbub Ali II
1911–50	Osman Ali Khan, Asaf Jah VII

21 *Wallajahs of Arcot (1749–1867) (Chapters 6–7)*

1749–95	Muhammad Ali Wallajah

22 *Sultanate of Mysore (1761–1799) (Chapter 7)*

1761–82	Haidar Ali
1782–99	Tipu Sultan

23 *Begums of Bhopal (1819–1926) (Chapter 9)*

1844–60 & 1868–1901	Sultan Shahjahan Begum
1901–26	Sultan Jahan Begum

GLOSSARY

The following list is a selection of some common non-English terms frequently used in Muslim South Asian history. All others have been explained within the text.

adab	manners, behaviour or conduct
ashraf	nobles; Muslims claiming descent from Sayyids, Shaikhs, Mughals and Pathans who had migrated to India from Arabia, Iran, Central Asia and Afghanistan
badshah	emperor
burqa	veil
dargah	shrine at a Sufi saint's tomb
dhimmi	the people of the Book, e.g. Christians or Jews
fakir	an ascetic
fatwa	religious ruling based on Islamic law
ghazi	an Islamic warrior
hadith	an established saying or story about Prophet Muhammad
hajj	pilgrimage to Mecca
hajji	one who has been on a pilgrimage to Mecca
hijrah	Muhammad's flight from Mecca to Medina
imam	a mosque prayer leader who preaches a sermon on Fridays
iqta	land granted by a Muslim ruler to an individual
jagir	lands whose revenue substituted as salary for a nobleman's services to a ruler
jami	congregational; e.g. jami mosque
jizya	poll tax levied on non-Muslims for their protection
jihad	spiritual war
khanazad	the most loyal of Mughal nobles
khanqah	Sufi hospice
khutba	Friday sermon at the mosque
madad-i mash	assignment of revenue from land to a religious institution or a scholar
madrasa	religious school or college

mamluk	slave-soldier exercising political power
mansabdar	the holder of a title of nobility within the hierarchy of Mughal nobility
masjid	a mosque
maulana	learned man
muhajir	refugee
Muharram	the first month in Islamic calendar, the month of Husain's martyrdom for the Shia
mujahid	Islamic fighter
mulla	Muslim cleric
mushaira	spiritually enriching musical gathering
nawab	Mughal governor
nazr	gift
nizam	governor, but specifically the title of the rulers of Hyderabad
pir	Sufi master
purdah	seclusion of women
qasbah	in the South Asian context, a distinctively Muslim settlement, usually a large village or a small town
Sayyid	a descendant of the Prophet
shahid	a martyr
sheikh	a Sufi spiritual leader
Shariah	Islamic legal and religious code
subah	a Mughal province
sultan	the title of a Muslim ruler
ulama	experts in religious law (singular: **alim**)
urs	commemoration of a Muslim saint
wahabi	a follower of a strict Islamic puritan ethos, traditionally practised in Saudi Arabia
wazir	chief minister
zakat	alms or charity
zamindar	landholder, and occasionally a tax collector
zenana	women's quarters

NOTES

PREFACE

1 Powicke 1955: 21

1 INTRODUCTION

1 Glasse 2001: 332, 464
2 Denny 1994: 45–50
3 Guillaume 1956: 6–10
4 Denny 1994: 60–1; Elias 2010: 326–7
5 Schimmel 1992: 29–34; Rahman 1979: 30–3; Guillaume 1956: 55–62
6 Denny 1994: 72–3
7 Glasse 2001; Denny 1994: 76
8 Rahman 1979: 33–7; Denny 1994: 118–37; Schimmel 1992: 34–42
9 Lapidus 2002: 31–44
10 Ibid. 40–4, 197–206
11 Denny 1994: 92–6, 206–7
12 Rahman 1979: 175–80; Schimmel 1992: 91–100
13 Glasse 2001: 255–7
14 Ernst 2003: 112–16; Schimmel 1992: 101–20
15 Lapidus 2002: 67–80
16 Rahman 1979: 50–67; Denny 1994: 158–68
17 Denny 1994: 168–71
18 Rahman 1979: 68–84
19 Denny 1994: 195–206
20 Pew Research Centre – http://en.wikipedia.org/wiki/list_of_countries_by_Muslim_population
21 Puniyani 2003: 146–9
22 Hardy 1961a: 115–27; Hardy 1961b: 294–309
23 Elliott & Dowson: 1867–77/2001: xv–xxvii
24 Rashid 1961: 139–51
25 Quoted in Lal 2003: 89
26 Smith 1961: 319–31
27 Lal 2003: 88–97
28 Aziz 1998: ix–xvii

2 THE ERA OF ARAB PREDOMINANCE (*c.* 600 TO 1000 CE)

1 Wink 1990: 192–218
2 Khan, M.A.S. 1997: 165–83; Eaton 2000: 32–9, 42–4
3 Hourani 1979: 61–84
4 Wink 1990: 63–4
5 Wink 1990: 69
6 Ibid.
7 Goitein 1955: 116–8; Wink 1990: 101–2
8 Heffening 1993: 757–9; Wink 1990: 71–2
9 Wink 1990: 72–5
10 Wink 1990: 73–5; Miller 1991: 458–9, 463
11 Wink 1990: 80
12 Ibid. 81–2
13 Ibid. 82
14 Majumdar & Pusalker 1954: 167
15 Al-Baladhuri in Elliot & Dowson 1867–1877: 116
16 Bosworth 1996: 3–5
17 Hodgson 1977a: 223–6: Nizami 1981: 5–6, 19–22
18 Wink 1990: 164–6
19 Ibid. 171–5
20 Ibid. 51–2
21 Maclean 1989: 65–6
22 Wink 1990: 202–3
23 Chachnamah in Elliot & Dowson 1867–1877: 170–4
24 Fatimi 1987: 97–105
25 Maclean 1989: 50ff.
26 Ibid. 68
27 Hodgson 1977a: 280–4; Lapidus 2002: 58–60
28 Wink 1990: 210–12
29 Ibid. 215
30 Ibid. 56, 175, 216
31 For a balanced evaluation of this subject, see Maclean 1989: 22–33
32 Wink 1990: 172
33 Ibid. 183
34 Ibid. 155ff.
35 Ibid. 171–2
36 Maclean 1989: 61–5
37 Ibid. 37
38 Ibid. 40
39 Ibid. 40–9
40 Ibid. 39
41 Ibid. 43
42 Ikram 1964: 10–11
43 Nizami 1981: 23
44 Hardy 1987: 111–7
45 Maclean 1989: 54–7
46 Ibid. 61–77
47 Ibid. 35–6
48 Sharma 1972: 26–8
49 Maclean 1989: 77–82
50 Ibid. 1989: 83–95

51 Ibid. 108–10
52 Ritter 1960: 162
53 Maclean 1989: 110–18
54 Massignon 1971: 100–1
55 Maclean 1989: 148–53
56 Daftary 1998: 200–202
57 Goodman 1990: 484
58 For a comprehensive survey, refer to Young *et al.* 1990
59 Nasr 1976: 36–48
60 Khan, M.A.S. 1997: 189–93
61 Hopkins 1990: 308–10; Khan, M.A.S. 1997: 187–9
62 Nasr 1976: 40
63 Khan, M.A.S. 1997: 208–29
64 Elliot & Dowson 1867–77/2001: 18
65 Bose *et al.* 1971: 95–7, 133–5; Nasr 1976: 97
66 Hopkins 1990: 302–3; Pingree 1997: 640–1
67 Joseph 2011: 462
68 Joseph 2011: 455–8; Hill 1993: 15–22
69 Nasr 1976: 79; Joseph 2011: 468
70 Nasr 1976: 155
71 Bose *et al.* 1971: 260
72 Nasr 1976: 173; Bose *et al.* 1971: 259–60
73 Cahen 1990: 188–9, 197
74 Ibid. 192–3
75 Elliot & Dowson 1867–1877: 113–30
76 Ibid. 114
77 Baloch 1983: 4–6
78 Hardy 1987: 111–17

3 TURKISH POWER AND PERSIAN CULTURE IN THE AGE OF TRANSITION (1000–1206)

1 Kwanten 1979: 18–26
2 Bosworth 1990: 155–67; Lapidus 2002: 125–32
3 Khurasan/Khorasan, etc., means East.
4 Bosworth 1973: 146–8, 206–9
5 Lapidus 2002: 38, 67, 112–17, 345–51
6 Bosworth 1973: 208–10
7 Wink 1997: 76–8
8 Frye 1975: 136–61
9 Wink 1997: 122–3
10 Bosworth 1973: 76
11 Nazim 1931: 96–9
12 Ibid. 115–21; 209–24
13 Bosworth 1977: 117
14 Ibid. 111–20
15 Wink 1997: 141
16 Ibid.
17 Ibid. 146
18 Ibid. 146–9
19 Sykes 1940: 213–15
20 Ibid. 213

21 Wink 1997: 149
22 Rabie 1975: 154–6
23 Ayalon 1975: 55–8; Bosworth 1975: 62–5
24 Bosworth 1975: 68
25 Bosworth 1968: 36–7
26 Lapidus 2002: 76–7
27 Bosworth 1973: 18–20
28 Ibid. 48–50; Lambton 1968: 210–11
29 Bosworth 1973: 91–3
30 Ibid. 62–3, 72
31 Ibid. 92
32 Ibid. 93–6
33 Ibid. 68–70
34 Wink 1997: 212–13; Bosworth 1977: 57–8
35 Bosworth 1973: 79–80
36 Ibid. 262
37 Lapidus 2002: 284–7
38 Busse 1975: 272–8
39 Bosworth 1966: 87–8
40 Bosworth 1962: 217–24
41 Bosworth 1973: 53
42 Thapar 2004: 72
43 Wink 1997: 301–8
44 Wheeler 2002: 87–90
45 Thapar 2004: 48–50
46 Ibid. 76–104, 136–8
47 Ibid. 27–35
48 Wink 1997: 4–5
49 Lazard 1975: 596–605
50 Frye 1975: 145
51 Nazim 1931: 156–9; Bosworth 1973: 131–5
52 Lane-Poole 1903: 30
53 Ibid. 29–30
54 Bosworth 1973: 132
55 Rypka 1968: 175–7
56 Joseph 2011: 463
57 Saliba 1990: 405–6, 416–17; Lawrence 1990: 285–7
58 Sachau 1888: 22
59 Ghani 1941: 175–7
60 Ibid. 195
61 Alam 2003: 178
62 Lazard 1975: 611–28
63 Schimmel 1973: 11; Alam 2003: 135–6
64 Schimmel 1973: 11–12; Alam 2003: 136–7
65 Rypka 1968: 442
66 Yusofi 1989: 891
67 Schimmel 1975b: 62–77
68 Masse 1971: 546
69 Schimmel 1975b: 88
70 Schimmel 1973: 10
71 Ghani 1941: 273ff.
72 Schimmel 1975b: 345

73 Ghani 1941: 288–330
74 Ghani 1941: 289; Schimmel 1975b: 360

4 THE DELHI SULTANATE AT ITS ZENITH (1206–1351)

1 Kumar 2007: 120–5; Jackson 1999: 26–9
2 Kumar 2007: 138–43; Jackson 1999: 34–41
3 Jackson 1999: 32–4, 37–8
4 Kumar 2007: vi
5 Jackson 1999: 47
6 Wink 1997: 158–61; Jackson 1999: 70–81
7 Nizami 1961: 104–5
8 Jackson 1999: 83–4
9 Jackson 1999: 85
10 Ibid. 171–6
11 Nizami 1981: 271–85
12 Jackson 1999: 273–7
13 Nizami 1961: 97–103
14 Kumar 2007: 288–9
15 Jackson 1999: 57–8; Nizami 1981: 369
16 Nizami 1961: 121–2; Titus 1979: 55–6; Jackson 1999: 37–8
17 Kumar 2007: 78–87
18 Wink 1997: 194–5; Jackson 1999: 65–6;
19 Wink 1997: 157
20 Kumar 2007: 143–88, 198–9; Jackson 1999: 61–5
21 Jackson 1999: 76–82
22 Ibid. 82–5, 171–7
23 Ibid. 182–6
24 Nizami 1961: 124
25 Ibid. 24–8
26 Kumar 2007: 213
27 Ikram 1964: 91–3
28 Hodgson 1977b: 49–52
29 Jackson 1999: 95
30 Nizami 1961: 184
31 Jackson 1999: 143–5
32 Wink 1997: 59–60, 111, 143–9
33 Jackson 1999: 124–6
34 Ibid. 208–13
35 Wink 2004: 153–5; Digby 1971: 11–22, 23–49
36 Morgan 1986: 44–50
37 Morgan 1986: 67–73
38 Brent 1976: 240–50; Spuler 1971: 6–7
39 Wink 1997: 202–11; Wink 2004: 119–21
40 Jackson 1999: 237
41 Ibid. 227–31
42 Moreland 1929: 32–5; Habib 1982: 60–3
43 Jackson 1999: 265–6
44 Dunn 1986: 183–212
45 Habib 1982: 57–8
46 Ibid. 75–6
47 Wink 1997: 215–6

48 Habib 1982: 78, 81–2
49 Ibid. 90
50 Ibid. 92
51 Abu-Lughod 1989: 270–2
52 Ibid. 284–6
53 Digby 1982: 97
54 D'Souza 2004: 72
55 Chaudhuri 1990: 83
56 Kumar 2007: 212–5; Nizami 1961: 150–73
57 Kumar 2007: 217–20
58 Ibid. 202–9; Habibullah 1961: 301–3
59 Wink 1997: 190–4
60 Digby 2003: 234–62
61 Nizami 1961: 174–229
62 Ibid. 230–40
63 Lawrence 1986: 104–28
64 Hodgson 1977b: 536–7
65 Jackson 1999: 289–94
66 Srivastava 1980: 134–50, 164–85; Habibullah 1961: 328–9
67 Habibullah 1961: 325–7
68 Hardy 1998: 228–9
69 Titus 1979: 27–30
70 Jackson 1999: 284–7
71 Srivastava 1980: 89
72 Wink 1997: 4–5, 123–4
73 Wink 1997: 274–6, 282–3
74 Srivastava 1980: 121
75 Titus 1979: 23
76 Srivastava 1980: 102
77 Ibid. 103–4, 106–7
78 Eaton 2004: 31–49, 81–93
79 Misra 1982: 68–9; Jackson 1999: 287–8
80 Page 2008: 43–6; Smith (no date): 152–3
81 Mitter 2001: 89–90
82 Harle 1986: 421; Dehejiya 1997: 255–8; Page 2008: 5
83 Dehejiya 1997: 252
84 Page 2008: 4
85 Dahejiya 1997: 255–8
86 Tomory 1982: 146; Page 2008: 5
87 Jackson 1999: 259–60
88 Davies 1989: 37; Page 2008: 123–4
89 Shokoohy & Shokoohy 2007; Davies 1989: 38
90 Ibid. 119, 124
91 Jackson 1999: 258–60
92 Ikram 1964: 45
93 Kumar 2007: 362–3
94 Kumar 2007: 366; Hardy 1960: 12–19
95 Jackson 1999: 283–4
96 Kumar 2007: 367–9; Jackson 1999: 7, 41, 45, 48, 51, 88
97 Jackson 1999: 21, 156, 162–4; Hardy 1960: 94–110
98 Kumar 2007: 370–3; Hardy 1960: 20–39; Jackson 1999: 21, 162–4, 191–2, 242, 292–3

99 Nizami 1961: 267–80
100 Ikram 1964: 102–3
101 Schimmel 1973: 3–4
102 Rizvi 2003: 154–68
103 Kumar 2007: 373
104 Ibid. 373–6
105 Mirza 1935: 136–7; Pinto 1989: 113
106 Schimmel 1973: 16
107 Ibid. 16–17
108 Mirza 1935: 152–73
109 Mirza 1950: xvii–xxxviii
110 Nath & Gwaliari 1981: 35–8
111 Mirza 1950: xxx

5 NEW CENTRES OF MUSLIM POWER AND CULTURE (1351–1556)

1 Habib & Nizami 1970: 569
2 Ibid. 580–1, 606–9; Jackson 1999: 304–5
3 Habib & Nizami 1970: 609–10; Jackson 1999: 286–9
4 Manz 1991: 41–65; Wink 2004: 122–5
5 Manz 1991: 72
6 Lal 1963: 30–1
7 Haig 1928: 206
8 Habib & Nizami 1970: 630
9 Lal 1963: 144–53
10 Wink 2004: 136
11 Dale 2004: 74–84, 135–71
12 Beveridge 1922: 320 n.2; Dale 2004: 136–48
13 Dale 2004: 177–9
14 Dale 2004: 291ff.
15 Beveridge 1922: 472–5
16 Anooshahr 2009: 42
17 Ibid. 20–6
18 Beveridge 1922: 344; Ross 1937: 13–14
19 Burn 1937: 21–35
20 Ibid. 66–8
21 Qanungo 1921: 347–60
22 Ibid. 370–80
23 Moreland 1929: 74
24 Qanungo 1921: 395–9
25 Ibid. 388–95
26 Ibid. 386–8
27 Ibid. 380–5
28 Ibid. 409, 415–18
29 Mujeeb 1967: 12–13
30 Titus 1979: 46; Schimmel 1980: 4–6, 72–4
31 Khan, M.I. 2003: 342–62; Schimmel 1980: 43–7
32 Eaton 1993: 47–50
33 Ibid. 55–63
34 Ibid. 63–9
35 Lal 1963: 9

36 Ibid. 144–53
37 Schimmel 1980: 40–3
38 Wink 2004: 195–201
39 Harle 1986: 431; Marshall 1928: 617–20
40 Eaton 2005: 37ff.
41 Ibid. 43
42 Ibid. 63–7
43 Ibid. 67–73
44 Ibid. 59–77
45 Eaton 1993: 134
46 Wink 2004: 125–9
47 Ibid. 134–8
48 Wink 2004: 119–25
49 Ibid. 201
50 Ibid. 138–40, 143–4, 146–7
51 Goyal 2002: 116; Lal 1963: 267–8; Stein 1998: 147
52 Stein 1998: 144–5; Rafiqi 2002: 29–41
53 Rafiqi 2002: 41–2; Lal 1963: 291–315
54 Risso 1995: 68–72
55 Davies 1989: 40
56 Blair & Bloom 1994: 154
57 Harle 1986: 425
58 Tomory 1982: 149
59 Harle 1986: 426
60 Tomory 1982: 151–2; Qanungo 1921: 399–403
61 Harle 1986: 427
62 Brown 1956: 42–4
63 Dani 1961: 8–25
64 Brown 1956: 35–41; Dani 1961: 55–70
65 Harle 1986: 429
66 Michell & Shah 1988: 32–45
67 Blair & Bloom 1994: 159
68 Michell & Shah 1988: 59–86
69 Michell & Zebrowski 1999: 71–2
70 Eaton 2005: 77
71 Dale 2004: 184–5
72 Crowe 1972: 16–21
73 Beveridge 1922: 486–7, 519
74 Moynihan 1980: 103–9; Dale 2004: 309–10, 315
75 Dale 2004: 186
76 Mitter 2001: 97–9
77 Blair & Bloom 1994: 162
78 Mitter 2001: 101
79 Joseph 2011: 410, 511–2; Bose *et al.* 1971: 99
80 Lal 1963: 246
81 Bose *et al.* 1971: 270
82 Alam 2003: 142–7
83 Ibid. 149
84 McGregor 1984: 60–73
85 Behl 2003: 180–208
86 Kaviraj 2003: 530

87 Schimmel 1975a: 132–3
88 Ibid. 133–4
89 Faruqi 2003: 821–4
90 Yashaschandra 2003: 591–3
91 Dale 2004: 23–7, 36–66
92 Anooshahr 2009: 20–37
93 Beveridge 1922: 480–521

6 THE MUGHAL ASCENDANCY: AKBAR AND HIS SUCCESSORS (1556–1689)

1 Richards 1993: 58–78
2 Wink 2009: 86–8
3 Richards 1993: 94–5
4 Beach 1992: 78ff.
5 Richards 1993: 102–3
6 Ibid. 113–18
7 Ibid. 121–3, 138–50
8 Sarker 2007: 154, 162–4
9 Richards 1993: 151–64
10 Raychaudhuri 1982a: 173–4
11 Richards 1993: 13
12 Ziegler 1978: 168–9, 209–10; Zaidi 1997: 15–24
13 Richards 1993: 25–8
14 Ibid. 95–6
15 Ibid. 183–4
16 Ibid. 32–3
17 Ibid. 202
18 Burn 1937: 159–60; Gordon 1993: 42–5
19 Richards 1993: 54, 220–4
20 Gordon 1993: 10–36
21 Ibid. 46–58
22 Ibid. 70–84
23 Ibid. 94
24 Richards 1993: 137, 167–8, 247
25 Ibid. 132–5
26 Richards 1978: 127
27 Ibid. 128–9, 139–50
28 Mukhia 2004: 42–3, 45–6
29 Richards 1978: 142
30 Ibid. 130–1
31 Ibid. 157–8
32 Khan, I.A. 2003: 121
33 Mukhia 2004: 100–6
34 Richards 1993: 58
35 Ibid. 63–6
36 Ibid. 145 (Table 3)
37 Maddison 2007: 117 (Table 3.4b)
38 Raychaudhuri 1965: 259–62
39 Moreland 1929: 110–23; Wink 2009: 78–80; Richards 1993: 84–6
40 Raychaudhuri 1982b: 348

41 Ibid. 331
42 Ibid. 340
43 Ibid. 325
44 Dasgupta 1982: 408–13
45 Risso 1995: 68–72
46 Ibid. 99–106
47 Chaudhuri 1982: 382–395
48 Dasgupta 1982: 428–30
49 Chaudhuri 1982: 400–2
50 Richards 1993: 76 (Table 1), 186
51 Moosvi 1987: 222 (Table 9.4), 247 (Table 10.6), 268 (Table 11.4), 270 (Table 11.5)
52 Eraly 2008: 165–81
53 Brown 1994: 131 (Table O)
54 Mukhia 2004: 14–71
55 Anooshahr 2009: 20ff.
56 Wink 2009: 90–1
57 Ibid. 97
58 Richards 1993: 38–9
59 Prasad 1997: 97–108
60 Wink 2009: 102
61 Mistree 2002: 420
62 Mukhia 2004: 46–7
63 Richards 1993: 37
64 Sharma 1972: 60–3; Wink 2009: 97–107
65 Prawdin 1963: 160–7; Mukhia 2004: 43–4, 58, 153; Khan, I.A. 1997: 79–96
66 Richards 1978: 150–3
67 Qureshi 1972: 62–9
68 Ibid. 91–3
69 Ibid. 94–6
70 Mukhia 2004: 24
71 Sharma 1972: 112
72 Ibid. 132
73 Bhatia 2006: 70–85
74 Chandra 2003: 142
75 Eaton 2004: 90–3 (Appendix 3)
76 Bhatia 2006: 50–2
77 Sharma 1972: 178
78 Richards 1993: 96–8
79 Ibid. 178
80 Eraly 2008: 44–9
81 Ibid. 50–2
82 Wade 2002: 232–42
83 Eraly 2008: 125–30
84 Lal 2005: 1–3, 24–9
85 Findly 2002: 37–43
86 Ibid. 27–31
87 Lal 2005: 61–5, 120–8, 193–207
88 Findly 2002: 32–3, 35–6, 52–3
89 Ibid. 55–6
90 Asher 1992: 1–18

91 Ibid. 49
92 Rezavi 1997: 173–87; Asher 1992: 52, 66
93 Asher 1992: 200
94 Naqvi 1986: 143–51; Sarker 2007: 132–43
95 Brown 1937: 545–7; Asher 1992: 53–8
96 Sarker 2007: 95–131
97 Asher 1992: 257–9
98 Koch 2002: 74–5; Tomory 1998: 159–60
99 Brown 1937: 552
100 Koch 2006: 83–102
101 Ibid. 22–81
102 Michell & Zebrowski 1999: 38–41, 84
103 Ibid. 90–2
104 Ibid. 92–4
105 Ibid. 51–3
106 Beach 1992: 16
107 Ibid. 39
108 Ibid. 21–5
109 Ibid. 25–9
110 Ibid. 40–1, 48
111 Tomory 1982: 252–3; Beach 1992: 52–5, 76, 86–8, 104–7
112 Beach 1992: 48–50, 60–1, 62–4
113 Ibid. 82–6
114 Ibid. 90–3
115 Ibid. 99–100
116 Ibid. xviii, 100–4
117 Ibid. 128–38
118 Michell & Zebrowski 1999: 145–7
119 Ibid. 162–77
120 Beach 1992: 110–3
121 Michell & Zebrowski 1999: 199
122 e.g. Francois Bernier, Thomas Coryat, Niccolao Manucci, John-Baptiste Tavernier among others
123 Alam 2003: 159–60
124 Ibid. 162–4
125 Ibid. 162–7
126 Ibid. 174–86
127 Schimmel 1980: 81–5; Alam 2003: 172–3
128 Alam 2003: 169–70, 173, 179; Schimmel 1973: 29–30; Schimmel 1980: 84–5
129 Prasad 1955: 381–3
130 Rashid 1961: 143–7
131 Hardy 1965: 921–3
132 Sarker 2007: 28
133 Schimmel 1980: 102
134 Schimmel 1973: 44–5
135 For scientific knowledge, see Khan, I.G. 1997: 121–8; for statistics, see Moosvi 1987: 3–35
136 Thackston 2002: 104–6
137 Qureshi 1972: 84–96
138 Armstrong 2001: 107–9; Schimmel 1980: 90–5; Richards 1993: 98–100
139 Schimmel 1980: 96–101

7 THE AGE OF MUGHAL DISINTEGRATION (1689–1765)

1 Quoted in Wolpert 2009: 191
2 Ali 1986/7: 90–1
3 Marshall 2003: 3–13
4 Prakash 2007: 140–1
5 Alam & Subrahmanyam 1998: 55–71
6 Sarkar 1937a: 230, 240–4, 244–6
7 Richards 1993: 244–5
8 Parthasarathi 1996: 209ff.
9 Richards 1993: 227–45
10 Ibid. 244–5
11 Stein 1998: 185
12 Richards 1993: 242
13 Ibid. 244
14 Stein 1998: 188–9
15 Raychaudhuri 1965: 268ff.
16 Alam 1986: 20–4
17 Sarkar 1937b: 320
18 Ibid. 325–6
19 Ibid. 327–30
20 Ibid. 339
21 Haig 1937a: 341–8
22 Haig 1937b: 437, 444
23 Ibid. 448
24 Gray 1991: 74
25 Ibid. 128
26 Bhadra 1984: 474–90
27 Singh 1988: 325–9
28 Alam 1986: 139–44
29 Ibid. 134ff.
30 Richards 1993: 256–8
31 Alam 1986: 13–14, 31–5
32 Gordon 1993: 37–40, 129–30, 189–94
33 Ibid. 114–31
34 Haig 1937a: 361
35 Ibid. 363
36 Ibid. 371–6; Haig 1937b: 438ff.
37 Gordon 1993: 151–3
38 Marshall 2003: 3–13
39 Eaton 1993: 142–5
40 Ali 1985: 534ff.; Prakash 2007: 138
41 Ali 1985: 570–6
42 Bayly 1988: 49; Ali 1985: 575–6, 592–5, 965–70, 969
43 Barnett 1980: 23–34
44 Alam 1986: 204–10, 219–24
45 Barnett 1980: 34–41
46 Ibid. 58–66
47 Sarkar 1937c: 377–8
48 Haig 1937a: 356–7; Sarkar 1937c: 379–82
49 Bayly 1988: 14–23
50 Leonard 1979: 403–13

51 Bayly 1988: 57–61
52 Dodwell 1929: 117–24, 157–64
53 Martineau 1929: 128–30; Dodwell 1929: 159–61
54 Bayly 1988: 50
55 Ibid. 50–1
56 Ibid. 53
57 Cole 1988: 24–7
58 Robinson 2001: 24–7
59 Cole 1988: 27–31
60 Ibid. 31–4
61 Ibid. 127–37
62 Leonard 1979: 403–13; Alavi 2007: 16
63 Mohan 1997: 65
64 Cole 1988: 115–17
65 Mohan 1997: 66
66 Bayly 1989a: 210
67 Ibid. 212–13
68 Ibid. 213
69 Rizvi 1969: 197–202
70 Kaul 1985: 353–66, 372–5
71 Ansari 1965: 254–5
72 Rizvi 1969: 294–308
73 Qureshi 1985: 111–13
74 Ahmad 1969: 40–2; Rizvi 1969: 264–71
75 Ahmad 1969: 41
76 Rizvi 1969: 261–77
77 Ahmad 1969: 8; Rizvi 1969: 245–9
78 Lambton 1978: 947–50; Rizvi 1969: 288–94
79 Alam 2003: 162–4
80 Ibid. 163–7
81 Alam 2003: 175–7; Schimmel 1973: 47–8; Ahmad 1969: 85–6
82 Schimmel 1973: 42–4
83 Alam 2003: 183–5
84 Schimmel 1975a: 127–32
85 Ibid. 153–5
86 Faruqi 2003: 839–40
87 Russell & Islam 1969a: 4
88 Haque 1992: 40–2; Schimmel 1975a: 166–7
89 Russell & Islam 1969a: 37–69; Haque 1992: 37–40; Schimmel 1975a: 174–8
90 Haque 1992: 42–4; Schimmel 1975a: 169–73
91 Russell & Islam 1969a: 232–70; Haque 1992: 30–7
92 Pritchett 2003: 866–70
93 Russell & Islam 1969a: 69–94
94 Robinson 2001: 42
95 Nasr 1976: 3–24; Nasr 1987: 29–40, 92–125
96 Robinson 2001: 218–22, 244–8
97 Ibid. 42, 107, 219
98 Alam 1986: 110–22, 306–7
99 Robinson 2001: 44–6
100 Ibid. 46–55, 213–15, 249–51
101 Ibid. 70
102 Bayly 1983: 123–4

103 Asher 1992: 295–8
104 Ibid. 298
105 Ibid. 300
106 Bayly 1983: 68–71
107 Asher 1992: 302–4
108 Ibid. 305–6
109 Ibid. 317–18
110 Ibid. 328–30

8 MUSLIMS UNDER THE EAST INDIA COMPANY (1765–1858)

1 Lawson 1993: 17
2 Marshall 1998: 271–9
3 Losty 1990: 71; Baron 2001: 179–81
4 Markovits 2002: 233–6
5 Fisher 1996: xvi–1; Fisher 1984: 394–8
6 Porter 2004: 380–1, n. 10
7 Bowen 1996: 37–8; Marshall 2007: 244–52, 261; Hyam 1976: 49–52, 206–9
8 Carter & Harlow 2003: 182–90
9 Fisher 1996: 20, 28–9
10 Ibid. 27
11 Bayly 1988: 58; Fisher 1996: 16–20
12 Fisher 1984: 393–422
13 Bayly 1988: 89–90
14 Fisher 1996: 249–52; Bayly 1988: 134
15 Mason 1974: 138–49; Black 2004: 209–17
16 Marshall 2007: 259–63
17 Baron 2001: 82–105
18 Ibid. 17–22
19 Bayly 1988: 94–5
20 Fisher 1996: 94–8
21 Bowring 1893: 27, 106–7
22 Bayly 1988: 95–7
23 Fisher 1996: 175–82
24 Ibid. 260–96
25 Baqir 1960: 127–44
26 Fisher 1996: 224–7
27 Ibid. 227–48
28 Chaudhuri 1983: 174–7; Bhattacharya 1983: 324–32
29 Beinhart & Hughes 2007: 111–15
30 Bayly 1988: 138–42
31 Ibid. 145–55
32 Bose 1993: 68–79
33 Malik 1980: 114
34 Bayly 1988: 150
35 Spear 1990: 95–6
36 Maddison 1971: 54–5, 70
37 Malik 1980: 122
38 Dalrymple 2002: 49–54; Collingham 2001: 51–9
39 Brown 1994: 131, Table O
40 Malik 1980: 111

41 Ibid. 121
42 Bayly 1988: 169–78
43 Bose 1993: 148–9
44 Ibid. 148–53
45 e.g. Gove 2006: 11–12, 92–3
46 Qureshi 1972: 139
47 Singh 2004: 271–4
48 Qureshi 1972: 140–52
49 Bayly 1988: 180; Mason 1974: 247–68
50 Mason 1974: 268–78
51 Qureshi 1972: 186–8
52 Jalal 2000: 60–8
53 Qureshi 1972: 188–96
54 Hardy 1972a: 62–70
55 Bayly 1988: 185
56 Jalal 2000: 29–31
57 Qureshi 1972: 204–6
58 Saul 2003: 199–200, 215–20
59 Metcalf 1965: 289–97; Saul 2003: 258–9
60 Hardy 1972a: 70–9
61 Metcalf 1965: 294
62 Bose 1993: 17–24, 68–79, 114–19
63 Malik 1980: 26–8
64 Ray 1998: 510
65 Asher 1992: 320–4
66 Malik 1980: 31–2
67 Bayly 1983: 350–9
68 Bayly 1996: 79–80
69 Ibid. 80
70 Ibid. 73–8, 229–35
71 Ibid. 264–7, 273, 277–8
72 Ibid. 248–50
73 Malik 1980: 39
74 Ibid. 44–5
75 Bayly 1989b: 106–50, 179–86, 233–5
76 Bayly 1996: 180–96
77 Faruqi 2003: 854–5; Pritchett 2003: 900
78 Pritchett 2003: 894–5
79 Ibid. 869, 886
80 Sadiq 1964: 119
81 Pritchett 2003: 879
82 Sadiq 1964: 120–3
83 Ibid. 142
84 Ibid. 145–63
85 Hardy 1972b: 63–5; Sadiq 1964: 178
86 Kaul 1985: 402–3; Hardy 1972b: 66–9
87 Saksena 2002: 96–7; Schimmel 1973: 25; Sadiq 1964: 208–9
88 Visram 2002: 33, 107
89 Ashraf 1982: 207–14; Bayly 1996: 86–7
90 Ashraf 1982: 223
91 Ibid. 218
92 Malik 1980: 48

93 Ibid. 49
94 Ashraf 1982: 226–31
95 Malik 1980: 55
96 Frykenberg 2003: 47–54
97 Ibid. 54–9
98 Oddie 2003: 164–77; Kulke 1974: 93–4
99 Powell 1993: 132–57
100 Ibid. 238–42
101 Ibid. 12–20
102 Ibid. 221–5, 234–8
103 Ibid. 242–62

9 STIRRINGS OF A MUSLIM MODERNITY UNDER THE RAJ (1858–1924)

1 Philips 1962: 10–11
2 Hardy 1972a: 75–9
3 Philips 1962: 11
4 Brown 1994: 128
5 Hardy 1972a: 70–1, 79ff.
6 Lelyveld 1982: 98–112; Metcalf 1978: 113–34
7 Basu 1974: 72ff.
8 Ibid. 7, 14, 61, 74
9 Siqueira 1960: 53–7; Basu 1974: 14–15
10 Visram 2002: 87
11 Belmekki 2010: 19–21; Hardy 1972a: 93–4, 122
12 Hunter 1871: 172ff.
13 Basu 1974: 147–8
14 Malik 1980: 143 (Table 4.2)
15 Hardy 1972a: 93
16 Lelyveld 1978: 68–92
17 Malik 1980: 84–90, 189–93
18 Lelyveld 1978: 35–56
19 Hardy 1972a: 102–3; Malik 1980: 198–99
20 Malik 1980: 222
21 Hashmi 1989: 89–96
22 Hardy 1972a: 103–4; Lelyveld 1978: 217–27; Hashmi 1989: 97–105
23 Lelyveld 1978: 115–18
24 Ibid. 188, 211–12, 218–20, 224–9, 231–3, 275–8
25 Basu 1974: 161
26 Jalal 2000: 68; Lelyveld 1982: 102–10; Belmekki 2010: 109–13
27 Schimmel 1980: 209–11
28 Ahmad 1967: 105; Zaman 2009: 225–8; Hashmi 1989: 48–57
29 Jones 1989: 59–60; Ahmad 1967: 106–7
30 Robinson 2000: 39, 78
31 Metcalf 1978: 113–18
32 Hashmi 1989: 66
33 Hardy 1972a: 71; Jones 1989: 61–2
34 Hasan 2008: 126
35 Brown 1994: 139–42
36 Ibid. 172–7
37 Belmekki 2010: 128–30; Brown 1994: 136

38 Sitaramayya 1969: 5–22
39 Brown 1994: 188
40 Ibid. 186 (Table R)
41 Ibid. 187 (Table S)
42 Ibid. 185 (Table Q)
43 Hasan 2000: 27
44 Pandey 1979: 14–15
45 Hardy 1972a: 130
46 Ibid. 125
47 Jaffrelot 2007: 6–10
48 Jalal 2000: 83–4; Thursby 1975: 76–88
49 Belmekki 2010: 119–26; Jalal 2000: 133–8
50 Cashman 1975: 78–9, 90–3; Bhatt 2001: 32–5; Thursby 1975: 89–102
51 Smith 1946: 157–94
52 Philips 1962: 151
53 Hardy 1972a: 148–51
54 Robinson 1974–5: 133–74
55 Philips 1962: 194
56 Robinson 1974–5: 151–9
57 Brown 1994: 149–51; Gopal 1959: 105–14
58 Aydin 2007: 31–4, 60–3, 93–5; Lapidus 2002: 489–93; Hardy 1972a: 175–80
59 Lapidus 2002: 283–98
60 Habib 2000: 65–74; Ahmad 1960: 86–90
61 Malik 1980: 235–8
62 Gopal 1959: 121–2
63 Hasan 2000: 68–80
64 Gopal 1959: 129–35; Hasan 2000: 81–7
65 Sarkar 1989: 168ff.
66 Smith 1946: 198–203
67 Ibid. 206
68 Hashmi 1989: 147–87; Hasan 2008: 145–8
69 Rai 2004: 26–30
70 Ibid. 174–82
71 Ibid. 149–60
72 Ibid. 162–3
73 Ibid. 145–7
74 Ibid. 137–9
75 Ibid. 165–8
76 Wood 1992: 130–6, 150–1
77 Miller 1976: 60–3; Dale 1977: 43–7
78 Wood 1992: 127–30, 136–44
79 Miller 1976: 85–100
80 Ibid. 109
81 Dale 1977: 41–2, 50–5
82 Hardgrave 1977: 57–99
83 Ibid. 65ff; Miller 1976: 128–33
84 Miller 1976: 148
85 Ibid. 149
86 Pandey 2006: 72–9
87 Ibid. 81
88 Pandey 1992: 128–32
89 Guha 2003: 155–67

90 Pinney 1990: 252–63
91 Pandey 2006: 88–9
92 Troll 1972: 43–5
93 Forward 1999: 15–28
94 Ahmad 1967: 87
95 Forward 1999: 49–69
96 Ahmad 1967: 61–2
97 Hasan 2005: 200–8; Sadiq 1964: 286; Saksena 2002: 295
98 Hasan 2005: 224–5, 227–31
99 Hasan 2005: 219–24
100 Schimmel 1975a: 226–7
101 Ibid. 238–9
102 Ibid. 237–8
103 Ibid. 238
104 Hassan 1971: 139–40; Schimmel 1975a: 244
105 Schimmel 1975a: 244
106 Malik 1971: 27–8
107 Sadiq 1964: 401–2
108 Saksena 2002: 307–8
109 Smith 1946: 196–7
110 Hasan 2005: 141–4; Sadiq 1964: 316–22; Schimmel 1975a: 231–2
111 Sadiq 1964: 339–42
112 Sharar 1989: 36–40, 60–75, 145–7, 155–9
113 Saksena 2002: 342; Sadiq 1964: 355–6
114 Saksena 2002: 341
115 Sadiq 1964: 394–7; Saksena 2002: 347–9
116 Schimmel 1975a: 236
117 Saksena 2002: 364–7
118 Amin 1996: 4–13
119 Ibid. 36–7, 67–72
120 Ibid. 215–16
121 Ibid. 219–26
122 Ibid. 262–3

10 UNITY OR SEPARATISM?: MUSLIM DILEMMA AT THE END OF THE RAJ (1924–1947/8)

1 Wells 2005: 14–18
2 Ibid. 32
3 Jalal 1985: 8
4 Aziz 1987a: 376–89
5 Smith 1946: 209–15
6 Hasan 2000: 177–210, 248–81
7 Smith 1946: 220–4
8 Banerjee 2000: 84–91
9 Smith 1946: 224–32
10 Hasan 1997: 63
11 Jalal 2000: 453–60
12 Hasan 1997: 65, 94–6; Hasan 2008: 204–5
13 Robinson 1991: 872–4
14 Alavi 1987: 68–73
15 Hasan 1997: 73–7

16 Hardy 1972a: 235; Hasan 1997: 70–3
17 Hasan 1997: 78–9
18 Nawaz 2002: 153–5, 170–2, 195–7, 205–7
19 Sayeed 1970: 276–93
20 Wolpert 1984: 92–102
21 Pandey 1979: 79; Hardy 1972a: 212
22 Sarkar 1989: 262–3
23 Hasan 2000: 267; Wolpert 1984: 105–7
24 Wells 2005: 185–94; Hardy 1972a: 213; Pandey 1979: 82–3
25 Wells 2005: 194–202
26 Philips & Wainwright 1970: 13–17; Manseragh 1970: 46–7; Moore 1970: 54–5; Wylie 1970: 517–19
27 Philips 1967: 1
28 Hasan 1988: 198–222
29 Hardy 1972a: 230; Panigrahi 2004: 121–31
30 Aziz 1987b: 652–8
31 Zaidi 1970: 264–5; Rao 1970: 426–43
32 Jalal 1985: 4
33 Ibid. 121–32
34 Ibid. 149–51, 160–3
35 Nanda 1970: 175–82
36 Wolpert 1984: 282–7; Hasan 2008: 156–8
37 Dalton 1970: 222–44
38 Moon 1998: 64
39 Collins & Lapierre 1975: 106–8
40 Ibid. 125–7
41 Roberts 1994: 55
42 Menon 1957: 350–86; Jalal 1985: 282–93; Pandey 1979: 215–17
43 Jalal 1985: 255; Moon 1998: 65–70
44 Talbot & Singh 2009: 7–24; also consult Kaul 2001; Khan 2007; Zamindar 2007
45 Roberts 1994: 91
46 Khan 2007: 125
47 Ibid. 109
48 Roberts 1994: 91–100
49 Khan 2007: 118
50 Ibid. 115
51 Zamindar 2007: 1–2, 230–4; Khan 2007: 119–20
52 Khan 2007: 124
53 Ibid. 156
54 Zamindar 2007: 120–57; Khan 2007: 156–7
55 Zamindar 2007: 45–76
56 For example, Korbel 1954; Puri 1995; Jha 1996; Schofield 2000; Hewitt 2001; Zutshi 2003
57 Ramusack 2004: 245–74
58 Talbot & Singh 2009: 53–7; Schofield 2000: 31; Korbel 1954: 48
59 Hewitt 2001: 65; Schofield 2000: 28–33
60 Schofield 2000: 33–41; Jha 1996: 12, 74ff.
61 Jha 1996: 124–6; Zutshi 2003: 259–311
62 Schofield 2000: 44
63 Ibid. 41–3, 49–52
64 Ibid. 61–2

65 Korbel 1954: 97–164
66 Puri 1995: 9
67 Hewitt 2001: 70–2
68 Minault 1998: 30
69 Ibid. 101
70 Ibid. 216–28
71 Ibid. 228–49
72 Ibid. 256–62; Forbes 1996: 55–7
73 Minault 1998: 111–22
74 Ibid. 129–48
75 Anjum 1992: 112–8
76 De Souza 2004: xi
77 Minault 1998: 47–9, 84–7
78 Lateef 1990: 74–94
79 Minault 1998: 182–7
80 Ibid. 178–80
81 Forbes 1996: 198–9
82 Minault 1998: 273–8
83 Lambert-Hurley 2007: 121–2
84 Minault 1998: 289–90
85 Forbes 1996: 83–90
86 Minault 1998: 303–5; Nawaz 2002: 92; Lateef 1990: 66–73
87 Iqbal 2004: 9–31
88 Ibid. 129–57
89 Hussain 1987: 50–106; Ahmad 1967: 163–7
90 Pandey 1979: 83–4
91 Ahmad 1967: 156–60
92 Douglas 1988: 147–50
93 Ibid. 158–62
94 Mujeeb 1967: 460–3
95 Azad 1988: 246–8; Douglas 1988: 224–5
96 Douglas 1988: 238–46
97 Hasan 1997: 114
98 Hasan 2008: 145–8
99 Hasan 1997: 176–7; Hasan 2008: 191
100 Hasan 1997: 91–7
101 Zbavitel 1976: 278–9
102 Coppola 1987: 49–55
103 Ibid. 43–4
104 Hasan 1997: 60–1, 131, note 96

11 EPILOGUE: NEW CHALLENGES IN A FRACTURED SUBCONTINENT (1947–2011)

1 Hasan 1997: 173
2 Ibid. 145–58
3 Ibid. 161–3
4 Ibid. 140–2
5 Alam 2007: 276–8
6 Ibid. 278–80
7 Hasan 1997: 194–6
8 Ibid. 196–214

9 Ibid. 278
10 Ibid. 261–3, 272
11 Mullally 2004: 671–92
12 Varadarajan 2002: 135–76
13 Engineer 2007: Internet Report
14 Sachar Committee 2006: Internet Report
15 Sikand 2001: 296–308; Engineer 2006: Internet Report
16 Bose 1997: 55–62, 95–6, 195–8; Schofield 2000: 156–7, 198–9
17 e.g. Jaffrelot (Ed.) 2002a; Jaffrelot 2002b; Talbot 2005; Nawaz 2008; Lieven 2011
18 Talbot 2005: 127–8
19 Jaffrelot 2002a: 63–4; Talbot 2005: 136–7
20 Talbot 2005: 139–47; Nawaz 2008: 86–9
21 Jaffrelot 2002a: 64–8
22 Nawaz 2008: 92–118; Talbot 2005: 153–6
23 Talbot 2005: 180–2
24 Ibid. 200
25 Nawaz 2008: 282–319
26 Talbot 2005: 229–30; Jaffrelot 2002a: 76–8
27 Lieven 2011: 235–6
28 Talbot 2005: 239–44
29 Nawaz 2008: 411–505
30 Lieven 2011: 69
31 Jaffrelot 2002a: 17–20
32 Ibid. 25–8
33 Lieven 2011: 309–27
34 Jaffrelot 2002a: 28–31
35 Siddiqqa 2007: 4–8, 219–42
36 Ibid. 13–24, 174–218
37 Ibid. 242–4
38 Kurin 1985: 2
39 Talbot 2005: 237–8
40 Nasr 2000: 139–80
41 Ibid. 84–91
42 Kamran 2009: 11–37
43 Gaborieau 2002: 248
44 Mohammad 2002: 228–9
45 Roy 2002a: 138–9
46 Roy 2002b: 149–59; Lieven 2011: 405–14
47 Lieven 2011: 477–81
48 Van Schendel 2009: 109–15; Bhattacharya 2002: 43–7
49 Van Schendel 2009: 118–43
50 Talbot 2005: 194–213
51 Jack 2011: 35; Van Schendel 2009: 173
52 Gerlach 2010: 167–8; Van Schendel 2009: 180–1
53 Ziring 1992: 96–103
54 Ziring 1992: 139
55 Ibid. 146–7, 164
56 Ibid. 130–4; Van Schendel 2009: 206, 241
57 Hussain 2007: 216
58 Ibid. 233, note 2
59 Gerlach 2010: 171–2

60 Van Schendel 2009: 211–5
61 Gerlach 2010: 172
62 Datta 2007: 241–3; Van Schendel 2009: 207
63 Datta 2007: 245–50
64 Ibid. 244
65 Ibid. 253–4
66 Ibid. 253
67 Coppola 1987: 55–62
68 Mazrui 1990: 116–39
69 Zaman 2000: 81–96
70 Hasan 1995: 88–99; Zaman 2000: 20–1
71 Zaman 2000: 127–56
72 Ibid. 238–49
73 Upstone 2010: 37–61; Nyman 2009: 109–25
74 Chughtai 1991: vii–xix, 7–19
75 Russell 1995: 302
76 Mitter 2001: 211–17, 235–6
77 Ibid. 217–19, 236–8

BIBLIOGRAPHY

Abu-Lughod, J.L. (1989) *Before European Hegemony: The World System AD 1250–1350*, Oxford: Oxford University Press.

Ahmad, A. (1960) 'Sayyid Ahmad Khan, Jamal Al-Din Al-Afghani and Muslim India', pp. 81–97, in D. Taylor (Ed.) (2011) *Islam in South Asia: Critical Concepts in Islamic Studies,* Vol. 2, London: Routledge.

Ahmad, A. (1967) *Islamic Modernism in India and Pakistan 1857–1964*, London: Oxford University Press.

Ahmad, A. (1969) *An Intellectual History of Islam in India*, Edinburgh: Edinburgh University Press.

Alam, A. (2007) 'Political Management of Islamic Fundamentalism: A View from India', pp. 266–95, in D. Taylor (Ed.) (2011) *Islam in South Asia: Critical Concepts in Islamic Studies,* Vol. 3, London: Routledge.

Alam, M. (1986) *The Crisis of Empire in Mughal North India: Awadh and the Punjab 1707–48*, Delhi, Oxford University Press.

Alam, M. (2003) 'The Culture and Politics of Persian in Pre-Colonial Hindustan', pp. 131–98, in S. Pollock (Ed.) (2003), *Literary Cultures in History: Reconstructions from South Asia*, Berkeley: University of California Press.

Alam, M. and Subrahmanyam, S. (Eds) (1998) *The Mughal State 1526–1750*, Delhi: Oxford University Press.

Alavi, H. (1987) 'Pakistan and Islam: Ethnicity and Ideology', p. 65, in F. Halliday and H. Alavi (Eds) (1987) *State and Ideology in the Middle East and Pakistan*, New York: Monthly Review Press.

Alavi, S. (Ed.) (2007) *The Eighteenth Century in India*, New Delhi: Oxford India Paperbacks.

Ali, M. (1985) *History of the Muslims of Bengal*, Riyadh: Ibn Saud Islamic University Publications.

Ali, M.A. (1986) 'Capital of the Sultans: Delhi during the 13th and 14th Centuries', pp. 34–44, in R.E. Frykenberg (Ed.) (1986) *Delhi Through the Ages: Essays in Urban History*, Delhi: Oxford University Press.

Ali, M.A. (1986–7) 'Recent Theories of Eighteenth Century India', pp. 90–9, in P.J. Marshall (Ed.) (2003) *The Eighteenth Century in Indian History: Evolution or Revolution?*, New Delhi: Oxford University Press.

Amin, S.N. (1996) *The World of Muslim Women in Colonial Bengal*, Leiden: E.J. Brill.

Anjum, M. (1992) *Muslim Women in India*, New Delhi: Radiant Publishers.

Anooshahr, A. (2009) *The Ghazi Sultans and the Frontiers of Islam*, London: Routledge.

Ansari, A.S.B. (1965) 'Al-Dihlawi, Shah Wali Allah', pp. 254–5, in *The Encyclopaedia of Islam*, Vol. 2, Leiden: E.J. Brill.

Armstrong, K. (2001) *Islam: A Short History*, London: Phoenix.

Arnold, T. (1896/1979) *The Preaching of Islam*, Lahore: Sharif Muhammad Ashraf.

Asher, C.B. (1992) *Architecture of Mughal India: The New Cambridge History of India,* Vol. I.4, Cambridge: Cambridge University Press.

Ashraf, M. (1982) *Muslim Attitudes Towards British Rule and Western Culture in India in the First Half of the Nineteenth Century*, Delhi: Idarah-i Adabiyat-i Delli.

Ayalon, D. (1975) 'Preliminary Remarks on the Mamluk Military Institution in Islam', pp. 44–58, in V.J. Parry and M.E. Yapp (Eds) (1975) *War, Technology and Society in the Middle East*, London: Oxford University Press.

Aydin, C. (2007) *The Politics of Anti-Westernism in Asia: Visions of World Order in Pan-Islamic and Pan-Asian Thought*, New York: Columbia University Press.

Azad, A.K. (1988) *India Wins Freedom*, Lahore: Vanguard Books.

Aziz, K.K. (1987a) *Rahmat Ali: A Biography*, Lahore: Vanguard Books.

Aziz, K.K. (1987b) *A History of the Idea of Pakistan*, Vol. 3, Lahore: Vanguard Books.

Aziz, K.K. (1998) *The Murder of History: A Critique of History Textbooks Used in Pakistan*, New Delhi: Renaissance Publishing House.

Baljon, J.M.S. (1986) *Religion and Thought of Shah Wali Allah Dihlawi, 1703–1762*, Leiden: E.J. Brill.

Baloch, N.A. (Ed.) (1983) *Fathnamah-i-Sind*, Islamabad: Institute of Islamic History, Culture and Civilization, Islamic University.

Banerjee, M. (2000) *The Pathan Unarmed: Opposition and Memory in the North West Frontier*, Karachi: Oxford University Press.

Baqir, M. (1960) 'The End of Muslim rule in Oudh', pp. 127–44, in Pakistan Historical Society, *A History of the Freedom Movement, Vol. 2, 1831–1905*, Karachi.

Barnett, R.B. (1980) *North India between Empires: Awadh, the Mughals and the British 1720–1801*, Berkeley: University of California Press.

Baron, A. (2001) *An Indian Affair; From Riches to Raj*, London: Channel Four Books.

Basu, A. (1974) *The Growth of Education and Political Development in India 1898–1920*, Delhi: Oxford University Press.

Bausani, A. (1968) 'Religion in the Saljuq Period', pp. 283–302, in J.A. Boyle (Ed.) (1968) *The Cambridge History of Iran*, Vol. 5, Cambridge: Cambridge University Press.

Bayly, C.A. (1983) *Rulers, Townsmen and Bazaars: North Indian Society in the Age of British Expansion 1770–1870*, Cambridge: Cambridge University Press.

Bayly, C.A. (1988) *Indian Society and the Making of the British Empire: The New Cambridge History of India,* Vol. II.1, Cambridge: Cambridge University Press.

Bayly, C.A. (1996) *Empire and Information: Intelligence Gathering and Social Communication in India, 1780–1870*, Cambridge: Cambridge University Press.

Bayly, S. (1989a) 'The South Indian State and the Creation of Muslim Community', pp. 203–39, in P.J. Marshall (Ed.) (2003) *The Eighteenth Century in Indian History: Evolution or Revolution?*, New Delhi: Oxford University Press.

Bayly, S. (1989b) *Saints, Goddesses and Kings: Muslims and Christians in South Indian Society, 1700–1900*, Cambridge: Cambridge University Press.

Beach, M.C. (1992) *Mughal and Rajput Painting: The New Cambridge History of India*, Vol. I.3, Cambridge: Cambridge University Press.

Behl, A. (2003) 'The Magic Doe: Desire and Narrative in a Hindavi Sufi Romance, circa 1503', pp. 180–208, in R.M. Eaton (Ed.) (2003) *India's Islamic Traditions 711–1750*, New Delhi: Oxford University Press.

Beinhart, W. and Hughes, L. (2007) *Environment and Empire*, Oxford: Oxford University Press.

Belmekki, B. (2010) *Sir Sayyid Ahmad Khan and the Muslim Cause in British India*, Berlin: Klaus Schwarz Verlag.

Beveridge, A.S. (1922) *The Baburnama in English*, Vols. 1 and 2, London: Luzac and Co.

Bhadra, G. (1984) 'Two Frontier Uprisings in Mughal India', pp. 474–90, in M. Alam and S. Subrahmanyam (Eds) (2000) *The Mughal State 1526–1750*, New Delhi: Oxford University Press.

Bhatia, M.L. (2006) *The Ulama, Islamic Ethics and Courts under the Mughals*, New Delhi: Manak Publications.

Bhatt, C. (2001) *Hindu Nationalism: Origins, Ideologies and Modern Myths*, Oxford: Berg.

Bhattacharya, F. (2002) 'East Bengal: Between Islam and a Regional Identity', pp. 39–60, in C. Jaffrelot (Ed.) (2002a) *A History of Pakistan and its Origins*, London: Anthem Press.

Bhattacharya, S. (1983) 'Regional Economy in Eastern India', pp. 270–332, in D. Kumar and M. Desai (Eds) (1983) *The Cambridge Economic History of India*, Vol. 2, Cambridge: Cambridge University Press.

Black, J. (2004) *The British Seaborne Empire*, New Haven, CT: Yale University Press.

Blair, S.S. and Bloom, J.M. (1994) *The Art and Architecture of Islam 1250–1800*, New Haven, CT: Yale University Press.

Blochmann, H. (1873) *The Ain-i-Akbari of Abul Fazl Allami*, Vol. 1, Book 1, Calcutta: Asiatic Society of Bengal.

Bose, D.M., Sen, S.N. and B.V. Subbarayappa (1971) *A Concise History of Science in India*, New Delhi: Indian National Academy of Science.

Bose, S. (1993) *Peasant Labour and Colonial Capital: Rural Bengal since 1770: The New Cambridge History of India*, Vol. III.2, Cambridge: Cambridge University Press.

Bose, S. (1997) *The Challenge in Kashmir: Democracy, Self-determination and a Just Peace*, New Delhi: Sage Publications.

Bosworth, C.E. (1962) 'The Titulature of the Early Ghaznavids', pp. 210–33, in *Oriens XV*, Leiden.

Bosworth, C.E. (1966) 'Mahmud of Ghazna in Contemporary Eyes and in Later Persian Literature', pp. 85–92, in *Iran IV*, London.

Bosworth, C.E. (1968) 'The Development of Persian Culture under the Early Ghaznavids', pp. 33–44, in *Iran VI*, London.

Bosworth, C.E. (1973) *The Ghaznavids: Their Empire in Afghanistan and Eastern Iran 994–1040*, Beirut: Librairie du Liban.

Bosworth, C.E. (1975) 'Recruitment, Muster and Review in Medieval Islamic Armies', pp. 59–77, in V.J. Parry and M.E. Yapp (Eds) (1975) *War, Technology and Society in the Middle East*, London: Oxford University Press.

Bosworth, C.E. (1977) *The Later Ghaznavids: Splendour and Decay – The Dynasty in Afghanistan and Northern India 1040–1186*, Edinburgh: Edinburgh University Press.

Bosworth, C.E. (1990) 'Administrative Literature', pp. 155–67, in M.J.L. Young, J.D. Latham and R.B. Sargeant (Eds) (1990) *Religion, Learning and Science in the Abbasid Period*, Cambridge: Cambridge University Press.

Bosworth, C.E. (1996) *The New Islamic Dynasties*, Edinburgh: Edinburgh University Press.

Bowen, H.V. (1996) *Elites, Enterprise and the Making of Britain's Overseas Empire 1688–1775*, Basingstoke: St. Martin's Press.

Bowring, L.R. (1893) *Haider Ali and Tipu Sultan*, Oxford: Clarendon Press.

Brent, P. (1976) *The Mongol Empire – Genghis Khan: His Triumph and Legacy*, London: Weidenfeld and Nicolson.

Brice, W. (Ed.) (1981) *An Historical Atlas of Islam*, Leiden: E.J. Brill.

Brittlebank, K. (1997) *Tipu Sultan's Search for Legitimacy: Islam and Kingship in a Hindu Domain*, Delhi: Oxford University Press.

Brown, J. (1994) *Modern India: The Origins of an Asian Democracy*, Oxford: Oxford University Press.

Brown, P. (1937) 'Monuments of the Mughul period', pp. 523–76, in R. Burn (Ed.) (1937) *The Cambridge History of India,* Vol. 4, Cambridge: Cambridge University Press.

Brown, P. (1956) *Indian Architecture: Islamic Period*, Bombay: Taraporewalla and Co.

Burn, R. (Ed.) (1937) *The Cambridge History of India,* Vol. 4, Cambridge: Cambridge University Press.

Busse, H. (1975) 'Iran under the Buyids', pp. 250–304, in R.N. Frye (Ed.) (1975) *The Cambridge History of Iran,* Vol. 4, Cambridge: Cambridge University Press.

Busteed, H.E. (1908/1972 reprint) *Echoes from Old Calcutta*, Shannon: Irish University Press.

Cahen, C. (1990) 'History and Historians', pp. 188–233, in M.J.L. Young, J.D. Latham and R.B. Sargent (Eds) (1990), *Religion, Learning and Science in the Abbasid Period*, Cambridge: Cambridge University Press.

Carter, M. and Harlow, B. (2003) *Archives of Empire: The East India Company to the Suez Canal, Selected Letters between Tipu and Company's Governor-General 1798–99*, Durham, NC: Duke University Press.

Cashman, R. (1975) *The Myth of the Lokmanya: Tilak and Mass Politics in Maharashtra*, Berkeley: University of California Press.

Chandra, S. (2003) 'Jizya and the State during the Seventeenth Century', pp. 133–49, in R.M. Eaton (Ed.) (2003) *India's Islamic Traditions 711–1750*, New Delhi: Oxford University Press.

Chatterjee, A. (1998) *Representations of India, 1740–1840: The Creation of India in Colonial Imagination*, Basingstoke: Macmillan Press.

Chaudhuri, B. (1983) 'Agrarian Relations in Eastern India', pp. 86–177, in D. Kumar and M. Desai (Eds) (1983) *The Cambridge Economic History of India,* Vol. 2, Cambridge: Cambridge University Press.

Chaudhuri, K.N. (1982) 'European Trade with India', pp. 382–407, in T. Raychaudhuri and I. Habib (Eds) (1982) *The Cambridge Economic History of India,* Vol. 1, Cambridge: Cambridge University Press.

Chaudhuri, K.N. (1990) *Asia before Europe: Economy and Civilisation of the Indian Ocean from the Rise of Islam to 1750*, Cambridge: Cambridge University Press.
Chughtai, I. (1991) *The Quilt and Other Stories*, New Delhi: The Women's Press.
Cole, J.R.I. (1988) *Roots of North Indian Shiism in Iran and Iraq: Religion and State in Awadh 1722–1859*, Berkeley: University of California Press.
Collingham, E.M. (2001) *Imperial Bodies: The Physical Experience of the Raj c. 1800–1947*, Cambridge: Polity Press.
Collins, I. and D. Lapierre (1975) *Freedom at Midnight*, London: Collins.
Coppola, C. (1987) 'Nationalism and Urdu Literature in India and Pakistan', pp. 35–68, in M.C. Hillmann (Ed.) (1987) *Essays on Nationalism and Asian Literatures*, Austin: University of Texas Press.
Crowe, S. (1972) *The Gardens of Mughal India: A History and a Guide*, London: Thames and Hudson.
Daftary, F. (1998) *A Short History of the Ismailis: Traditions of a Muslim Community*, Edinburgh: Edinburgh University Press.
Dale, S.F. (1977) 'The Islamic Frontier in South West India: The Shahid as a Cultural Ideal among the Mappilas of Malabar', pp. 41–55, in *Modern Asian Studies*, 11(1).
Dale, S.F. (2004) *The Garden of the Eight Paradises: Babur and the Culture of Empire in Central Asia, Afghanistan and India 1483–1530*, Leiden: E.J. Brill.
Dalrymple, W. (2002) *White Mughals: Love and Betrayal in Eighteenth Century India*, London: Flamingo.
Dalton, D.G. (1970) 'Gandhi during Partition: A Case Study in the Nature of Satyagraha', pp. 222–44, in C.H. Philips and M.D. Wainwright (Eds) (1970) *The Partition of India: Policies and Perspective 1935–1947*, London: George, Allen and Unwin.
Dani, A.H. (1961) *Muslim Architecture in Bengal*, Dacca: Asiatic Society of Pakistan.
Darke, H. (1960) *The Book of Government, or Rules for Kings: The Siyasatnama of Nizam ul-Mulk*, London: Routledge and Kegan Paul.
Dasgupta, A. (1982) 'Indian Merchants and the Trade in the Indian Ocean', pp. 407–33, in T. Raychaudhuri and I. Habib (Eds) (1982) *The Cambridge Economic History of India*, Vol. 1, Cambridge: Cambridge University Press.
Datta, S. (2007) 'Islamic Militancy in Bangladesh: The Threat from Within', pp. 240–65, in D. Taylor (Ed.) (2011) *Islam in South Asia: Critical Concepts in Islamic Studies*, Vol. 3, London: Routledge.
Davies, P. (1989) *Monuments of India*, Vol. 2, London: Penguin.
De Souza, E. (2004) *Purdah: An Anthology*, New Delhi: Oxford University Press.
Dehejiya, V. (1997) *Indian Art*, London: Phaidon Press.
Denny, F.M. (1994) *An Introduction to Islam*, New York: Macmillan.
Digby, S. (1971) *War Horse and Elephant in the Delhi Sultanate: A Study of Military Supplies*, Oxford: Orient Monographs.
Digby, S. (1982) 'The Currency System', pp. 93–101, in T. Raychaudhuri and I. Habib (Eds) (1982) *The Cambridge Economic History of India*, Vol. 1, Cambridge: Cambridge University Press.
Digby, S. (2003) 'The Sufi Shaikh as a Source of Authority in Medieval India', pp. 234–62, in R.M. Eaton (Ed.) (2003) *India's Islamic Traditions 711–1750*, New Delhi: Oxford University Press.

Dodwell, H. (Ed.) (1929) *The Cambridge History of India,* Vol. 5, Cambridge: Cambridge University Press.

Douglas, I.J. (1988) *Abul Kalam Azad: An Intellectual and Religious Biography*, Oxford: Oxford University Press.

D'Souza, E. (2004) *Medieval India*, Mumbai: Manan Prakashan.

Dunn, R.E. (1986) *The Adventures of Ibn Battuta: A Muslim Traveler of the 14th Century*, London: Croom Helm.

Durant, W. (1935) *Our Oriental Heritage*, New York: Simon and Schuster.

Eaton, R.M. (1993) *The Rise of Islam and the Bengal Frontier 1204–1760*, Berkeley: University of California Press.

Eaton, R.M. (2000) *Essays on Islam and Indian History*, Delhi: Oxford University Press.

Eaton, R.M. (2003) *India's Islamic Traditions 711–1750*, New Delhi: Oxford University Press.

Eaton, R.M. (2004) *Temple Desecration and Muslim States in Medieval India*, Gurgaon: Hope India Publications.

Eaton, R.M. (2005) *A Social History of the Deccan, 1300–1761, The New Cambridge History of India,* Vol. I.8, Cambridge: Cambridge University Press.

Elias, J.J. (Ed.) (2010) *Key Themes for the Study of Islam*, Oxford: One World.

Elliott, H.M. (1871/2004) *Tuzuk-I Timuri, The Autobiography of Timur*, Lahore: Sang-e-Meel.

Elliott, H.M. and J. Dowson (1867–1877/2001) *A History of India as Told by its Own Historians: The Muhammadan Period,* Vols. 1 and 7, New Delhi: Low Price Publications.

Engineer, A.A. (2006) 'Indian Muslims – Reservation or No Reservation?', Centre for Study of Society and Secularism, Mumbai: http://www.csss-isla.com/aboutus.htm

Engineer, A.A. (2007) 'Identity and Social Exclusion-Inclusion: A Muslim Perspective', Centre for Study of Society and Secularism, Mumbai: http://www.csss-isla.com/aboutus.htm

Eraly, A. (2008) *The Mughal World: India's Tainted Paradise*, London: Phoenix.

Ernst, C.W. (2003) 'Between Orientalism and Fundamentalism: Problematising the teaching of Sufism', pp. 112–16, in M.W. Brannon (Ed.) (2003) *Teaching Islam*, Oxford: Oxford University Press.

Esposito, J.L. (2003) *The Oxford Dictionary of Islam*, Oxford: Oxford University Press.

Falconer, A.E.I. (1991) *Sufi Literature and the Journey to Immortality*, Delhi: Motilal Banarsidass.

Faruqi, S.R. (2003) 'A Long History of Urdu Culture, Part 1', pp. 805–63, in S. Pollock (Ed.) (2003) *Literary Cultures in History: Reconstructions from South Asia*, Berkeley: University of California Press.

Fatimi, S.Q. (1987) 'The Twin Ports of Daybul: A Study in the Early Maritime History of Sind', pp. 97–195, in H. Khuhro (Ed.) (1987) *Sind through the Centuries*, Karachi: Oxford University Press.

Findly, E.B. (2002) 'The Lives and Contributions of Mughal Women', pp. 27–58, in Z. Ziad, (Ed.) (2002) *The Magnificent Mughals*, Oxford: Oxford University Press.

Fisher, M. (1984) 'Indirect Rule in the British Empire: The Foundations of the Residency System in India (1764–1858)', *Modern Asian Studies*, 18(3): 393–428.

Fisher, M. (1996) *The Politics of the British Annexation of India 1757–1857*, New Delhi: Oxford India Paperbacks.

Forbes, G. (1996) *Women in Modern India*, Cambridge: Cambridge University Press.

Forward, M. (1999) *The Failure of Islamic Modernism: Sayyid Ameer Ali's Interpretation of Islam*, Bern: Peter Lang.

Fredunbeg, M.K. (1900/1985 reprint) *The Chachnama, An Ancient History of Sind*, Lahore: Vanguard Books.

Frembgen, J.W. (2008) *Journey to God: Sufis and Dervishes in Islam*, Oxford: Oxford University Press.

Frye, R.N. (1975) 'The Samanids', pp. 136–61, in R.N. Frye (Ed.) (1975) *The Cambridge History of Iran*, Vol. 4, Cambridge: Cambridge University Press.

Frykenberg, R. (2003) *Christians and Missionaries in India: Cross-Cultural Communication since 1500*, London: Routledge Curzon.

Gaborieau, M. (2002) 'Islam and Politics', pp. 235–49, in C. Jaffrelot (Ed.) (2002a) *A History of Pakistan and its Origins*, London: Anthem Press.

Gerlach, C. (2010) *Extremely Violent Societies*, Cambridge: Cambridge University Press.

Ghani, M.A. (1941) *Pre-Mughal Persian in Hindustan*, Allahabad: The Allahabad Law Journal Press.

Gibb, H.A.R. (1971) *The Travels of Ibn Battuta 1325–1354*, Vol. 3, London: Hakluyt Society.

Glasse, C. (2001) *The Concise Encyclopaedia of Islam*, London: Stacey International.

Goitein, S.D. (1955) *Jews and Arabs: Their Contacts through the Ages*, New York: Schocken Books.

Goodman, L.E. (1990) 'The Translation of Greek Materials into Arabic', pp. 477–97, in M.J.L. Young *et al.* (Eds) (1990) *Religion, Learning and Science in the Abbasid Period*, Cambridge: Cambridge University Press.

Gopal, R. (1959) *Indian Muslims: A Political History*, Bombay: Asia Publishing House.

Gordon, S. (1993) *The Marathas 1600–1818, The New Cambridge History of India*, Vol. II.4, Cambridge: Cambridge University Press.

Gove, M. (2006) *Celsius 7/7: How the West's Policy of Appeasement has Provoked Yet more Fundamentalist Terror – and What has to be Done Now*, London: Weidenfeld and Nicolson.

Goyal, D.R. (2002) 'Indian Response to Islam', pp. 106–17, in A.A. Engineer (Ed.) (2002) *Islam in India: The Impact of Civilizations*, Delhi: Shipra Publications.

Gray, J. (Ed.) (1991) *John Stuart Mill: On History and Other Essays*, Oxford: Oxford University Press.

Guha, R. (Ed.) (2011) *Makers of Modern India*, London: Harvard University Press.

Guha, S. (2003) 'The Politics of Identity and Enumeration in India *c.* 1600–1990', 148–67, in *Society for Comparative Study of Science and History*, 2003.

Guillaume, A. (1956) *Islam*, London: Penguin.

Habib, I. (1982) '"Agrarian economy" and "Non-Agricultural Production and the Urban Economy"', pp. 48–93, in T. Raychaudhuri and I. Habib (Eds) (1982) *The Cambridge Economic History of India*, Vol. 1, Cambridge: Cambridge University Press.

Habib, I. (Ed.) (1997) *Akbar and His India*, New Delhi: Oxford University Press.

Habib, I. (2000) 'Reconciling Science with Islam in 19th Century India', pp. 53–80, in D. Taylor (Ed.) (2011) *Islam in South Asia: Critical Concepts in Islamic Studies*, Vol. 2, London: Routledge.

Habib, M. and K.A. Nizami (1970) *A Comprehensive History of India*, Vol. 5, *The Delhi Sultanate*, New Delhi: People's Publishing House.

Habibullah, A.B.M. (1961) *The Foundation of Muslim Rule in India*, Allahabad: Central Book Depot.

Haig, W. (Ed.) (1928) *The Cambridge History of India*, Vol. 3, Cambridge: Cambridge University Press.

Haig, W. (1937a) 'Muhammad Shah', pp. 341–76, in R. Burn (Ed.) (1937) *The Cambridge History of India*, Vol. 4, Cambridge: Cambridge University Press.

Haig, W. (1937b) 'Ahmad Shah, Alamgir II and Shah Alam', pp. 428–48, in R. Burn (Ed.) (1937) *The Cambridge History of India*, Vol. 4, Cambridge: Cambridge University Press.

Haq, S.M. (1960) 'The Last Days of the Mughal Dynasty', pp. 1–26, in Pakistan Historical Society (1960) *A History of the Freedom Movement*, Karachi: Pakistan Historical Society.

Haque, I. (1992) *Glimpses of Mughal Society and Culture: A Study based on Urdu Literature in the Second Half of the Eighteenth Century*, New Delhi: Concept Publishing Company.

Hardgrave, R.L. (1977) 'The Mappila Rebellion: Peasant Revolt in Malabar', *Modern Asian Studies*, 1(1): 57–99.

Hardy, P. (1960) *Historians of Medieval India*, London: Luzac and Co.

Hardy, P. (1961a) 'Some Studies in Pre-Mughal Muslim Historiography', pp. 115–27, in C.H. Philips (Ed.) (1961) *Historians of India, Pakistan and Ceylon*, London: Oxford University Press.

Hardy, P. (1961b) 'Modern Muslim Historical Writing on Medieval Muslim India', pp. 294–309, in C.H. Philips (Ed.) (1961) *Historians of India, Pakistan and Ceylon*, London: Oxford University Press.

Hardy, P. (1965) 'Firishta', pp. 921–3, in *Encyclopaedia of Islam*, Vol. 2, Leiden: E.J. Brill.

Hardy, P. (1972a) *The Muslims of British India*, Cambridge: Cambridge University Press.

Hardy, P. (1972b) 'Ghalib and the British', pp. 54–69, in R. Russell (Ed.) (1972) *Ghalib, The Poet and His Age*, London: George, Allen and Unwin.

Hardy, P. (1987) 'Is the Chachnama Intelligible to the Historian as Political Theory?', pp. 111–17, in H. Khuhro (Ed.) (1987) *Sind through the Centuries*, Karachi: Oxford University Press.

Hardy, P. (1998) 'Growth of Authority Over a Conquered Political Elite: Early Delhi Sultanate as a Possible Case Study', pp. 216–41, in J.F. Richards (Ed.) (1998) *Kingship and Authority in South Asia*, Delhi: Oxford University Press.

Harle, J.C. (1986) *The Art and Architecture of the Indian Sub-Continent*, London: Penguin.

Hasan, M. (1988) 'The Muslim Mass Contacts Campaign: Analysis of a Strategy of Political Mobilization', pp. 198–222, in R. Sisson and S. Wolpert (Eds) (1988) *Congress and Indian Nationalism: The Pre-Independence Phase*, Berkeley: University of California Press.

Hasan, M. (Ed.) (1995) *India Partitioned: The Other Face of Freedom*, Vol. 1, New Delhi: Lotus collections, Roli Books.

Hasan, M. (1997) *Legacy of a Divided Nation: India's Muslims since Independence*, Oxford: Oxford University Press.

Hasan, M. (2000) *Nationalism and Communal Politics in India 1885–1930*, Delhi: Manohar.

Hasan, M. (2005) *A Moral Reckoning: Muslim Intellectuals in Nineteenth Century Delhi*, New Delhi: Oxford University Press.

Hasan, M. (2008) *Moderate or Militant: Images of India's Muslims*, New Delhi: Oxford University Press.

Hashmi, M.A. (1989) *Muslim Response to Western Education: A Study of Four Pioneer Institutions*, New Delhi: Commonwealth Publishers.

Hassan, R. (1971) 'The Development of (Iqbal's) Political Philosophy', pp. 136–58, in H. Malik, (Ed.) (1971) *Iqbal: Poet-Philosopher of Pakistan*, New York: Columbia University Press.

Heffening, W. (1993) 'Muta', pp. 757–9, in *The Encyclopaedia of Islam,* Vol. 7, Leiden: E.J. Brill.

Hewitt, V. (2001) *Towards the Future: Jammu and Kashmir in the Twenty First Century*, Cambridge: Granta Editions.

Hill, D.R. (1993) *Islamic Science and Engineering*, Edinburgh: Edinburgh University Press.

Hodgson, M.G. (1977a) *The Venture of Islam:* Vol. 1, Chicago: The University of Chicago Press.

Hodgson, M.G. (1977b) *The Venture of Islam,* Vol. 2, Chicago: The University of Chicago Press.

Hopkins, J.F.P. (1990) 'Geography and Navigational Literature', pp. 301–27, in M.J.L. Young, J.D. Latham and R.B. Sargent (Eds) (1990), *Religion, Learning and Science in the Abbasid Period*, Cambridge: Cambridge University Press.

Hourani, G.F. (1979) *Arab Seafaring in the Indian Ocean in Ancient and Early Medieval Times*, Princeton, NJ: Princeton University Press.

Hunter, W.W. (1871/1969 reprint) *The Indian Musulmans*, Delhi: Indological Book House.

Hussain, A. (1999) 'Speak Up' (tr. from Urdu), www.poemhunter.com/poem/speak-4/

Hussain, I. (2007) 'Fundamentalism and Bangladesh: No Error, No Terror', pp. 215–39, in D. Taylor (Ed.) (2011) *Islam in South Asia: Critical Concepts in Islamic Studies,* Vol. 3, London: Routledge.

Hussain, R. (1987) *Iqbal: Poet and his Politics*, New Delhi: Uppal Publishing House.

Hyam, R. (1976) *Britain's Imperial Century 1815–1914: A Study of Empire and Expansion*, London: Batsford.

Ikram, S. (1964) *Muslim Civilization in India*, New York: Columbia University Press.

Iqbal, M. (1911) *Tarana-i Hind*, www.enotes.com/topic/Saare_Jahan_Se_Achcha

Iqbal, M. (2004 edition) *The Reconstruction of Religious Thought in Islam*, Lahore: Sang-e-Meel Publications.

Islam, R. (1982) *A Calendar of Documents on Indo-Persian Relations,* Vol. II, 1500–1750, Karachi: Institute of Central and West Asian Studies.

Jack, I. (2011) 'It's not the Arithmetic of Genocide That's Important. It's that We Pay Attention', *The Guardian*, 21 May, p. 35.

Jackson, P. (1999) *The Delhi Sultanate: A Political and Military History*, Cambridge: Cambridge University Press.

Jaffrelot, C. (Ed.) (2002a) *A History of Pakistan and its Origins*, London: Anthem Press.

Jaffrelot, C. (Ed.) (2002b) *Pakistan: Nationalism without a Nation?*, New Delhi: Manohar.

Jaffrelot, C. (2007) *Hindu Nationalism: A Reader*, Princeton, NJ: Princeton University Press.

Jalal, A. (1985) *The Sole Spokesman: Jinnah, the Muslim League and the Demand for Pakistan*, Cambridge: Cambridge University Press.

Jalal, A. (2000) *Self and Sovereignty: Individual and Community in South Asian Islam since 1850*, London: Routledge.

Jha, P.S. (1996) *Kashmir 1947: Rival Versions of History*, Delhi: Oxford University Press.

Jones, K.W. (1989) *Socio-Religious Reform Movements in British India: The New Cambridge History of India,* Vol. III.1, Cambridge: Cambridge University Press.

Joseph, G.G. (2011) *The Crest of the Peacock: Non-European Roots of Mathematics*, Princeton, NJ: Princeton University Press.

Kamran, T. (2009) 'Contextualising Sectarian Militancy in Pakistan: A Case Study of Jhang', pp. 11–37, in D. Taylor (Ed.) (2011) *Islam in South Asia: Critical Concepts in Islamic Studies*, Vol. 3, London: Routledge.

Kaul, H.K. (Ed.) (1985) *Historic Delhi: An Anthology*, New Delhi: Oxford India Paperbacks.

Kaul, S. (Ed.) (2001) *The Partitions of Memory: The Afterlife of the Division of India*, Delhi: Permanent Black.

Kaviraj, S. (2003) 'The Two Histories of Literary Culture in Bengal', pp. 503–66, in S. Pollock (Ed.) (2003) *Literary Cultures in History: Reconstructions from South Asia*, Berkeley: University of California Press.

Keay, J. (2000) *India: A History*, London: Harper Collins.

Khan, I.A. (1997) 'Akbar's Personality Traits and World Outlook: A Critical Reappraisal', pp. 79–96, in I. Habib (Ed.) (1997) *Akbar and his India*, New Delhi: Oxford University Press.

Khan, I.A. (2003) 'The Nobility under Akbar and the Development of his Religious Policy 1560–80', pp. 120–31, in R.M. Eaton (Ed.) (2003) *India's Islamic Traditions 711–1750*, New Delhi: Oxford University Press.

Khan, I.G. (1997) 'Scientific Concepts in Abul Fazl's Ain-i Akbari', pp. 121–8, in I. Habib (Ed.) (1997) *Akbar and his India*, New Delhi: Oxford University Press.

Khan, M.A.S. (1997) *Early Muslim Perception of India and Hinduism*, New Delhi: South Asian Publishers.

Khan, M.I. (2003) 'The Impact of Islam on Kashmir in the Sultanate Period, 1320–1556', pp. 342–62, in R.M. Eaton (Ed.) (2003) *India's Islamic Traditions 711–1750*, New Delhi: Oxford University Press.

Khan, Y. (2007) *The Great Partition: The Making of India and Pakistan*, New Haven, CT: Yale University Press.

Khilnani, S. (1997) 'India's Mapmaker: Cyril Created Two Nations in 36 days. On the 37th Day He Rested. Then He Got out Quick', *The Observer* (London) (Review Section), 22 June.

Koch, E. (2002) *Mughal Architecture: An Outline of its History and Development*, New Delhi: Oxford University Press.

Koch, E. (2006) *The Complete Taj Mahal*, London: Thames and Hudson.
Korbel, J. (1954) *Danger in Kashmir*, Princeton, NJ: Princeton University Press.
Kulke, E. (1974) *The Parsees in India: A Minority as Agent of Social Change*, Munich: Weltforum Verlag.
Kumar, S. (2007) *The Emergence of the Delhi Sultanate*, New Delhi: Permanent Black.
Kurin, R. (1985) 'Islamization in Pakistan: A View from the Countryside', pp. 1–10, in D. Taylor (Ed.) (2011) *Islam in South Asia: Critical Concepts in Islamic Studies*, Vol. 3, London: Routledge.
Kwanten, L. (1979) *The Imperial Nomads*, Leicester: Leicester University Press.
Lafrance, P. (2002) 'Between Caste and Tribe', pp. 189–218, in C. Jaffrelot (Ed.) (2002a) *A History of Pakistan and its Origins*, London: Anthem Press.
Lal, K.S. (1963) *The Twilight of the Sultanate*, Delhi: Asia Publishing House.
Lal, R. (2005) *Domesticity and Power in the Early Mughal World*, Cambridge: Cambridge University Press.
Lal, V. (2003) *The History of History: Politics and Scholarship in Modern India*, Oxford: Oxford University Press.
Lambert-Hurley, S. (2007) *Muslim Women, Reform and Princely Patronage: Nawab Sultan Jahan Begum of Bhopal*, London: Routledge.
Lambton, A.K.S. (1968) 'The Internal Structure of the Saljuq Empire', pp. 203–82, in J.A. Boyle, (Ed.) (1968) *The Cambridge History of Iran*, Vol. 5, Cambridge: Cambridge University Press.
Lambton, A.K.S. (1978) 'Khalifa in Political Theory', pp. 947–50, in *The Encyclopaedia of Islam*, Vol. 4, Leiden: E.J. Brill.
Lane-Poole, S. (1903) *Medieval India*, New York: Haskell House Publishers.
Lapidus, I.M. (2002) *A History of Islamic Societies*, Cambridge: Cambridge University Press.
Lateef, S. (1990) *Muslim Women in India: Political and Private Realities 1890s to 1980s*, New Delhi: Kali for Women.
Lawrence, B. (1986) 'The Earliest Chishtiya and Shaikh Nizam ud-Din Awliya', pp. 104–28, in R.E. Frykenberg (Ed.) (1986) *Delhi Through the Ages: Essays in Urban History, Culture and Society*, Delhi: Oxford University Press.
Lawrence, B. (1990) 'Abu Rayhan Biruni: Indology', pp. 285–87, in *Encyclopaedia Iranica*, Vol. 4, London: Routledge and Kegan Paul.
Lawrence, B. (1992) *Nizam ud-Din Auliya: Morals for the Heart*, New York: Paulist Press.
Lawrence, W. (1895) *The Valley of Kashmir*, London: Oxford University Press.
Lawson, P. (1993) *The East India Company: A History*, London: Longman.
Lazard, G. (1975) 'The Rise of the New Persian Language', pp. 595–657, in R.N. Frye (Ed.) (1975) *The Cambridge History of Iran*, Vol. 4, Cambridge: Cambridge University Press.
Lelyveld, D. (1978) *Aligarh's First Generation: Muslim Solidarity in British India*, Princeton, NJ: Princeton University Press.
Lelyveld, D. (1982) 'Disenchantment at Aligarh: Islam and the Realm of the Secular in Late Nineteenth Century India', pp. 98–112, in D. Taylor (Ed.) (2011) *Islam in South Asia: Critical Concepts in Islamic Studies,* Vol. 2, London: Routledge.
Leonard, K. (1979) 'The "Great Firm" Theory of the Decline of the Mughal Empire', pp. 398–418, in M. Alam and S. Subrahmanyam (Eds) (2000) *The Mughal State*, New Delhi: Oxford University Press.

Lieven, A. (2011) *Pakistan: A Hard Country*, London: Allen Lane.
Losty, J.P. (1990) *Calcutta: City of Palaces*, London: The British Library.
Lowe, W.H. (1884/1976) *Muntakhab-ut-Tawarikh by Al-Badaoni*, Vol. 2, Karachi: Karimsons.
Maclean, D. (1989) *Religion and Society in Arab Sind*, Leiden: E.J. Brill.
Maddison, A. (1971) *Class Structure and Economic Growth: India and Pakistan since the Moghuls*, London: George Allen and Unwin.
Maddison, A. (2007) *Contours of the World Economy, 1–2030 AD*, Oxford: Oxford University Press.
Majumdar, R.C. and A.D. Pusalker (Eds) (1954) *The Classical Age*, Bombay: Bharatiya Vidya Bhavan.
Malik, H. (Ed.) (1971) *Iqbal: Poet-Philosopher of Pakistan*, New York: Columbia University Press.
Malik, H. (1980) *Sir Sayyid Ahmad Khan and Muslim Modernization in India and Pakistan*, New York: Columbia University Press.
Manseragh, P.N.S. (1970) 'Some Reflections on the Transfer of Power in Plural Societies', pp. 43–53, in C.H. Philips and M.D. Wainwright (Eds) (1970) *The Partition of India: Policies and Perspectives 1935–1947*, London: George Allen and Unwin.
Manz, B.F. (1991) *The Rise and Rule of Tamerlane*, Cambridge: Cambridge University Press.
Marek, J. (1971) '(Iqbal's) Perceptions of International Politics', pp. 159–73, in H. Malik (Ed.) (1971) *Iqbal: Poet-Philosopher of Pakistan*, New York: Columbia University Press.
Markovits, C. (Ed.) (2002) *A History of Modern India 1480–1950*, London: Anthem Press.
Marshall, J. (1928) 'The Monuments of Muslim India', pp. 568–640, in W. Haig (Ed.) (1928) *Cambridge History of India*, Vol. 3, Cambridge: Cambridge University Press.
Marshall, P.J. (1998) 'The English in Asia to 1700', pp. 264–85, in N. Canny (Ed.) (1998) *The Oxford History of British Empire*, Vol. 1, *The Origins of Empire*, Oxford: Oxford University Press.
Marshall, P.J. (2001) 'The British in Asia: Trade to Dominion, 1700–65', pp. 487–507, in P.J. Marshall (Ed.) (2001) *The Oxford History of the British Empire*, Vol. 2, *The Eighteenth Century*, Oxford: Oxford University Press.
Marshall, P.J. (Ed.) (2003) *The Eighteenth Century in Indian History: Evolution or Revolution?*, New Delhi: Oxford University Press.
Marshall, P.J. (2007) *The Making and Unmaking of Empires: Britain, India and America 1750–83*, Oxford: Oxford India Paperbacks.
Martineau, A. (1929) 'Dupleix and Bussy', pp. 125–40, in H. Dodwell (Ed.) (1929) *Cambridge History of India*, Vol. 5, Cambridge: Cambridge University Press.
Mason, P. (1974) *A Matter of Honour: An Account of the Indian Army, its Officers and Men*, London: Penguin.
Masse, H.H. (1971) 'Hudjwiri', p. 546, in *The Encyclopaedia of Islam*, Vol. 3, Leiden: E.J. Brill.
Massignon, L. (1971) 'Al-Halladj', pp. 99–104, in *The Encyclopaedia of Islam*, Vol. 3, Leiden: E.J. Brill.
Mazrui, A. (1990) '"Satanic Verses" or a Satanic Novel: Moral Dilemmas of the Rushdie Affair', *Third World Quarterly*, 12(1): 116–39.

McGregor, R.S. (1984) *Hindi Literature from its Beginnings to the 19th Century*, Wiesbaden: Otto Harrassowitz.

Menon, V.P. (1957) *The Transfer of Power in India*, Princeton, NJ: Princeton University Press.

Metcalf, B. (1978) 'The *Madrasa* at Deoband: A Model for Religious Education in Modern India', pp. 113–34, in D. Taylor (Ed.) (2011) *Islam in South Asia: Critical concepts in Islamic Studies*, Vol. 2, London: Routledge.

Metcalf, T.R. (1965) *The Aftermath of Revolt 1857–1870*, Princeton, NJ: Princeton University Press.

Michell, G and S. Shah (1988) *Ahmadabad*, Mumbai: Marg Publications.

Michell, G. and M. Zebrowski (1999) *Architecture and Art of the Deccan Sultanates: The New Cambridge History of India*, Vol. I.7, Cambridge: Cambridge University Press.

Miller, R.E. (1976) *Mappila Muslims of Kerala: A Study in Islamic Trends*, New Delhi: Orient Longman.

Miller, R.E. (1991), 'Mappila', pp. 458–66, in *The Encyclopaedia of Islam*, Vol. 6, Leiden: E.J. Brill.

Minault, G. (1998) *Secluded Scholars: Women's Education and Muslim Social Reform in Colonial India*, Oxford: Oxford University Press.

Minto, M. (1934) *India, Minto and Morley 1905–1910*, London: Macmillan.

Mirza, M.W. (1935) *The Life and Works of Amir Khusrau*, Delhi: Idara-i Adabiyat-i Delli (Delhi).

Mirza, M.W. (1950) *The Nuh Sipihr of Amir Khusraw*, London: Islamic Research Association and Oxford University Press.

Misra, S.C. (1982) *The Rise of Muslim Power in Gujarat*, Bombay: Munshiram Manoharlal.

Mistree, K. (2002) 'Parsi Arrival and Early Settlements in India', pp. 411–33, in P.J. Godrej and F.P. Mistree (Eds) (2002) *A Zoroastrian Tapestry: Art, Religion and Culture*, Ahmadabad: Mapin Publishing.

Mitter, P. (2001) *Indian Art*, Oxford: Oxford University Press.

Mohammad, A. (2002) 'The Many Faces of Islam', pp. 221–34, in C. Jaffrelot (Ed.) (2002a) *A History of Pakistan and its Origins*, London: Anthem Press.

Mohan, S. (1997) *Awadh under the Nawabs: Politics, Culture and Commercial Relations, 1722–1856*, New Delhi: Manohar.

Moon, P. (1998) *Divide and Quit: An Eye-Witness Account of the Partition of India*, Oxford: Oxford University Press.

Moore, R.J. (1970) 'The Making of India's Paper Federation', pp. 54–78, in C.H. Philips and M.D. Wainwright (Eds) (1970) *The Partition of India: Policies and Perspectives 1935–1947*, London: George Allen and Unwin.

Moosvi, S. (1987) *The Economy of the Mughal Empire, c. 1595, A Statistical Study*, Delhi: Oxford University Press.

Moreland, W.H. (1929) *The Agrarian System of Muslim India*, Delhi: Kanti Publications.

Morgan, D. (1986) *The Mongols*, Oxford: Basil Blackwell.

Moynihan, E.B. (1980) *Paradise as a Garden: In Persia and Mughal India*, London: Scholar Press.

Mujeeb, M. (1967) *The Indian Muslims*, London: George, Allen and Unwin.

Mukhia, H. (2004) *The Mughals of India*, Oxford: Blackwell.

Mullally, S. (2004) 'Feminism and Multicultural Dilemmas in India: Revisiting the Shah Bano Case', *Oxford Journal of Legal Studies*, 24(4): 671–92.

Nanda, B.R. (1970) 'Nehru, the Indian National Congress and the Partition of India 1945–1947', pp. 148–87, in C.H. Philips and M.D. Wainwright (Eds) (1970) *The Partition of India: Policies and Perspectives 1935–1947*, London: George Allen and Unwin.

Naqvi, H.A. (1986) 'Shahjahanabad: The Mughal Delhi 1638–1803: An Introduction', pp. 143–51, in R.E. Frykenberg (Ed.) (1986) *Delhi through the Ages: Essays in Urban History, Culture and Society*, Delhi: Oxford University Press.

Nasr, S.H. (1976) *Islamic Science: An Illustrated Study*, London: World of Islam Festival Publication.

Nasr, S.H. (1987) *Science and Civilization in Islam*, Cambridge: Islamic Texts Society.

Nasr, S.V.R. (2000) 'The Rise of Sunni Militancy in Pakistan: The Changing Role of Islamism and the Ulama in Society and Politics', *Modern Asian Studies*, 34(1): 139–80.

Nath, R. and F. Gwaliari (1981) *India as Seen by Amir Khusrau*, Jaipur: Historical Research Documentation Programme.

Nawaz, J. (2002) *Father and Daughter: A Political Autobiography*, Oxford: Oxford University Press.

Nawaz, S. (2008) *Crossed Swords: Pakistan, Its Army and Its Wars Within*, Oxford: Oxford University Press.

Nazim, M. (1931) *The Life and Times of Sultan Mahmud of Ghazna*, Lahore: Khalil and Co.

Nizami, K.A. (1961) *Some Aspects of Religion and Politics in India during the Thirteenth Century*, Aligarh: Department of History Muslim University.

Nizami, K.A. (Ed.) (1981) *Politics and Society During the Early Medieval Period – Collected Works of Professor Mohammed Habib*, Vol. 2, New Delhi: People's Publishing House.

Nyman, J. (2009) *Home, Identity and Mobility in Contemporary Diasporic Fiction*, Amsterdam: Rodopi B.V.

Oak, P.N. (1969) *Taj Mahal: The True Story – The Tale of a Temple Vandalised*, Houston: A. Ghosh.

Oddie, G.A. (2003) 'Constructing Hinduism: The Impact of Protestant Missionary Movement on Hindu Self-understanding', pp. 155–82, in R.E. Frykenberg (Ed.) (2003) *Christians and Missionaries in India: Cross-Cultural Communications since 1500*, London: Routledge Curzon.

Page, J.B. (2008 edition by George Michell) *Indian Islamic Architecture*, Leiden: E.J. Brill.

Pakistan Government (1989) *Quaid-i-Azam: Speeches and Statements 1947–48*, Islamabad: Government of Pakistan Publications.

Pandey, B.N. (Ed.) (1979) *The Indian Nationalist Movement, 1885–1947: Select Documents*, London: Macmillan.

Pandey, G. (1992) 'The Colonial Construction of Communalism: British Writings on Banaras in the Nineteenth Century', pp. 94–134, in V. Das (Ed.) (1992) *Mirrors of Violence: Communities, Riots and Survivors in South Asia*, Delhi: Oxford University Press.

Pandey, G. (2006) *The Construction of Communalism in Colonial North India*, New Delhi: Oxford University Press.

Panigrahi, D.N. (2004) *The Partition of India: The Story of Imperialism in Retreat*, London: Routledge.

Parthasarathi, P. (1996) 'Merchants and the Rise of Colonialism', pp. 199–224, in S. Alavi. (Ed.) (2007) *The Eighteenth Century in India*, Oxford: Oxford India Paperbacks.

Pew Research Centre – http://en.wikipedia.org/wiki.list_of_countries_by_Muslim_population

Philips, C.H. (Ed.) (1961) *Historians of India, Pakistan and Ceylon*, London: Oxford University Press.

Philips, C.H. (1962) *The Evolution of India and Pakistan 1858 to 1947: Select Documents*, London: Oxford University Press.

Philips, C.H. (1967) *The Partition of India – Montague Burton Lecture*, Leeds: Leeds University Press.

Philips, C.H. and M.D. Wainwright (Eds) (1970) *The Partition of India: Policies and Perspectives 1935–1947*, London: George, Allen and Unwin.

Pingree, D. (1997), 'SindHind', pp. 640–1, in *The Encyclopaedia of Islam*, Vol. 9, Leiden: E.J. Brill.

Pinney, C. (1990) 'Colonial Anthropology in the Laboratory of Mankind', pp. 252–63, in C. Bayly (Ed.) (1990) *The Raj: India and the British 1600–1947*, London: National Portrait Gallery Publications.

Pinto, D. (1989) 'The Mystery of the Nizam ud-Din's Dargah: The Accounts of Pilgrims', pp. 112–24, in C. Troll (Ed.) (1989) *Muslim Shrines in India: Their Character, History and Significance*, Delhi, Oxford University Press.

Pollock, S. (Ed.) (2003) *Literary Cultures in History: Reconstructions from South Asia*, Berkeley: University of California Press.

Porter, B. (2004) *The Absent-Minded Imperialists: Empire, Society and Culture in Britain*, Oxford: Oxford University Press.

Powell, A. (1993) *Muslims and Missionaries in Pre-Mutiny India*, London: Curzon Press.

Powicke, F.M. (1955) *Modern Historians and the Study of History*, London: Odhams Press.

Prakash, O. (2007) 'Trade and Politics in Eighteenth Century Bengal', pp. 136–64, in S. Alavi (Ed.) (2007) *The Eighteenth Century in India*, Oxford: Oxford India Paperbacks.

Prasad, I. (1955) *The Life and Times of Humayun*, Calcutta: Orient Longmans.

Prasad, P. (1997) 'Akbar and the Jains', pp. 97–108, in I. Habib (Ed.) (1997) *Akbar and his India*, New Delhi: Oxford University Press.

Prawdin, M. (1963) *The Builders of the Moghul Empire*, London: George Allen and Unwin.

Pritchett, F.W. (2003) 'A Long History of Urdu Literary Culture, Part 2: Histories, Performances and Masters', pp. 864–911, in S. Pollock (Ed.) (2003) *Literary Cultures in History: Reconstructions from South Asia*, Berkeley: University of California Press.

Puniyani, R. (2003) *Communal Politics: Facts versus Myths*, New Delhi: Sage.

Puri, B. (1995) *Kashmir: Towards Insurgency*, New Delhi: Orient Longman.

Qanungo, K. (1921) *Sher Shah: A Critical Study Based on Original Sources*, Calcutta: Kar Majumdar and Co.

Qureshi, I.H. (1972) *Ulema in Politics*, Karachi: Maaref.
Qureshi, I.H. (1985) *The Muslim Community in the Indo-Pakistan Sub-continent, 610–1947: A Brief Historical Analysis*, New Delhi: Renaissance Publishing House.
Rabie, H. (1975) 'The Training of the Mamluk Faris', pp. 153–63, in V.J. Parry and M.E. Yapp (Eds) (1975) *War, Technology and Society in the Middle East*, London: Oxford University Press.
Racine, J. (2002) 'Pakistan in the Game of the Great Powers', pp. 97–111, in C. Jaffrelot (Ed.) (2002a) *A History of Pakistan and its Origins*, London: Anthem Press.
Rafiqi, A.Q. (2002) 'India's Interface with Islam', pp. 36–45, in A. Engineer (Ed.) (2002) *Islam in India: The Impact of Civilizations*, Delhi: Shipra Publications.
Rahman, F. (1979) *Islam*, Chicago: University of Chicago Press.
Rai, M. (2004) *Hindu Rulers, Muslim Subjects: Islam, Rights and the History of Kashmir*, London: Hurst.
Ramusack, B.M. (2004) *The Indian Princes and their States*, Cambridge: Cambridge University Press.
Rao, B.S. (1970) 'India 1935–1947', pp. 415–67, in C.H. Philips and M.D. Wainwright (Eds) (1970) *The Partition of India: Policies and Perspectives 1935–1947*, London: George Allen and Unwin.
Rashid, A. (1961) 'The Treatment of History by Muslim Historians in Mughal Official and Biographical Works', pp. 139–51, in C.H. Phillips (Ed.) (1961) *Historians of India, Pakistan and Ceylon*, London: Oxford University Press.
Ray, R.K. (1998) 'Indian Society and the Establishment of British Supremacy 1765–1818', pp. 508–29, in P.J. Marshall (Ed.) (1998) *Oxford History of British Empire,* Vol. 2, Oxford: Oxford University Press.
Raychaudhuri, T. (1965) 'The Agrarian System of Mughal India: A Review Essay', pp. 259–83, in M. Alam and S. Subrahmanyam (Eds) (1998) *The Mughal State 1526–1750*, New Delhi: Oxford University Press.
Raychaudhuri, T. (1982a) 'The Mughal Empire', pp. 172–93, in T. Raychaudhuri and I. Habib (Eds) (1982) *The Cambridge Economic History of India,* Vol. 1, *c.* 1200–1750, Cambridge: Cambridge University Press.
Raychaudhuri, T. (1982b) 'Inland Trade', pp. 325–59, in T. Raychaudhuri and I. Habib. (Eds) (1982) *The Cambridge Economic History of India,* Vol. 1, *c.* 1200–1750, Cambridge: Cambridge University Press.
Rezavi, S.A.N. (1997) 'Revisiting Fatehpur Sikri: An Interpretation of Certain Buildings', pp. 173–87, in I. Habib (Ed.) (1997) *Akbar and his India*, New Delhi: Oxford University Press.
Richards, J.F. (1978) 'The Formulations of Imperial Authority under Akbar and Jahangir', pp. 126–67, in M. Alam and S. Subrahmanyam (Eds) (1998) *The Mughal State 1526–1750*, New Delhi: Oxford University Press.
Richards, J.F. (1993) *The Mughal Empire*, *The New Cambridge History of India,* Vol. I.5, Cambridge: Cambridge University Press.
Risso, P. (1995) *Merchants and Faith: Muslim Commerce and Culture in the Indian Ocean*, Boulder, CO: Westview Press.
Ritter, H. (1960) 'Abu Yazid (Bayazid) al-Bistam', pp. 162–3, in *The Encyclopaedia of Islam*, Vol. 1, Leiden: E.J. Brill.
Rizvi, S.A.A. (1969) *Shah Wali Allah and His Times*, Canberra: Marijat Publishing House.

Rizvi, S.A.A. (2003) *A History of Sufism in India*, Vol. 1: *Early Sufism and its History in India to ad 1600*, Delhi: Munshiram Manoharlal.

Roberts, A. (1994) *Eminent Churchillians*, London: Weidenfeld and Nicholson.

Robinson, F. (1974–5) *Separatism Among Indian Muslims: The Politics of the United Provinces' Muslims 1860–1923*, Cambridge: Cambridge University Press.

Robinson, F.C.R. (1991) 'Mawdudi', pp. 872–4, in *The Encyclopaedia of Islam*, Vol. 6, Leiden: E. J. Brill.

Robinson, F. (2000) *Islam and Muslim History in South Asia*, New Delhi: Oxford University Press.

Robinson, F. (2001) *The Ulama of Farangi Mahall and Islamic Culture in South Asia*, London: Hurst and Co.

Rogers, A. and H. Beveridge (1909/1989) *The Tuzuk-i-Jahangiri*, Vols. 1 and 2, New Delhi: Atlantic Publishers.

Ross, D. (1937) 'Babur', pp. 1–20, in R. Burn (Ed.) (1937) *The Cambridge History of India*, Vol. 4, Cambridge: Cambridge University Press.

Roy, O. (2002a) 'Islam and Foreign Policy: Central Asia and the Arab-Persian World', pp. 134–47, in C. Jaffrelot (Ed.) (2002a) *A History of Pakistan and its Origins*, London: Anthem Press.

Roy. O. (2002b) 'The Taliban: A Strategic Tool for Pakistan', pp. 149–59, in C. Jaffrelot (Ed.) (2002b) *Pakistan: Nationalism without a Nation*, New Delhi: Manohar.

Russell, R. (1995) *Hidden in the Lute: An Anthology of Two Centuries of Urdu Literature*, Manchester: Carcanet Press.

Russell, R. and K. Islam (1969a) *Three Mughal Poets: Mir, Sauda, Mir Hasan*, London: George Allen and Unwin.

Russell, R. and K. Islam, K. (1969b) *Ghalib 1797–1869: Life and Letters*, Vol. 1, London: George Allen and Unwin.

Rypka, J. (1968) *History of Iranian Literature*, Dordrecht: D. Reidel Publishing Co.

Sachar Committee Report (2006): http://en.wikipedia.org/wiki,Sachar_Committee_Report

Sachau, E. (1888) *Alberuni's India*, Vols. 1 and 2, London: Trubner.

Sadiq, M. (1964) *A History of Urdu Literature*, London: Oxford University Press.

Saksena, R. (2002) *A History of Urdu Literature*, New Delhi: Indigo Books.

Saliba, G. (1990) 'Al-Biruni and the Science of his Time', pp. 405–23, in M.J.L. Young, J.D. Latham and R.B. Sargent (Eds) (1990) *Religion, Learning and Science in the Abbasid Period*, Cambridge: Cambridge University Press.

Sarkar, J. (1937a) 'Aurangzib 1658–1681', pp. 222–59, in R. Burn (Ed.) (1937) *The Cambridge History of India*, Vol. 4, Cambridge: Cambridge University Press.

Sarkar, J. (1937b) 'From Bahadur Shah to Rafi ud-Daula', pp. 319–40, in R. Burn (Ed.) (1937) *The Cambridge History of India*, Vol. 4, Cambridge: Cambridge University Press.

Sarkar, J. (1937c) 'The Hyderabad State: 1724–62', pp. 377–91, in R. Burn (Ed.) (1937) *The Cambridge History of India*, Vol. 4, Cambridge: Cambridge University Press.

Sarkar, J. (1973) *The History of Bengal: Muslim Period, 1200–1757*, Patna: Academica Asiatica.

Sarkar, S. (1989) *Modern India 1885–1947*, London: Macmillan.

Sarker, K. (2007) *Shah Jahan and his Paradise on Earth*, Kolkata: Bagchi and Co.

Saul, D. (2003) *The Indian Mutiny 1857*, London: Penguin.
Sayeed, K.B. (1970) 'The Personality of Jinnah and his Political Strategy', pp. 276–93, in C.H. Philips and M.D. Wainwright (Eds) (1970) *The Partition of India: Policies and Perspectives 1935–1947*, London: George Allen and Unwin.
Schimmel, A. (1973) *Islamic Literatures of India*, Wiesbaden: Otto Harrassowitz.
Schimmel, A. (1975a) *Classical Urdu Literature from the Beginning to Iqbal*, Wiesbaden: Otto Harrassowitz.
Schimmel, A. (1975b) *Mystical Dimensions of Islam*, Chapel Hill: University of North Carolina Press.
Schimmel, A. (1980) *Islam in the Indian Sub-Continent*, Leiden: E.J. Brill.
Schimmel, A. (1992) *Islam: An Introduction*, New York: State University of New York Press.
Schofield, V. (2000) *Kashmir in Conflict: India, Pakistan and the Unfinished War*, London: I.B. Tauris.
Schwartzberg, J. (Ed.) (1992) *A Historical Atlas of South Asia*, Oxford: Oxford University Press.
Shaida, K.H. (2008) *Khusro, the Indian Orpheus: A Hundred Odes*, Charleston, SC: Bookserge Publishing Co. Available at: www.booksurge.com
Sharar, A.H. (1989) *Lucknow: The Last Phase of an Oriental Culture*, Delhi: Oxford University Press.
Sharma, B.N. (1972) *Social and Cultural History of Northern India 1000–1200*, New Delhi: Abhinav Publications.
Sharma, R. (1972) *The Religious Policy of the Mughal Emperors*, London: Asia Publishing House.
Sharma, S. (2000) *Persian Poetry at the Indian Frontier: Masud Saad Salman of Lahore*, Delhi: Permanent Black.
Shokoohy, M. and N. Shokoohy (2007) *Tughluqabad: A Paradigm for Indo-Islamic Urban Planning and its Architectural Components*, London: British Academy and Araxus Books.
Siddiqqa, A. (2007) *Military Inc.: Inside Pakistan's Military Economy*, London: Pluto Press.
Sikand, Y. (2001) 'A New Indian Muslim Agenda: The Dalit Muslims and the All-India Backward Muslim *Morcha*', pp. 296–308, in D. Taylor (Ed.) (2011) *Islam in South Asia: Critical Concepts in Islamic Studies,* Vol. 3, London: Routledge.
Singh, C. (1988) 'Conformity and Conflict: Tribes and the "Agrarian System" of Mughal India', *Indian Economic and Social History Review,* 23(3): 319–40.
Singh, J. (2010) *Jinnah: India-Partition-Independence*, Oxford: Oxford University Press.
Singh, K. (2004) *A History of the Sikhs,* Vol. 1, 1469–1839, Delhi: Oxford University Press.
Siqueira, T. (1960) *Modern Indian Education*, Calcutta: Oxford University Press.
Sitaramayya, B.P. (1969) *History of the Indian National Congress,* Vol. 1, 1885–1935, Delhi: S. Chand and Co.
Smith, V. (no date) *A History of Fine Arts in India and Ceylon*, 3rd edition ed. by Karl Khandalavala, Bombay: D.B. Taraporewala and Sons.
Smith, W.C. (1946) *Modern Islam in India: A Social Analysis*, London: Victor Gollancz.

Smith, W.C. (1961) 'Modern Muslim Historical Writing in English', pp. 319–31, in C.H. Philips (Ed.) (1961) *Historians of India, Pakistan and Ceylon*, London: Oxford University Press.

Spear, P. (1990) *A History of India,* Vol. 2, London: Penguin.

Sprenger, A. (1851) *Al-Masudi's Historical Encyclopaedia Entitled 'Meadows of Gold and Mines of Gems',* Vol. 1, London: Oriental Translation Fund.

Spuler, B. (1971) *The Mongols in History*, London: Pall Mall Press.

Srivastava, K.L. (1980) *The Position of Hindus under the Delhi Sultanate 1206–1526*, New Delhi: Munshiram Manoharlal.

Stein, B. (1998) *A History of India*, Oxford: Blackwell.

Suri, P. (1968) 'Babur', pp. 98–105, in M. Hasan (1968) *Historians of Medieval India*, Meerut: Meenakshi Prakashan.

Swarup, D. (1986) *Politics of Conversion*, Delhi: Deen Dayal Research Institute.

Sykes, P. (1940) *A History of Afghanistan,* Vol. 1, London: Macmillan.

Talbot, I. (2002) 'Does the Army shape Pakistan's Foreign Policy?', pp. 311–35, in C. Jaffrelot (Ed.) (2002b) *Pakistan: Nationalism without a Nation*, New Delhi: Manohar.

Talbot, I. (2005) *Pakistan: A Modern History*, London: Hurst and Co.

Talbot, I. and G. Singh (2009) *The Partition of India*, Cambridge: Cambridge University Press.

Taleb, A. (1814) *Travels in Asia, Africa and Europe during the Years 1799, 1800, 1801, 1802 and 1803*, Translated from Persian by Charles Stewart, London: Longman Hurst, Rees, Orme and Brown Co.

Taylor, D. (Ed.) (2011) *Islam in South Asia: Critical Concepts in Islamic Studies,* Vol. 3, London: Routledge.

Thackston, W.M. (2002) 'Literature', pp. 83–111, in Z. Ziad (Ed.) (2002) *The Magnificent Mughals*, Oxford: Oxford University Press.

Thapar, R. (2004) *Somanatha: The Many Voices of a History*, New Delhi: Penguin/Viking.

Thursby, G.R. (1975) *Hindu–Muslim Relations in British India*, Leiden: E.J. Brill.

Titus, M. (1979) *Indian Islam – A Religious History of Islam in India*, New Delhi: Oriental Books Reprint.

Tomory, E. (1982) *A History of Fine Arts in India and the West*, Hyderabad: Orient Longman.

Troll, C.W. (1972) 'A Nineteenth Century Restatement of Islam', pp. 43–5, in D. Rothermund (1975) *Islam in Southern Asia: A Survey of Current Research*, Wiesbaden: Franz Steiner Verlag.

Upstone, S. (2010) *British Asian Fiction: Twenty First Century Voices*, Manchester: Manchester University Press.

Van Schendel, W. (2009) *A History of Bangladesh*, Cambridge: Cambridge University Press.

Varadarajan, S. (2002) *Gujarat: The Making of a Tragedy*, New Delhi: Penguin Books.

Visram, R. (2002) *Asians in Britain: 400 Years of History*, London: Pluto Press.

Wade, B.C. (2002) 'Music and Dance', pp. 229–68, in Z. Ziad (Ed.) (2002) *The Magnificent Mughals*, Oxford: Oxford University Press.

Wells, I.B. (2005) *Jinnah, Ambassador of Hindu–Muslim Politics*, London: Seagull Books.

Wheeler, B.M. (2002) *Prophets in the Quran: An Introduction to the Quran and Muslim Exegesis*, London: Continuum.

Wink, A. (1990) *Al-Hind: The Making of the Indo-Islamic World,* Vol. 1, *Early Medieval India and the Expansion of Islam 7th –11th Centuries*, Leiden: E.J. Brill.

Wink, A. (1997) *Al-Hind: The Making of the Indo-Islamic World,* Vol. 2, *The Slave Kings and the Islamic Conquest 11th to 13th Centuries*, Leiden: E.J. Brill.

Wink, A. (2004) *Al-Hind: The Making of the Indo-Islamic World,* Vol. 3, *Indo-Islamic Society 14th–15th Centuries*, Leiden: E.J. Brill.

Wink, A. (2009) *Makers of the Muslim World: Akbar*, Oxford: Oneworld.

Wolpert, S. (1984) *Jinnah of Pakistan*, Oxford: Oxford University Press.

Wolpert, S. (2009 edition) *A New History of India*, Oxford: Oxford University Press.

Wood, C. (1992) 'Peasant Revolt: An Interpretation of Moplah Violence in the Nineteenth and Twentieth Centuries', pp. 126–52, in D. Hardiman (Ed.) (1992) *Peasant Resistance in India 1858–1914*, Delhi: Oxford University Press.

Wylie, Sir F. (1970) 'Federal Negotiations in India 1935–37 and After', pp. 517–26, in C.H. Philips and M.D. Wainwright (Eds) (1970) *The Partition of India: Policies and Perspectives 1935–1947*, London: George Allen and Unwin.

Yashaschandra, S. (2003) 'From Hemcandra to Hind Swaraj: Religion and Power in Gujarati Literary Culture', pp. 567–611, in S. Pollock (Ed.) (2003) *Literary Cultures in History: Reconstructions from South Asia*, Berkeley: University of California Press.

Young, M.J.L., J.D. Latham and R.B. Sargeant (Eds) (1990) *Religion, Learning and Science in the Abbasid Period*, Cambridge: Cambridge University Press.

Yusofi, G.H. (1989) 'Bayhaqi, Abul-Fazl', pp. 889–94, in E. Yashater (Ed.) (1989) *Encyclopaedia Iranica,* Vol. 3, London: Routledge and Kegan Paul.

Zaidi, S.I.A. (1997) 'Akbar and the Rajput Principalities: Integration into Empire', pp. 15–24, in I. Habib (Ed.) (1997) *Akbar and his India*, Oxford: Oxford University Press.

Zaidi, Z.H. (1970) 'Aspects of the Development of Muslim League Policy 1937–47', pp. 145–75, in C.H. Philips and M.D. Wainwright (Eds) (1970) *The Partition of India: Policies and Perspectives 1935–1947*, London: George Allen and Unwin.

Zaman, M.Q. (2009) 'Studying Hadith in a Madrasa in the Early Twentieth Century', pp. 225–39, in B. Metcalf (2009) *Islam in South Asia: in Practice*, Princeton, NJ: Princeton University Press.

Zaman, N. (2000) *A Divided Legacy: The Partition in Selected Novels of India, Pakistan and Bangladesh*, Oxford: Oxford University Press.

Zamindar, V. (2007) *The Long Partition and the Making of Modern South Asia: Refugees, Boundaries, Histories*, New York: Columbia University Press.

Zbavitel, D. (1976) *Bengali Literature*, Wiesbaden: Otto Harrassowitz.

Ziad, Z. (2002) *The Magnificent Mughals,* Oxford: Oxford University Press.

Ziegler, N.P. (1978) 'Some Notes on Rajput Loyalties during the Mughal Period', pp. 168–210, in M. Alam and S. Subrahmanyam (Eds) (1998) *The Mughal State 1526–1750*, New Delhi: Oxford University Press.

Ziring, L. (1992) *Bangladesh from Mujib to Ershad; An Interpretative Study*, Dhaka: Dhaka University Press.

Zutshi, C. (2003) *Languages of Belonging: Islam, Regional Identity and the Making of Kashmir*, Delhi: Permanent Black.

INDEX

While the conventional method of having surnames before personal names of people is followed here, this is not always possible with many South Asian, particularly Muslim names. Such names therefore appear in this list in the same order as they are written in the text.

INDEX OF PERSIAN/URDU/BENGALI TEXTS